T0331893

Trace Amines

Experimental and Clinical Neuroscience

Trace Amines: *Comparative and Clinical Neurobiology,*
    edited by *A. A. Boulton, A. V. Juorio,*
    and *R. G. H. Downer,* 1988
Neuropsychopharmacology of the Trace Amines,
    edited by *A. A. Boulton, P. R. Bieck, L. Maitre,*
    and *P. Riederer,* 1985
Neurobiology of the Trace Amines, edited by
    A. A. Boulton, G. B. Baker, W. G. Dewhurst,
    and *M. Sandler,* 1984
Neural Membranes, edited by *Grace Y. Sun,*
    Nicolas Bazan, Jang-Yen Wu, Giuseppe Porcellati,
    and *Albert Y. Sun,* 1983

Trace Amines

Comparative and Clinical Neurobiology

Edited by

A. A. Boulton and A.V. Juorio

University of Saskatchewan, Saskatoon, Saskatchewan, Canada

and

R. G. H. Downer

University of Waterloo, Waterloo, Ontario, Canada

Humana Press • Clifton, New Jersey

Library of Congress Cataloging-in-Publication Data

Trace amines: comparative and clinical neurobiology / edited by A.A. Boulton, A.V. Juorio, and R.G.H. Downer
     760 p. cm.—(Experimental and Clinical Neuroscience)
     Based on the proceedings of a conference held May 28–30, 1987 in Isla Margarita, Venezuela as a satellite of the 11th International Society for Neurochemistry Meeting.
     Sponsored by the International Society for Neurochemistry.
     Includes bibliographies and index.
     ISBN 0-89603-144-6
     1. Biogenic amines—Physiological effect—Congresses. 2. Biogenic amines—Metabolism—Congresses. 3. Neurobiology—Congresses.
I. Boulton, A. A. (Alan A.) II. Juorio, A. V. (Augusto V.) III. Downer, Roger G. H. IV. International Society for Neurochemistry. Meeting. (11th: 1987: La Guaira, Distrito Federal, Venezuela) V. International Society for Neurochemistry. VI. Series.
     [DNLM: 1. Amines—congresses. 2. Neurobiology—congresses.
QU 60 T759 1987]
     QP801.B66T73 1988
     591'.0188—dc19   88-6847
                                        CIP

Preface

This is the third Proceedings book to arise from biennial conferences on the Trace Amines. Since our first meeting in 1983 in Edmonton, Canada, progress has been brisk and, as will be seen from the ensuing pages, it is now possible to include major contributions from invertebrate neurobiologists as well as receptorologists. In the opening session we heard about the distribution of the trace amines—now clearly a misnomer—in insects and the pharmacological, receptor, and synaptic characteristics of octopamine and tryptamine as well as the possibility of monoamines in general being targets for insecticide discovery.

In mammalian brain the distribution and characterization of the tryptamine receptor has proceeded to the point where two types have been described as well as novel agonists and antagonists, and, for the first time, a binding site for p-tyramine has been described. The combination of lesions and pharmacological and metabolic manipulations now permits the mapping of trace aminergic pathways, and the rapidly accumulating evidence from releasing drugs, *in situ* microdialysis, iontophoresis, and second messenger systems lends credence to the claim that the trace amines possess neuromodulatory functions. Although drugs with specific actions on trace aminergic systems are still required, it is interesting to note that prodrugs of phenylethylamine have now been synthesized and that intriguing new condensation adducts have been identified following interaction between trace and other amines and cigaret smoke constituents and alcohol metabolites. Some of these latter substances are capable of interacting with some receptors and enzymes.

As biological markers of stress, migraine, depression, schizophrenia, and tardive dyskinesia, the trace amines or their acidic metabolites or their associated enzymes seem particularly good candidates.

Alan A. Boulton
Augusto V. Juorio
Roger G. H. Downer

Acknowledgments

This book is based on the Proceedings of the Trace Amines: Their Comparative and Clinical Neurobiology conference which was held May 28th to 30th, 1987 at the Hotel Bella Vista, Isla Margarita, Venezuela. The meeting was also arranged as a Satellite of the 11th International Society for Neurochemistry Meeting and we are grateful to the ISN for their sponsorship and also for a financial contribution. In addition we are grateful to the following pharmaceutical companies for their financial support:

American Cyanamid Company, Princeton, New Jersey, USA
Boehringer Ingelheim (Canada) Ltd., Burlington, Ontario, Canada
Ciba-Geigy AG, Basel, Switzerland
Ciba-Geigy Canada Ltd., Mississauga, Ontario, Canada
Ciba-Geigy Ltd., Summit, New Jersey, USA
Hoechst Roussel Pharmaceuticals Inc., Somerville, New Jersey, USA
Lilly Research Laboratories, Indianapolis, Indiana, USA
Pfizer Canada Inc., Pointe Claire, Dorval, Quebec, Canada
Rhône-Poulenc Pharma Inc., Montreal, Quebec
Stauffer Chemical Company, Richmond, California, USA
Upjohn Company of Canada, Don Mills, Ontario, Canada
Warner-Lambert Company (Parke-Davis Canada Inc.), Scarborough, Ontario, Canada

Contributors

BAKER, GLEN B. • Department of Psychiatry, University of Alberta, 1st Floor, Clinical Sciences Building, Edmonton, Alberta, Canada T6G 2G3

BIECK, PETER R. • Humanpharmakologisches Institut Ciba-Geigy GMBH, Ob dem Himmelreich 9, D-7400 Tübingen, Germany

BOULTON, ALAN A. • Neuropsychiatric Research Division, A114 CMR Building, University of Saskatchewan, Saskatoon, Saskatchewan, Canada S7N OWO

CELUCH, STELLA M. • Neuropsychiatric Research Division, CMR Building, University of Saskatchewan, Saskatoon, Saskatchewan, Canada S7N OWO

CIPRIAN-OLLIVIER, JORGE • Francisco de Vittoria 2322/24, Buenos Aires, Argentina 1425

COLLINS, MICHAEL A. • Department of Biochemistry, Loyola University Stritch School of Medicine, Maywood, IL 60153, USA

COOPER, THOMAS B. • Analytical Psychopharmacology Lab, Nathan S. Kline Institute, Orangeburg, NY 10962, USA

DIAMOND, BRUCE I. • Department of Psychiatry, Medical College of Georgia, 1515 Pope Avenue, Augusta, GA 30912, USA

DOWNER, ROGER G. H. • Department of Biology, University of Waterloo, Waterloo, Ontario, Canada N2L 3G1

DURDEN, DAVID A. • Neuropsychiatric Research Division, CMR Building, University of Saskatchewan, Saskatoon, Saskatchewan, Canada S7N OWO

DYCK, LILLIAN E. • Neuropsychiatric Research Division, CMR Building, University of Saskatchewan, Saskatoon, Saskatchewan, Canada S7N OWO

EVANS, PETER D. • Department of Zoology, University of Cambridge, Downing Street, Cambridge, England CB1 3EJ, UK

FARAJ, BAHJAT • Department of Radiology, 467 Woodruff Memorial Building, Emory University School of Medicine, Atlanta, GA 30322, USA

FLETCHER, PAUL J. • Neuropsychiatric Research Division, CMR Building, University of Saskatchewan, Saskatoon, Saskatchewan, Canada S7N OWO

GREENSHAW, ANDREW J. • Department of Psychiatry, University of Alberta, 1st Floor, Clinical Sciences Building, Edmonton, Alberta, Canada T6G 2G3

HARRIS, JOSEPH • Department of Chemistry, Arizona State University, Tempe, AZ 85287, USA

HENRY, DAVID P. • Lilly Laboratory for Clinical Research, University of Indiana, School of Medicine, Indianapolis, IN 46202 USA 46202

JENNINGS, KENT R. • Insecticide Discovery Department, American Cyanamid Co., PO Box 400, Princeton, NJ 08540, USA

JUORIO, AUGUSTO V. • Neuropsychiatric Research Division, CMR Building, University of Saskatchewan, Saskatoon, Saskatchewan, Canada S7N OWO

KIENZL, ELIZABET • LBI for Clinical Neurobiology, Lainz Hospital, Vienna, Austria

LANGE, ANGELA B. • Department of Zoology, University of Toronto, Toronto, Ontario, Canada M5S 1A8

LANGER, SALOMON Z. • Departement de Recherche Biologique, Laboratoires d'Etudes et de Recherches Synthelabo, 58 rue de la Glaciere, 75013 Paris, France

LARSON, ALICE A. • Department of Vet Biology, University of Minnesota, College of Veterinary Medicine, 295 AnSci/Vet Med. Building, 1988 Fitch Avenue, St. Paul, MN 55108, USA

MIDGLEY, JOHN • Nuclear Medicine Center, Veterans Administration Medical Center, 1601 Archer Road, Gainesville, FL 32602, USA

NATHANSON, JAMES A. • Department of Neurology, Massachusetts General Hospital, Boston, MA 02114, USA

NGUYEN, TUONG VAN • Neuropsychiatric Research Division, CMR Building, University of Saskatchewan, Saskatoon, Saskatchewan, Canada S7N OWO

ORIKASA, SHUZO • Neuropsychiatric Research Division, CMR Building University of Saskatchewan, Saskatoon, Saskatchewan, Canada S7N OWO

OSBORNE, NEVILLE N. • Nuffield Laboratory of Ophthalmology, University of Oxford, Walton Street, Oxford, OX2 6AW, UK

PATERSON, I. ALICK • Neuropsychiatric Research Division, CMR Building, University of Saskatchewan, Saskatoon, Saskatchewan, Canada S7N OWO

PERRY, DAVID C. • Department of Pharmacology, George Washington University School of Medicine, 2300 Eye Street, NW, Washington, DC 20037, USA

READER, TOMAS A. • Centre de Recherche en Sciences Neurologiques, Universite de Montreal, Case Postale 6128, succursale A, Montreal, Quebec, Canada H3C 3J7

RICHARDSON, MARY ANN • Clinical Research Division, The Nathan S. Kline Institute for Psychiatric Research, Orangeburg, NY 10962, USA

SANDLER, MERTON • Department of Chemical Pathology, Queen Charlotte's Hospital, Goldhawk Road, London W6 OXG, UK

SARPER, RAUF • Nuclear Medicine, VA Medical Center, 1670 Clairmont Road, Decatur, GA 30033, USA

SATOH, NOBUNORI • Department of Pharmacology, Tohoku College of Pharmacy, 4-4-1, Komatsushima, Sendai 983, Japan

SLOLEY, B. DUFF • Neuropsychiatric Research Division, CMR Building, University of Saskatchewan, Saskatoon, Saskatchewan, Canada S7N OWO

TIPTON, KEITH F. • Biochemistry Department, Trinity College, University of Dublin, Dublin 2, Ireland, UK

VACCARI, ANDREA • Instituto di Farmacologia, via Porcell, 90100 Cagliari, Italy

WHIM, MATTHEW • Department of Zoology, University of Cambridge, Downing Street, Cambridge, CB2 3EJ, UK

WILSON, JOHN X. • Department of Physiology, Health Sciences Center, University of Western Ontario, London, Ontario, Canada N6A 5C1

WOOD, PAUL L. • Department of Neuropharmacology, Neuroscience Research, Ciba-Geigy Ltd., 556 Morris Avenue, Summit, NJ 07901, USA

YU, PETER • Neuropsychiatric Research Division, CMR Building, University of Saskatchewan, Saskatoon, Saskatchewan, Canada S7N OWO

Contents

Neurobiology: Invertebrates

Invited Contributions

Submitted Contributions

Neurobiology: Vertebrates

Invited Contributions

Submitted Contributions

Clinical and Metabolic

Invited Contributions

Submitted Contributions

List of Abbreviations

A	Adrenaline, epinephrine
ACh	Acetylcholine
AChe	Acetylcholinesterase
ADP, etc.	5' (Pyro)-Diphosphates of adenosine, etc.
AMP, CMP, GMP, IMP, UMP	5'-Phosphates of adenosine, cytidine, guanosine, inosine, and uridine
AMPH	Amphetamine
ATP, etc.	5' (Pyro)-triphosphates of adenosine, etc.
ATPase	Adenosine triphosphatase
BBB	Blood brain barrier
BP	Blood pressure
Bz	Benzylamine
ChAT	Choline acetyltransferase
Ci	Curie
CNS	Central nervous system
CoA and acyl-CoA	Coenzyme A and its acetyl derivatives (e.g., acetyl-CoA)
COMT	Catecholamine-O-methyltransferase
cpm	Counts per minute
CSF	Cerebrospinal fluid
Cyclic AMP	3',5'-Cyclic adenosine monophosphate
Cyclic GMP	3',5'-Cyclic guanosine monophosphate
DMT	N,N-Dimethyltryptamine
DNA	Deoxyribonucleic acid
DNase	Deoxyribonuclease
DOPA	3,4-Dihydroxyphenylalanine
DOPAC	3,4-Dihydroxyphenylacetic acid
Dopamine or DA	3,4-Dihydroxyphenylethylamine
dpm, dps	Disintegrations per minute, disintegrations per second
DSM III	Diagnostic and statistical manual of mental disorders
EDTA	Ethylene-diaminetetraacetate
EGTA	Ethyleneglycol bis(aminoethylether)tetraacetate
g	Average gravity
GABA	γ-Aminobutyric acid
GC	Gas chromatography
GC–MS	Gas chromatography/mass spectrometry
GLC	Gas–liquid chromatography
h	Hour
HAMDS	Hamilton depression rating scale

5-HIAA	5-Hydroxyindoleacetic acid
m-HMA	meta-Hydroxymandelic acid
p-HMA	*para*-Hydroxymandelic acid
m-HPA	*meta*-Hydroxyphenylacetic acid
p-HPA	*para*-Hydroxyphenylacetic acid
HPLC	High performance liquid chromatography
5-HT	5-Hydroxytryptamine
HVA	Homovanillic acid
IAA	Indoleacetic acid
ip	Intraperitoneally
iv	Intravenously
μm	Micron
MA	Mandelic acid
MAO	Monoamine oxidase
MAO-A	Monoamine oxidase, type A
MAO-B	Monoamine oxidase, type B
MAOI	Monoamine oxidase inhibitor
5-MeODMt	5-Methoxy-*N,N*-dimethyltryptamine
MHPG	3-Methoxy-4-hydroxyphenylethylene glycol
min	Minute
MOPEG	*See* MHPG
MS	Mass spectrometry
3-MT	3-Methoxytyramine
NA	Noradrenaline, norepinephrine
NCI	Negative chemical ionization
m-OA	*meta*-Octopamine
p-OA	*para*-Octopamine
PAA	Phenylacetic acid
PE	2-Phenylethylamine
PEH	Phenelzine, β-Phenylethylhydrazine
Pi	Orthophosphate (inorganic)
po	Per os (by mouth)
PPi	Pyrophosphate (inorganic)
PST	Phenosulfotransferase
RNA	Ribonucleic acid
RNase	Ribonuclease
rpm	Revolutions per minute
s	Second
sc	Subcutaneous
SD	Standard deviation
SEM	Standard error of mean
T	Tryptamine
m-TA	*meta*-Tyramine
o-TA	*ortho*-Tyramine

p-TA	*para*-Tyramine
TCP	Tranylcypromine
THBC	Tetrahydro-beta-carbolines
TLC	Thin-layer chromatography
Tris	2-Amino-2-hydroxymethylpropane-1,3-diol
VMA	3-Methoxy-4-hydroxymandelic acid

Neurobiology: Invertebrates

PHARMACOLOGICAL CHARACTERISTICS OF OCTOPAMINE-SENSITIVE ADENYLATE CYCLASE AND N-ACETYL OCTOPAMINE TRANSFERASE IN INSECTS

R.G.H. DOWNER, J.W.D. GOLE, R.J. MARTIN, G.L. ORR

Department of Biology
University of Waterloo
Waterloo, Ontario
Canada, N2L 3G1

INTRODUCTION

p-Octopamine (OA) occurs in the nervous systems of many invertebrates at much higher concentrations than in vertebrate tissues (Klemm, 1985). A variety of physiological effects have been ascribed to OA including modulation of the activity of excitable tissues in annelids (Tanaka and Webb, 1983), crustacea (Evans et al, 1987) and insects (Hoyle, 1975; Evans and O'Shea, 1978) and the induction of specific behavioural responses in nematodes (Horitz et al, 1982) and crustacea (Livingston et al, 1980; Bevengut and Clarac, 1982; Kravitz et al, 1981). In insects, OA has also been implicated in excitation of the firefly lantern (Carlson, 1986), stimulation of glycogenolysis in nerve cord and fat body (Robertson and Steele, 1972; Downer, 1979a,b). Stimulation of cardiac contraction (Collins and Miller, 1977), enhancement of substrate utilisation by flight muscle (Goosey and Candy, 1980), release of diacylglycerol from locust fat body (Orchard et al. 1982) and regulation of release of peptide hormones from the corpus cardiacum (Orchard et al., 1983; Downer et al., 1984). The physiological effects of OA in insects are those that might be associated with a generalised response to excitation and suggest that OA may serve a sympathomimetic role in insects.

MECHANISMS OF OA ACTION

At least some of the actions of OA are expressed through interaction of the amine with OA-specific receptors that are coupled to adenylate cyclase. An OA-sensitive adenylate cyclase was first described in a thoracic ganglion preparation of the American cockroach, Periplaneta americana (Nathanson and Greengard, 1973) and similar complexes have subsequently been reported in a variety of tissues and species (Bodnaryk, 1982; David and Coulon, 1986). The insect adenylate cyclase resembles that of other eukaryotic cells in having a requirement for magnesium ions and GTP, demonstrating elevated production of cyclic AMP in the presence of sodium fluoride and forskolin and in showing a positive dose-dependent response to calmodulin (Orr et al., 1985).

Some actions of OA may not be mediated through the adenylate cyclase-cyclic AMP system. Evidence in support of this possibility is provided by the demonstration that OA elevates intracellular calcium levels in cultured haemocytes of Malacosoma disstria. The OA-mediated increase in calcium is blocked by an OA-antagonist whereas this compound has no effect on the rise in calcium levels caused by treatment of the cells with the hypertrehalosemic peptides, HT-I and HT-II (Jahagirdar et al., 1987). Although these preliminary observations do not provide definitive evidence of a second mechanism of OA action (see Evans, 1984), they strongly suggest involvement of an agonist-dependent calcium gating mechanism similar to that described for the action of 5-HT on salivary glands of adult Calliphora (Litosch et al. 1985).

PHARMACOLOGY OF OA RECEPTORS

Aspects of the pharmacology of OA-receptors have been studied in a variety of insect species and tissues using various physiological responses or cyclic AMP production as criteria for assessment of pharmacological effects. Studies with derivatives of phenylamines, phenolamines and catecholamines indicate that phenolamines are the most effective agonists of OA-mediated effects and that the D(-) stereoisomer is more potent than the L(+) or DL forms (Evans, 1980). Furthermore, optimal activity is achieved when the β-carbon is hydroxylated and the phenolic-OH is in the para-position (Dougan and Wade, 1978; Harmar and

TABLE 1

Effect of Various Antagonists on Octopamine-Mediated Processes

INSECT PREPARATION	CHLORPROMAZINE ANTI EMETIC slight ANTI HISTAMINIC ANTI ADRENERGIC	CYPROHEPTADINE ANTI HISTAMINE ANTI PRURITIC	METOCHLOPROMIDE ANTI EMETIC	MIANSERIN ANTI SEROTONIN ANTI HISTAMINE
Locusta migratoria lateral oviduct; cAMP production (Orchard & Lange, 1986)	D	C	B	B
Locusta migratoria flight muscle; cAMP production (Lafon-Cazal & Bockaert, 1985)	C	B	B	A
Periplaneta americana thoracic ganglion; cAMP production (Harmer & Horn, 1977)	B	A		
Periplaneta americana haemocytes; cAMP production (Orr et al., 1985)	B	B	B	A
Periplaneta americana corpus cardiacum; cAMP production (Gole et al., 1987)		B		A
Periplaneta americana nerve cord; cAMP production (Downer et al., 1985)		A		A
Photinus pyralis light organ, cAMP production (Nathanson, 1985)	A			
Drosophila melanogaster head homogenate; ^{3}H-OA binding (Dudai, 1984)	A			
Schistocerca gregaria flight muscle; twitch amplitude (Evans, 1981)	B	B	B	B
Schistocerca gregaria flight muscle; relaxation rate (Evans, 1981)	C	B	B	B
Schistocerca gregaria flight muscle; myogenic rhythm (Evans, 1981)	A	A	D	B
% OF STUDIES IN WHICH COMPOUND IS A or B	66	90	83	100
% OF STUDIES IN BROKEN PREPARATIONS IN WHICH COMPOUND IS A or B	75	100	100	100
% OF STUDIES IN WHOLE PREPARATIONS IN WHICH COMPOUND IS A or B	50	83	75	100

Key: A, effective antagonist, K_1 or $IC_{50} \leq 1 \times 10^{-7}$ M; B, good antagonist, K_1 or $IC_{50} \leq 1 \times 10^{-5}$ M; C, poor antagonist, K_1 or $IC_{50} \leq 1 \times 10^{-3}$ M; D, no antagonism.

TABLE 1 (continued)

Effect of Various Antagonists on Octopamine-Mediated Processes

INSECT PREPARATION	PHENOXYBENZAMINE ANTI HYPERTENSIVE	PHENTOLAMINE α-ADRENERGIC BLOCKER	PROPRANOLOL β-ADRENERGIC BLOCKER	YOHIMBINE ADRENERGIC BLOCKER
Locusta migratoria lateral oviduct; cAMP production (Orchard & Lange, 1986)		A		B
Locusta migratoria flight muscle; cAMP production (Lafon-Cazal & Bockaert, 1985)		A		D
Periplaneta americana thoracic ganglion; cAMP production (Harmer & Horn, 1977)		A	D	D
Periplaneta americana haemocytes; cAMP production (Orr et al., 1985)	B	A	D	C
Periplaneta americana corpus cardiacum; cAMP production (Gole et al., 1987)		B	D	
Periplaneta americana nerve cord; cAMP production (Downer et al., 1985)		A		
Photinus pyralis light organ, cAMP production (Nathanson, 1985)		A	D	D
Drosophila melanogaster head homogenate; ^{3}H-OA binding (Dudai, 1984)	A	A	C	B
Schistocerca gregaria flight muscle; twitch amplitude (Evans, 1981)	C	B	C	D
Schistocerca gregaria flight muscle; relaxation rate (Evans, 1981)	B	B	C	D
Schistocerca gregaria flight muscle; myogenic rhythm (Evans, 1981)	A	A	A	A
% OF STUDIES IN WHICH COMPOUND IS A or B	66	100	9	33
% OF STUDIES IN BROKEN PREPARATIONS IN WHICH COMPOUND IS A or B	66	100	0	20
% OF STUDIES IN WHOLE PREPARATIONS IN WHICH COMPOUND IS A or B	66	100	20	50

Key: A, effective antagonist, K_1 or $IC_{50} \leq 1 \times 10^{-7}$ M; B, good antagonist, K_1 or $IC_{50} \leq 1 \times 10^{-5}$ M; C, poor antagonist, K_1 or $IC_{50} \leq 1 \times 10^{-3}$ M; D, no antagonism.

Table 2: Effect of aminergic agonists and antagonists on the N-acetylation of
p-octopamine by N-acetyltransferase from Malpighian tubules and
cerebral ganglia (BRAIN) of <u>Periplaneta</u> <u>americana</u>

	% INHIBITION	
TREATMENT	MALPIGHIAN TUBULES	BRAIN
AGONISTS:		
<u>Octopamine</u>		
p-Synephrine	n.s.	n.s.
Naphazoline	8.5*	10.6
Demethylchlordimeform	23.8	12.6
<u>Dopamine</u>		
ADTN	97.8	68.5
Epinine	55.3	22.9
<u>α-Adrenergic</u>		
Clonidine	n.s.	8.2
ANTAGONISTS:		
<u>Octopamine</u>		
Mianserin	23.4	20.7
Phentolamine	17.2	7.3
Cyproheptadine	38.2	25.0
Gramine	70.1	24.8
<u>Dopamine</u>		
+Butaclamol	17.1	20.7
cis-Flupenthixol	74.9	33.4
SKF-82526 (also agonist)	98.2	63.2
<u>β-Antagonist</u>		
Propranolol	n.s.	6.9

*Values indicate mean % inhibition of activity relative to control values for
4 determinations.
Standard errors of the means were less than 5% of the means.
All values presented are significant from control values, $p < 0.05$, model 1
ANOVA.
n.s. = not significant from control values
All compounds were included in the assay medium at equimolar concentrations
with octopamine (1.0 mM).

Horn, 1977).

Several adrenergic agonists have been tested for their ability to mimic OA-mediated processes. Evans (1981) used the differing responses of three OA-mediated processes in locust flight muscle to clonidine, naphazoline and tolazoline as partial criteria to propose three pharmacologically distinct OA-receptors.

Agonism and partial agonism towards OA-receptors have also been demonstrated with formamidine derivatives (Evans and Gee, 1980; Hollingworth and Murdock, 1980; Nathanson and Hunnicutt 1981; Gole et al., 1983) and derivatives of phenyliminoimidazolidines (Nathanson, 1985). However, the formamidine derivatives may interact also with dopamine- and 5-HT-receptors in the insect nervous system (Downer et al., 1985).

Studies on the antagonism of OA-receptors are limited to the use of antagonists that have been developed for use with defined vertebrate receptors. Thus, at the present time, the OA-receptors in insects are characterised according to their interaction with drugs that, in vertebrates, bind to aminergic receptors other than OA. Generalisations are further complicated by variations in the experimental procedures employed by different investigators and by inter-specific and inter-tissue differences. In spite of these constraints, it is apparent that OA-receptors are pharmacologically distinct from those of any other aminergic receptor that has been described in vertebrate tissues and that more than one OA-receptor occurs. The effects of some antagonists on some insect preparations are summarised in Table 1. The summary indicates that the most potent antagonists of OA-mediated processes are the α-adrenergic antagonists, phentolamine and mianserin, which is an anti-serotonergic and anti-histaminergic compound. Both phentolamine and mianserin were effective at A or B levels in all preparations tested (see Table 1 for details of classification). Chlorpromazine, cyproheptadine, metochlopromide and phenoxybenzamine were effective against most of the preparations tested; however, the poor antagonism of these compounds against some preparations suggests that they may be useful in distinguishing between different OA-receptor types.

There have been few studies on the binding of

octopamine to putative octopamine receptors in insects although preliminary investigations on whole head homogenates of <u>Drosophila</u> (Dudai, 1984) and firefly light organ (Hashemzadeh <u>et al</u>., 1985) using ^{3}H-OA suggest some specific binding of the ligand. Studies with ^{3}H-mianserin indicate that this ligand has a much greater affinity for OA-receptors than for any other aminergic receptor in the insect nervous system and preliminary observations suggest the presence of high and low affinity binding sites with negative co-operativity (N. Minhas, unpublished observations).

OCTOPAMINE DEGRADATION

A high affinity, sodium-dependent uptake mechanism for OA has been described in the nerve cord of cockroach (Evans, 1978) and larval firefly light organs (Carlson and Evans, 1986). These studies suggest that OA-uptake in insects may be analogous to the uptake of norepinephrine in vertebrates. The fate of the sequestered OA within the nervous system has not been fully resolved; however, there is strong evidence to suggest that the primary enzymatic pathway for OA-catabolism involves N-acetylation. N-acetyltransferase (NAT) activity against OA has been described in the nervous tissue of locust (Hayashi <u>et al</u>., 1977; Mir and Vaughan, 1981) and the European corn borer (Evans <u>et al</u>., 1980).

In addition to serving as a neurotransmitter and neuromodulator, OA also functions as a neurohormone and circulating levels in the haemolymph are elevated in response to excitation or physical activity (Bailey <u>et al</u>., 1984; Davenport and Evans, 1984). The excitation-induced levels of OA are rapidly restored to basal levels with NAT activity in the Malpighian tubules contributing appreciably to the process of OA inactivation (Downer and Martin, 1987).

The development of a sensitive analytical procedure involving high performance liquid chromatography with coulometric electrochemical detection has enabled the comparison of NAT activity from Malpighian tubules and cerebral ganglia of the American cockroach (R.J. Martin, unpublished observations). A variety of biogenic amines and related derivatives were tested for their effect on the N-acetylation of OA by NAT derived from the two tissue sources (R.J. Martin and R.G.H. Downer, in preparation). Some differences in the NAT from Malpighian tubules and cerebral ganglia are indicated by the actions of 4-methoxyphenylethylamine which has no effect on NAT from Malpighian tubules yet it inhibits the cerebral ganglion enzyme by 26%; also, N,N-dimethyltyramine inhibits Malpighian tubule NAT by 77% but the brain enzyme by only 24%. The results demonstrate also that greatest inhibition of NAT activity is observed when the competing compound has a ring hydroxyl moiety in the 4 position. Studies with aminergic agonists and antagonists confirm differences in the properties of NAT from Malpighian tubules and cerebral ganglia especially with regard to dopaminergic agonists and antagonists (Table 2). The NAT of insects is active against a variety of monoamines and, indeed, has greater activity against dopamine than octopamine, therefore, it is not surprising that compounds which interfere with dopamine binding to dopaminergic receptors inhibit the action of NAT against OA.

The formamidine, demethylchlordimeform, also inhibits NAT activity from both Malpighian tubules and cerebral ganglia. Thus, in addition to their interactions with aminergic receptors, it is likely that formamidines also perturb the inactivation of monoamines in the insect. The significance of this action in expressing the insecticidal action of formamidines has not been resolved.

CONCLUSIONS

The obvious physiological importance of OA in insects and other invertebrates indicates that, at least in some organisms, it is much more than a trace amine. The increasing availability of sensitive analytical techniques and appreciation of invertebrate systems is likely to result in major advances in our understanding of the biochemistry, pharmacology and physiology of this monoamine. It is hoped that this brief review of the subject will provide impetus for further study of OA.

ACKNOWLEDGEMENTS

Supported by operating grants from the Natural Sciences and Engineering Research Council of Canada and American Cyanamid.

REFERENCES

Bailey B.A., Martin R.J., and Downer R.G.H. (1984) Haemolymph octopamine levels during and following flight in the American cockroach, Periplaneta ameri- cana L. Can. J. Zool. 62, 19-22.

Bevengut M., and Clarac F. (1982) Contrôle de la posture du crabe Carcinus maenas par des amines biogènes. C.R. Acad. Sci., Paris, sér. III, 295, 23-28.

Bodnaryk R.P. (1982) Biogenic amine-sensitive adenylate cyclase in insects. Insect Biochem. 12, 1-6.

Carlson A.D. (1968) Effect of adrenergic drugs on the lantern of the larval Photinus firefly. J. Exp. Biol. 48, 381-187.

Carlson A.D., and Evans P.D. (1986) Inactivation of octopamine in larval firefly light organs by a high-affinity uptake mechanism. J. Exp. Biol. 122, 369-385.

Collins C., and Miller T.A. (1977) Studies on the action of biogenic amines on cockroach heart. J. Exp. Biol. 67, 1-15.

Davenport A.P., and Evans P.D. (1984) Stress induced changes in the octopamine levels of insect haemo- lymph. Insect Biochem. 14, 135-143.

David J-C., and Coulon J-F. (1985) Octopamine in inverte- brates and vertebrates. A review. Prog. Neurobiol. 24, 141-185.

Dougan D.F.H., and Wade D.N. (1978) Action of octopamine agonists and steroisomers at a specific octopamine receptor. Clin. exp. Pharmac. Physiol. 5, 333-339.

Downer R.G.H. (1979a) Induction of hypertrehalosemia by excitation in Periplaneta americana. J. Insect Physiol. 29, 55-63.

Downer R.G.H. (1979b) Trehalose production in isolated fat body of the American cockroach, Periplaneta americana. Comp. Biochem. Physiol. 62C, 31-34.

Downer R.G.H., and Martin R.J. (1987) N-acetylation of octopamine: a potential target for insecticide development. In The search for novel insecticides: toxicants affecting GABA, octopamine and other neuroreceptors in invertebrates (Green M., Holling- worth R.M., and Hedin P.A., eds) in press, American Chemical Society, New York.

Downer R.G.H., Gole J.W.D., Orr G.L., and Orchard I. (1984) The role of octopamine and cyclic AMP in regulating hormone release from corpora cardiaca of the American cockroach. J. Insect Physiol. <u>30</u>, 457–462.

Downer R.G.H., Gole J.W.D., and Orr G.L. (1985) Interaction of formamidines with octopamine-, dopamine- and 5-hydroxytryptamine-sensitive adenylate cyclase in nerve cord of <u>Periplaneta</u> <u>americana</u>. Pestic. Sci. <u>16</u>, 472–478.

Dudai Y. (1982) High-affinity octopamine receptors revealed in <u>Drosophila</u> by binding of (^{3}H)-octopamine. Neurosci. Lett. <u>28</u>, 163–167.

Evans P.D. (1978) Octopamine: a high-affinity uptake mechanism in the nervous system of the cockroach. J. Neurochem. <u>30</u>, 1015–1022.

Evans P.D. (1980) Octopamine receptors in insects. <u>In</u>: Receptors for Neurotransmitters, Hormones and Pheromones in Insects (Battelle D.B., Hall L.M., and Hildebrand J.E. eds) pp 245–258, Elsemer, Amsterdam.

Evans P.D. (1981) Multiple receptor types for octopamine in the locust. J. Physiol. <u>318</u>, 99–122.

Evans P.D. (1984) The role of cyclic nucleotides and calcium in the mediation of the modulatory effects of octopamine on locust skeletal muscle. J. Physiol. <u>345</u>, 325–340.

Evans P.D.,and Gee J.D. (1980) Action of formamidine pesticides on octopamine receptors. Nature <u>287</u>, 60–62.

Evans P.D., and O'Shea M. (1978) The identification of an octopaminergic neurone and the modulation of a myogenic rhythm in the locust. J. Exp. Biol. <u>73</u>, 235–260.

Evans P.D., Kravitz E.A., and Talamo B.R. (1976) Octopamine release at two points along lobster nerve trunks. J. Physiol. <u>262</u>, 71–89.

Evans P.M., Soderlund D.M., and Aldrich J.R. (1980) <u>In</u> <u>vitro</u> N-acetylation of biogenic amines by tissues of the European corn borer, <u>Ostrinia</u> <u>nubilalis</u> hübner. Insect Biochem. <u>10</u>, 375–380.

Gole J.W.D., Orr G.L., and Downer R.G.H. (1983) Interaction of formamidines with octopamine-sensitive adenylate cyclase receptor in the nerve cord of <u>Periplaneta</u> <u>americana</u> L. Life Sci. <u>32</u>, 2939–2947.

Gole J.W.D., Orr G.L., and Downer R.G.H. (1987) Pharmacology of octopamine-, dopamine- and 5-hydroxytryptamine-sensitive adenylate cyclase in the corpus cardiacum of the American cockroach, <u>Periplaneta</u> <u>americana</u> L. Archs. Insect Biochem. Physiol. in press.

Goosey M.W., and Candy D.J. (1980) Effects of D- and L-octopamine and of pharmacological agents on the metabolism of locust flight muscle. Biochem. Soc. Trans. 8, 532-533.

Harmar A.J., and Horn A.S. (1977) Octopamine-sensitive adenylate cyclase in cockroach brain: effects of agonists, antagonists and guanylyl nucleotides. Molec. Pharmacol. 13, 512-520.

Hashemzadeh H., Hollingworth R.M., and Voliva A. (1985) Receptors for ^{3}H-octopamine in the adult firefly light organ. Life Sci. 37, 433-440.

Hayashi S., Murdock L.L. and Florey E. (1977) Octopamine metabolism in invertebrates (Locusta, Astacus, Helix): evidence for N-acetylation in arthropod tissues. Comp. Biochem. Physiol. 58C, 183-191.

Hollingworth R.M., and Murdock L.L. (1980) Formamidine pesticides: octopamine-like action in a firefly. Science 208, 74-76.

Horvitz H.R., Chalfie M., Trent C., Sulston J.E., and Evans P.D. (1982) Serotonin and octopamine in the nematode Caenorhabditis elegans. Science 216, 1012-1014.

Hoyle G. (1975) Evidence that insect dorsal unpaired median (DUM) neurons are octopaminergic. J. Exp. Zool. 193, 433-439.

Jahagirdar A.P., Milton G., Viswanatha T., and Downer R.G.H. (1987) Calcium involvement in mediating the action of octopamine and hypertrehalosemic peptides on insect haemocytes. FEBS Letts. in press.

Klemm N. (1985) The distribution of biogenic monoamines in invertebrates. In: Neurobiology: current comparative approaches. (Gilles R., and Balthazart J. eds) pp 280-296, Springer-Verlag, Berlin.

Kravitz E.A., Beltz B.S., Glusman S., Goy M.F., Harris-Warwick R.M., Johnstone M.F., Livingstone M.S., Schwartz T.L., and Siwicki K. (1983) Neurohormones and lobsters: biochemistry to behaviour. Trends Neurosci. 6, 346-349.

Lafon-Cazal M., and Bockaert J. (1985) Pharmacological characterization of octopamine-sensitive adenylate cyclase in the flight muscle of Locusta migratoria L. Eur. J. Pharmacol. 119, 53-59.

Livingstone M.S., Harris-Warwick R.M., and Kravitz E.A. (1980) Serotonin and octopamine produce opposite postures in lobsters. Science 208, 76-79.

Litosch I., Wallis C., and Fain J.N. (1985) 5-hydroxy-
    tryptamine stimulates inositol phosphate production
    in a cell-free system from blowfly salivary glands.
    J. Biol. Chem. 260, 5464–5471.
Mir A.K., and Vaughan P.F.T. (1981) Biosynthesis of
    N-acetyldopamine and N-acetyloctopamine by Schisto-
    cerca gregaria nervous tissue. J. Neurochem. 36,
    441–446.
Nathanson J.A. (1985a) Phenyliminoimidazolines: charac-
    terization of a class of potent agonists of octo-
    pamine-sensitive adenylate cyclase and their use in
    understanding the pharmacology of octopamine recep-
    tors. Molec. Pharmacol. 28, 254–268.
Nathanson J.A. (1985b) Characterization of octopamine-
    sensitive adenylate cyclase: elucidation of a class
    of potent and selective octopamine-2-receptor
    agonists with toxic effects in insects. Proc. Nat.
    Acad. Sci. U.S.A. 82, 599–603.
Nathanson J.A., and Greengard P. (1973) Octopamine-
    sensitive adenylate cyclase: evidence for a biologi-
    cal role of octopamine in nervous tissue. Science
    180, 308–310.
Nathanson J.A., and Hunnicutt E.J. (1981) N-demethylchlor-
    dimeform. A potent partial agonist of octopamine-
    sensitive adenylate cyclase. Molec. Pharmacol. 20,
    68–75.
Orchard I., and Lange A. (1986) Pharmacological profile of
    octopamine receptors on the lateral oviducts of the
    locust, Locusta migratoria. J. Insect Physiol. 32,
    741–745.
Orchard I., Carlisle J.A., Loughton B.A., Gole J.W.D., and
    Downer R.G.H. (1982) In vitro studies on the effects
    of octopamine on locust fat body. Gen. Comp. Endo-
    crinol. 48, 7–13.
Orchard I., Loughton B.A., Gole J.W.D., and Downer R.G.H.
    (1983) Synaptic transmission elevates adenosine
    3',5'-monophosphate (cyclic AMP) in locust neuro-
    secretory cells. Brain Res. 258, 152–155.
Orr G.L., Gole J.W.D., and Downer R.G.H. (1985) Charac-
    terisation of an octopamine-sensitive adenylate
    cyclase in hemocyte membrane fragments of the Ameri-
    can cockroach, Periplaneta americana. Insect
    Biochem. 15, 695–701.
Robertson H.A., and Steele J.E. (1972) Activation of
    insect nerve cord phosphorylase by octopamine and
    adenosine 3',5'-monophosphate. J. Neurochem. 19,

1603–1606.
Tanaka K.R. and Webb R.A. (1983) Octopamine action on the spontaneous contractions of the isolated nerve cord of <u>Lumbricus</u> <u>terrestris</u>. Comp. Biochem. Physiol. <u>76C</u>, 113–120.

OCTOPAMINE RECEPTORS ON LOCUST SKELETAL MUSCLE

PETER D. EVANS.

A.F.R.C. Unit of Insect Neurophysiology and Pharmacology, Dept. of Zoology, University of Cambridge, Downing Street, Cambridge, CB2 3EJ, U.K.

Octopamine is a biogenic amine that was first discovered in the posterior salivary glands of the octopus (Erspamer and Boretti, 1951). Since that time it has been shown to be a ubiquitous constituent of all nervous systems examined, all the way from nematodes and earthworms up to vertebrates, including humans (see Evans, 1980; 1985a). The functional role of octopamine in the vertebrate nervous system, however, is at present enigmatic (Harmar, 1980; Talamo, 1980; Robertson, 1981). Octopamine occurs in three different structural isomeric forms, para-, meta- and ortho-octopamines, all of which have been shown to occur naturally in vertebrates (Williams, Couch and Midgley, 1984). The occurrence of p- and $\underline{m}$-octopamine in the sympathetic nervous system (Ibrahim, Couch, Williams, Fregley and Midgley, 1985) and brain (Danielson, Boulton and Robertson, 1977; David and Delacour, 1980) has led to the suggestion that they may function as co-transmitters or modulators of noradrenaline action. In addition the close structural similarities between $\underline{m}$- and $\underline{p}$-octopamine and noradrenaline suggest that many of the observed physiological effects of octopamine application on vertebrate tissues may be mediated via interactions with adrenergic receptors. Thus the actions of the stereoisomers of octopamine, and its N-methylated analogue synephrine, have been examined on both alpha adrenergic receptor subtypes (Brown, McGrath Midgley, Muir, O'Brien, Thonoor and Williams, 1987) and beta-adrenergic receptor subtypes (Williams, Jordan,

17

Thonoor and Midgley, 1987). In both cases, however, it was concluded that if m- and p-octopamine are coreleased with noradrenaline in amounts proportional to their concentration in tissue, then their activities on both alpha and beta-adrenergic receptors would be too low for them to be considered physiological. To data no specific octopaminergic neurones have been found in the vertebrate nervous system, although a differential distribution has been reported in rat brain (Buck, Murphy and Molinoff, 1977). However, experimental evidence for the existence of specific octopaminergic receptors in vertebrates is limited to a very few papers in which single neurones in spinal cord, cerebral cortex (Hicks and McLennan, 1978a,b) and thalamus (Dao and Walker, 1980) have been shown to respond differentially to iontophoretic application of p-octopamine and noradrenaline.

In invertebrate nervous systems, on the other hand, specific octopaminergic neurones have been identified, together with target sites for the action of their released octopamine (see Evans, 1980; 1985a). To date the majority of identified octopaminergic neurones occur in insects where octopamine functions as a neuromodulator, a neurohormone and a true neurotransmitter (Evans, 1980; 1985a; Orchard, 1982). The first octopaminergic neurone to be identified in insects was one of a group of neutral red staining cells on the dorsal surface of the metathoracic ganglion of the locust (Fig.1) (Evans and O'Shea, 1977; 1978). This cell has been designated the dorsal unpaired median cell to the extensor-tibiae muscle of the locust hindleg (DUMETi) (Hoyle et al., 1974). It is an unusual neurone since it is unpaired and sends axons to innervate the extensor-tibiae muscle in both the left and right legs (Fig.2). This preparation presents a very useful model system in which to study the pharmacology and mode of action of octopamine receptors on insect skeletal muscle (see Evans and Myers, 1986). The muscle is innervated by three physiologically identified motorneurones, a fast excitatory (FETi), a slow excitatory (SETi) and a branch of the common inhibitor (CI). In addition it is also innervated by DUMETI the modulatory octopaminergic neurone mentioned above. The latter cell does not make specific neuromuscular contacts in the muscle, but rather its terminals end as blindly ending neurosecretory terminals between the muscle fibres (Hoyle et al., 1980). The extensor muscle itself is a large muscle and can provide large quantities of material for

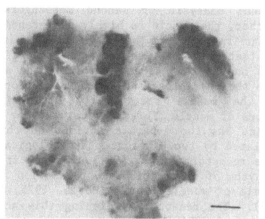

Fig.1   Light micrograph of the dorsal surface of a locust metathoracic ganglion showing the organization of the DUM neurone somata as revealed by neutral red staining (0.01 mg/ml in isotonic saline for 3h). Scale bar 100μm. (from Evans and O'Shea, 1978).

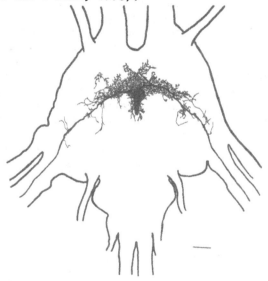

Fig.2   Anatomy of the DUMETi neuron from the metathoracic ganglion of the locust. The cell was filled intracellularly with cobaltous ions which were subsequently precipitated as the sulphide. The preparation was then subjected to whole-mount silver intensification using a modification of Timm's method. Scale bar, 100μm. (from Evans, 1982)

the biochemical analysis of second messengers (Evans, 1984a). Further, the muscle is highly differentiated such that different regions can be identified that contain exclusively, fast, slow or intermediate muscle fibre types (Hoyle, 1978) and it is thus easy to study the actions of octopamine on the physiology and biochemical properties of regions of the muscle specifically containing these different types of muscle fibres (Evans, 1985b).

Octopamine has two separate actions on this muscle that are mediated by two distinct pharmacological classes of octopamine receptor (Evans, 1981). First, it can slow a myogenic rhythm of contraction and relaxation found in a proximal bundle of muscle fibres (Hoyle, 1975, Evans and O'Shea, 1978). The receptors mediating this action have been designated OCTOPAMINE$_1$ class receptors. Second, octopamine can modulated neuromuscular transmission and muscular contraction in the bulk of the non-myogenic fibres in the muscle. The receptors mediating these effects have been designated the OCTOPAMINE$_2$ class receptors. These can be further subdivided into the OCTOPAMINE$_{2A}$ receptors located presynaptically on the terminals of the slow motorneurone and when stimulated by octopamine released from the neurosecretory terminals of the DUMETi neurone (Morton and Evans, 1984), they potentiate both the spontaneous and neuronally evoked release of neurotransmitter. The second subtype, the OCTOPAMINE$_{2B}$, receptors are located postsynaptically, or extrasynaptically since the DUMETi neurone does not form morphologically distinct neuromuscular junctions, on the muscle fibres themselves. Stimulation of the latter class of receptors increases the rate of relaxation of both fast and slow excitatory motorneurone generated twitch tension (O'Shea and Evans, 1979), tetanic and 'catch' tension (Evans and Siegler, 1982) in this muscle.

OCTOPAMINE$_1$ receptors can be distinguished from OCTOPAMINE$_2$ receptors on the basis of both antagonist and agonist studies (Evans, 1981, see Tables 1A and 1B). Chlorpromazine and yohimbine are much better blockers of the former receptors than is metoclopramide, whereas the converse is true of the latter receptors. Also clonidine is a more potent agonist than naphazoline at OCTOPAMINE$_1$ receptors, while the converse is true for OCTOPAMINE$_2$ receptors. Similarly OCTOPAMINE$_{2A}$ receptors can be distinguished from OCTOPAMINE$_{2B}$ receptors on the basis of both agonist and antagonist studies (see Table 1A and 1B). Cyproheptadine, mianserin and metoclopramide are better

Table 1A Action of agonists on octopamine receptors.

| Drug | Receptor Class | | |
	$Octopamine_1$ EC_{50} (M)	$Octopamine_{2A}$ EC_{50} (M)	$Octopamine_{2B}$ EC_{50} (M)
Clonidine	6.8×10^{-10}	6.4×10^{-6}	2.0×10^{-5}
Naphazoline	1.2×10^{-8}	1.3×10^{-8}	2.2×10^{-7}
Tolazoline	1.5×10^{-9}	3.2×10^{-6}	6.0×10^{-7}

EC_{50} for $Octopamine_1$ receptors is the concentration of a 5 min pulse of the drug required to reduce the frequency of the myogenic rhythm by 50%. EC_{50} for $Octopamine_{2A}$ and $Octopamine_{2B}$ receptors is concentration of a 30s pulse of drug required to produce 50% of maximal response to octopamine in SETi twitch amplitude and relaxation rate, respectively (SETi fired at 1Hz). For further details of experiments and additional drugs see Evans, P.D. (1981)

Table 1B Action of antagonists on octopamine receptors.

| Drug | Receptor Class | | |
	$Octopamine_1$ EC_{50} (M)	$Octopamine_{2A}$ EC_{50} (M)	$Octopamine_{2B}$ EC_{50} (M)
Metoclopro-mide	–	1.0×10^{-6}	9.5×10^{-6}
Yohimbine	2.8×10^{-7}	–	–
Chlorpro-mazine	2.6×10^{-8}	1.6×10^{-4}	7.0×10^{-5}
Cyprohep-tadine	3.7×10^{-8}	2.2×10^{-6}	5.1×10^{-5}
Mianserin	4.5×10^{-6}	1.2×10^{-6}	2.0×10^{-5}

EC_{50} for $Octopamine_1$ receptors is concentration of drug required to reduce response of myogenic rhythm to a 5 min pulse of 10^{-7} M DL-octopamine by 50%. EC_{50} for $Octopamine_{2A}$ and $Octopamine_{2B}$ receptors is concentration of a drug required to reduce response of SETi-induced twitch amplitude and relaxation rate, respectively, to a 30s pulse of 10^{-6}M DL-octopamine by 50% (SETi fired at 1Hz). For further details of experiments and additional drugs see Evans, P.D. (1981)

blockers of 2A than 2B receptors, whilst the converse is true for chlorpromazine. In addition naphazoline is a much better agonist than tolazoline at 2A receptors and tolazoline is much better than clonidine at 2B receptors. An important point to emphasise is that this receptor classification was evolved on the basis of pharmacological differences between the receptor subclasses. This is preferable to any classification based purely on a functional or mode of action basis since it avoids situations where pharmacologically identical receptors are classified differently if they are coupled to different transducing mechanisms for mediating their actions. However, in the case of the locust octopamine receptors the pharmacologically distinct $OCTOPAMINE_1$ and $OCTOPAMINE_2$ class receptors do in fact also have different modes of action (Evans, 1981, 1984a,b,c)

The drugs that are effective at distinguishing the different classes of octopamine receptor in locust skeletal muscle turn out to be a mixture of agents that have previously been described to be active at vertebrate α-adrenergic receptors, dopamine receptors, 5-HT receptors and histamine receptors (Evans, 1981, 1984a,b, 1985a,b). Thus at present there are no readily available specific octopaminergic agents available for such studies. However, the pharmacological differences described above suggest that it should be possible to synthesise such specific agonists and antagonists that would be capable of distinguishing octopamine receptors from other biogenic amine receptors and also of distinguishing between the different pharmacological classes of octopamine receptor.

The above conclusion is supported by recent evidence from studies on the actions of different octopamine isomers on the octopamine receptor subtypes on locust skeletal muscle. Octopamine occurs as three different positional isomeric forms i.e. para, meta and ortho, each of which occurs as the (+) and (−) enantiomorphs. Each of these six isomeric forms of octopamine has been synthesised and characterized by Prof. J. Midgley (University of Strathclyde, Glasgow) and we have examined their effects on the different subclasses of octopamine receptor in the locust extensor-tibiae muscle (Evans, Thonoor and Midgley, in preparation). The most effective form on the $OCTOPAMINE_{2B}$ receptors mediating the increase in relaxation rate of slow motorneurone induced twitch tension is the (−)para isomer which has a threshold of between 10^{-9} and 10^{-10} M and an ED_{50} of 5×10^{-8} M. The (+)

para isomer was around an order of magnitude less effective with an ED_{50} of 5×10^{-7} M. The (-) meta and (+) meta isomers were poor agonists that never maximally activated the receptors at concentrations below 10^{-3} M and needed between 10^{-6} and 10^{-5} M concentrations to demonstrate any significant effect. In contrast when the same isomers were tested on the OCTOPAMINE$_1$ receptors responsible for the slowing of the myogenic rhythm the meta isomers were relatively much more active. At the OCTOPAMINE$_1$ receptors (-) para octopamine was active causing a 50% reduction in the rhythm frequency at 10^{-9} M. Again the (+) para isomer was about an order of magnitude less effective. However, on this receptor both the (-) meta and (+) meta isomers were capable of causing a 100% reduction in rhythm frequency, with the (-) meta isomer causing a 50% reduction in frequency at 3×10^{-8} M and with a threshold for an observable slowing of the rhythm occuring between 10^{-10} and 10^{-11} M. Thus the (-) meta isomer of octopamine may well provide an important lead compound in the development of agonists and antagonists that are highly selective for the OCTOPAMINE$_1$ class receptors.

Recently a set of compounds, namely some substituted phenyliminoimidazolidines, NC5 and NC7, have been suggested to be potent agonists for OCTOPAMINE$_2$ receptors that bring about their actions by the activation of adenylate cyclase (Nathanson, 1985a,b). NC5 and NC7 are effective agonists of the OCTOPAMINE$_{2B}$ receptors mediating the increase in relaxation rate of slow motorneurone induced twitch tension. They also activate the OCTOPAMINE$_{2A}$ receptors in this preparation. In both cases the compounds have thresholds between 10^{-9} and 10^{-10} M and NC5 is slightly more effective than NC7 especially at higher concentrations (Evans 1987). However, both NC7 and NC5 are active on the OCTOPAMINE$_1$ receptors mediating the slowing of the myogenic rhythm, with NC7 being the most effective causing a 50% reduction in the rhythm frequency at 10^{-10} M. Thus the phenyliminoimidazolidines are not specific for OCTOPAMINE$_2$ receptors in the locust (Evans, 1987).

A considerable amount of evidence indicates that the OCTOPAMINE$_{2A}$ and OCTOPAMINE$_{2B}$ receptors in locust skeletal muscle mediate their actions via increased cyclic AMP levels generated by the activation of adenylate cyclase (Evans, 1984a,b). All the physiological actions of these receptors can be mimicked by elevating cyclic AMP levels by mechanisms that bypass the receptor activation process. Thus they can be mimicked by the addition of the phosphodiesterase inhibitor isobutylmethylxanthine (IBMX) to the preparation. IBMX can also potentiate the physiological actions of octopamine. In addition these effects are mimicked by addition of forskolin, the specific diterpene activator of adenylate cyclase and also by the addition of highly permeable and slowly metabolised analogues of cyclic AMP, such as 8 chlorophenylthio cyclic AMP. Further evidence comes from the fact that short pulses of octopamine introduced into the muscle superfusate increase cyclic AMP levels, but not cyclic GMP levels, in a dose dependent way. The pharmacological profile and the time courses of the biochemical changes in cyclic AMP level mirror those of the physiological responses to octopamine.

The dose-response curve for the octopamine mediated increase in cyclic AMP levels in the extensor-tibiae muscle is unusual in that the rising phase of the sigmoid curve extends over more than 2 log units of concentration before entering the linear portion (Evans, 1984a). This suggests that there may be more than one component to the response. This can be seen more clearly by replotting the data on a log-log plot (Fig.3). This clearly reveals that there are two proportionally related components to the response. Each component has an initial linear functional slope of unity where a ten fold increase in octopamine concentration produces a ten fold change in cyclic AMP accumulation. Thus in these regions of the curve there is no cooperativity between agonist molecules. The existence of these two linear components to the curve joined by a non-linear section could be explained in two ways. It

Fig.3. A log-log dose response curve for the action of octopamine and phenyliminoimidazoline derivations on cyclic AMP levels in the extensor-tibiae muscle. The results are expressed as the increase in cAMP levels in pmol mg^{-1} protein in the experimental muscle above that found in the contralateral control muscle. In experiments using 10^{-4}M isobutylmethylxanthine (IBMX) both experimental and control muscles were preincubated for 10 min in IBMX before exposure of the experimental muscle to agonists plus IBMX for 10 min and the control to a further 10 min incubation in IBMX. Each value is the mean of four determinations and the bars represent standard errors of the mean (from Evans, 1987).

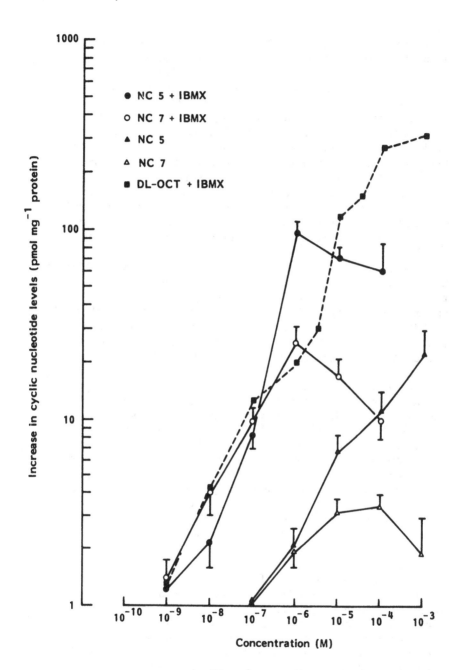

Fig. 3

could mean that the preparation has a single receptor type that increases cyclic AMP levels and that at higher agonist concentrations it may undergo some form of agonist induced configurational change that alters its affinity for octopamine. An alternative explanation is that there are two distinct independent receptor sites involved. However, the relationship between cyclic AMP levels and physiological responses is complex since the increases in cyclic AMP levels could be confined to a functionally distinct subcompartment of muscle fibres within the muscle. A direct test of the latter hypothesis reveals that this is indeed the case (Evans, 1985b) with the regions of the muscle that contain the largest proportions of slow and intermediate muscle fibre types exhibiting the largest octopamine dependent increases in cyclic AMP levels.

At present it is not possible to identify definitely how the two biochemically identified receptor sites relate to the OCTOPAMINE$_{2A}$ and OCTOPAMINE$_{2B}$ receptors identified pharmacologically. However, it seems unlikely that the OCTOPAMINE$_{2A}$ sites on the presynaptic terminals of the slow motorneurone could account for a substantial part of the cyclic AMP accumulation measured in the whole muscle. It thus seems very likely that the higher affinity component of the cyclic AMP accumulation will represent the actions of the OCTOPAMINE$_{2B}$ receptors, since they both have the same thresholds and peak in the same regions of the dose-response curve. It further seems likely that the lower affinity component of the cyclic AMP increases could represent an octopamine mediate increase in cyclic AMP for which we have, as yet, not identified any corresponding physiological effect. Indeed it could even be related to a purely biochemical effect, such as changes in carbohydrate metabolism, that do not have any directly corresponding physiological responses. The idea that the two components of the cyclic AMP increase mediated by octopamine are generated by independent receptor systems receives further support from biochemical studies on the actions of NC5 and NC7 on the extensor-tibiae muscle (Fig.2). These compounds give dose-response curves for cyclic AMP accumulation that superimpose upon the higher affinity component of the octopamine curve, but lack any second component corresponding to the lower affinity octopamine component. Thus NC5 and NC7 although not specific agonists of OCTOPAMINE$_2$ receptors, since they activate OCTOPAMINE$_1$ receptors in the locust, can

differentiate between the receptors mediating the two
components of the octopamine sensitive increase in cyclic
AMP levels in the extensor-tibiae muscle. These compounds
have also been suggested to be able to distinguish
differences in the OCTOPAMINE$_2$ receptors mediating
increases in cyclic AMP levels in a number of other insect
preparations including firefly light organ, cockroach
nerve cord and tobacco hornworm nerve cord (Nathanson,
1985a,b).

At present the exact mode of action of the
OCTOPAMINE$_1$ receptors mediating the slowing of the
myogenic rhythm remains to be elucidated. However, the
bulk of the evidence available to date suggests that they
do not mediate their actions via increases in cyclic AMP
levels. Rather, it is more likely that octopamine
mediates its actions by a mechanism involving the release
of Ca^{++} from intracellular stores perhaps by a mechanism
involving receptor activated breakdown of phospholipids
(see Evans, 1984c).

Thus the extensor-tibiae muscle of the hindleg of the
locust with its identified octopaminergic innervation from
DUMETi has proved extremely useful in the study of the
modulatory actions of octopamine. The octopamine content
of the cell has been measured and its actions shown to
mimic those of applying octopamine exogenously. In
addition the pharmacological profile of the receptor
subtypes mediating the effects of the octopamine released
from DUMETi have been characterised physiologically and
also in terms of the biochemistry of changes induced in
second messenger levels. The generality of the above
findings has been emphasised more recently by the finding
that octopaminergic DUM cells also innervate other
preparations in insects. These include the ovarioles of
the locust (Lange and Orchard, 1984; Orchard and Lange,
1985, 1986), the light organs of the firefly (Christensen
and Carlson, 1981; 1982; Christensen et al., 1983) and the
dorsal longitudinal muscles of the locust (see Whim and
Evans, this volume). In these preparations extensive
pharmacological data is also now available for the
octopamine receptors mediating the action of the released
octopamine.

In the vertebrate nervous system the study of the
functional role of octopamine is hindered by a lack of
identified octopaminergic neurones and a lack of a
demonstration of specific octopaminergic receptors. Here
octopamine is regarded as a trace amine since it is

present in much lower quantities than noradrenaline and only has weak effects on adrenergic receptors. Indeed from studies of octopamine stereoisomers on alpha and beta adrenergic receptor subtypes (Brown et al., 1987; Williams et al., 1987) it has been concluded that if m- and p-octopamine are coreleased with noradrenaline in amounts proportional to their concentration in tissue, then their activities on both alpha and beta adrenergic receptors would be too low for them to be considered physiological. A correlate of this conclusion must be that if octopamine is to play a physiological role in the vertebrate nervous system then it is likely to be on a specific class of octopaminergic receptors. In view of the pharmacological similarities pointed out in this article between insect octopamine receptors and other vertebrate amine receptors, notably alpha-adrenergic receptors, what is urgently required in this field are selective agonists and antagonists for the various subtypes of octopamine receptors demonstrated to be present in insects. It seems very likely that the development of such agents using insect preparations as bioassays may well be useful in the differentiation of specific octopaminergic recepotrs from adrenergic receptors in the vertebrate nervous system, a distinction which is essential for a better understanding of the functional role of octopamine in vertebrates.

Brown, C.M., McGrath, J.C., Midgley, J.M., Muir, A.G.B., O'Brien, J.W., Thonoor, C.M. and Williams, C.M. (1987) Alpha-adrenergic activities of octopamine and synephrine stereoisomers. Brit. J. Pharmacol. (submitted) .

Buck, S.H., Murphy, R.C. and Molinoff, P.B. (1977) The normal occurrence of octopamine in the central nervous system of the rat. Brain Res. 122, 281-297.

Christensen, T.A. and Carlson, A.D. (1981) Symmetrically organized dorsal unpaired median DUM neurones and flash control in the male firefly, Photuris versicolar. J. Exp. Biol. 93, 133-147.

Christensen, T.A. and Carlson, A.D. (1982) The neurophysiology of larval firefly luminescence direct activation through four bifurcating DUM neurons. J. comp. Physiol. 148, 503-514.

Christensen, T.A., Sherman, T.G., McCaman, R.E. and Carlson, A.D. (1983) Presence of octopamine in firefly photomotor neurons. Neuroscience 9, 183-189.

Danielson, T.J., Boulton, A.A. and Robertson, H.A. (1977) m-Octopamine, p-octopamine and phenylethanolamine in rat brain: a sensitive, specific assay and the effects of some drugs. J. Neurochem. 29, 1131-1135.

Dao, W.P.C. and Walker, R.J. (1980) Effect of cyproheptadine on the octopamine-induced responses in the mammalian central nervous system, Experientia 36, 584-585.

David, J.C., Coulon, J.F., Lafon-Cazal, M. and Vinson, D. (1981) Can L-dopa be a precursor of m-octopamine in the cephalic ganglions of the locust Locusta migratoria L? Experientia 37, 804-805.

David, J.C. and Delacour, J. (1980) Brain contents of phenylethanolamine, m-octopamine and p-octopamine in the roman strain of rats. Brain Res. 195, 231-235.

Erspamer, V. and Boretti, G. (1951) Identification and characterization, by paper chromatography, of enteramine, octopamine, tyramine, histamine and allied substances in extracts of posterior salivary glands of Octopoda and in other tissue extracts of vertebrates and invertebrates. Arch. int. Pharmacodyn. 88, 296-332.

Evans, P.D. (1980) Biogenic amines in the insect nervous system. Adv. Insect Physiol. 15, 317-473.

Evans, P.D. (1981) Multiple receptor types for octopamine in the locust J. Physiol. 318, 99-122.

Evans, P.D. (1982) Properties of modulatory octopamine receptors in the locust. Ciba Fdn. Symp. 88, 48-69.

Evans, P.D. (1984a) A modulatory octopaminergic neurone increases cyclic nucleotide levels in locust skeletal muscle. J. Physiol. (Lond.). 348, 307-324.

Evans, P.D. (1984b) The role of cyclic nucleotides and calcium in the mediation of the modulatory effects of octopamine on locust skeletal muscle. J. Physiol. (Lond.). 348, 325-340.

Evans, P.D. (1984c) Studies on the mode of action of octopamine, 5-hydroxytryptamine and proctolin on a myogenic rhythm in the locust. J. exp. Biol. 110, 231-251.

Evans, P.D. (1985a) Octopamine. In Comprehensive Insect Biochemistry, Physiology and Pharmacology. Eds. G.A. Kerkut and L. Gilbert pp499-530, Pergamon Press, Oxford.

Evans, P.D. (1985b) Regional differences in responsiveness to octopamine within a locust skeletal muscle. J. Physiol (Lond.) 366, 331-341.

Evans, P.D. (1987) Phenyliminoimidazolidine derivatives activate both OCTOPAMINE$_1$ and OCTOPAMINE$_2$ receptor subtypes in locust skeletal muscle. J. Exp. Biol. 129, 239-250.

Evans, P.D. and Myers, C.M. (1986) Peptidergic and aminergic modulation of insect skeletal muscle. J. exp. Biol. 124, 143-176.

Evans, P.D. and O'Shea, M. (1977) An octopaminergic neurone modulates neuromuscular transmission in the locust. Nature, Lond. 270, 257-259.

Evans, P.D. and O'Shea, M. (1978) The identification of an octopaminergic neurone and the modulation of a myogenic rhythm in the locust. J. exp. Biol. 73, 235-260.

Evans, P.D. and Siegler, M.V.S. (1982) Octopamine mediated relaxation of maintained and catch tension in locust skeletal muscle. J. Physiol. 324, 93-112.

Harmar, A.J. (1980) Neurochemistry of octopamine. In Modern Pharmacology-Toxicology, Vol 12: Noncatecholic Phenylethylamines, Part 2 Edited by A.D. Mosnaim and M.E. Wolff. pp97-149, Marcel Dekker, New York and Basal.

Hicks, T.P. and McLennan, H. (1978a) Actions of octopamine upon dorsal horn neurones of the spinal cord. Brain Res. 157, 402-406.

Hicks, T.P. and McLennan, H. (1978b) Comparison of the actions of octopamine and catecholamines on single neurones of the rat cerebral cortex. Brit. J. Pharmacol. 64, 485-491.

Hoyle, G. (1975) Evidence that insect dorsal unpaired median (DUM) neurons are octopaminergic. J. exp. Zool. 193, 425-431.

Hoyle, G., Colquhoun, W. and Williams, M. (1980) Fine structure of an octopaminergic neuron and its terminals. J. Neurobiol. 11, 103-126.

Hoyle, G., Dagan, D., Moberly, B. and Colquhoun, W. (1974) Dorsal unpaired median insect neurons make neurosecretory endings on skeletal muscle. J exp Zool 187, 159-165.

Ibrahim, K.E., Couch, M.W., Williams, C.M., Fregly, M.J. and Midgley, J.M. (1985) m-Octopamine: Normal occurrence with p-octopamine in mammalian sympathetic nerves. J. Neurochem. 44, 1862-1867.

Lange, A.B. and Orchard, I. (1984) Dorsal unpaired median neurons, and ventral bilaterally paired neurones, project to a visceral muscle in an insect. J. Neurobiol. 15, 441-453.

Morton, D.B. and Evans, P.D. (1984) Octopamine release from an identified neurone in the locust. J. exp. Biol. 113, 269-287.

Nathanson, J.A. (1985a) Characterization of octopamine sensitive adenylate cyclase: Elucidation of a class of potent and selective octopamine -2 receptor agonists with toxic effects in insects. Proc. Natl. Acad. sci. U.S.A. 82, 599-603.

Nathanson, J.A. (1985a) Characterization of octopamine sensitive adenylate cyclase: Elucidation of a class of potent and selective octopamine-2 receptor agonists with toxic effects in insects. Proc. Natl. Acad. Sci. U.S.A. 82, 599-603.

Orchard, I. (1982) Octopamine in insects: neurotransmitter, neurohormone and neuromodulator. Canad. J. Zool. 60, 659-669.

Orchard, I. and Lange, A.B. (1985) Evidence for octopaminergic modulation of an insect visceral muscle. J. Neurobiol. 16, 171-181.

Orchard, I. and Large, A.B. (1986) Pharmacological profile of octopamine receptors on the lateral oviducts of the locust, Locusta migratoria. J. Insect Physiol. 32, 741-745.

O'Shea, M. and Evans, P.D. (1979) Potentiation of neuromuscular transmission by an octopaminergic neurone in the locust. J. exp. Biol. 79, 169-190.

Robertson, H.A. (1981) Octopamine - after a decade as a putative neuroregulator. In Essays in Neurochemistry and Neuropharmacology, vol. 5 Edited by M.B.H. Youdim, W. Lovenberg, D.F. Sharman and J.R. Lagnado. pp47-73, Wiley, New York..

Talamo, B.R. (1980) Function of octopamine in the nervous system. In Modern Pharmacology-Toxicology, vol. 12, Noncatecholic Phenylethylamines. Part 2 Edited by A.D. Mosnaim and M.E. Wolf. pp261-292, Marcel Dekker, New York and Basel..

Williams, C.M., Couch, M.W. and Midgley, J.M. (1984) Natural occurrence and metabolism of the isomeric octopamines and synephrines. In Neurobiology of the Trace Amines Eds. A.A. Boulton, G.B. Baker, W.G. Dewhurst and M. Sandler pp.97-105, The Humana Press, Clifton, New Jersey.

Williams, C.M., Jordan, R., Thonoor, C.M. and Midgley, J.M. (1987) Beta-adrenergic activities of octopamine and synephrine stereoisomers. J. Pharm. Pharmacol. (in press).

THE FIREFLY LIGHT ORGAN: A MODEL FOR A TRACE AMINE SYNAPSE

James A. Nathanson

Department of Neurology
Harvard Medical School
Massachusetts General Hospital
Boston, Massachusetts 02114

Evidence presented during the past 15 years, from this and other laboratories, strongly supports a role for octopamine as a neurotransmitter and neuromodulator in insects and other invertebrates (Lingle, Marder and Nathanson, 1982; Evans, 1984). Much evidence for octopamine's function as a neurotransmitter has come from studies carried out with the firefly light organ, and it now appears that octopamine may be the major, if not the sole mediator of neural regulation of light emission in this species. As described below, studies of octopamine's action in the firefly have relevance, not only to the control of light emission in the firefly, but also to a greater understanding of octopamine and octopamine receptor pharmacology in general.

OCTOPAMINE RECEPTORS AND CYCLIC AMP

Shortly after the discovery of large amounts of octopamine in invertebrate nerve tissue (Molinoff and Axelrod, 1972; Jurio and Molinoff, 1974) and the demonstration that octopamine could cause electrophysiological effects on snail neurons (Walker et al, 1972), Robertson and Steele (1972; 1973) reported that octopamine could lead to the activation of glycogen phosphorylase in insect nerve cord. Because the analogous action of norepinephrine in stimulating glycogenolysis in vertebrates involved activation of adenylate cyclase, Nathanson and Greengard

(1973; 1974) postulated that certain actions of octopamine might involve the formation of cyclic AMP. They identified an octopamine-sensitive adenylate cyclase which was present in insect nerve cord and enriched in synaptic areas as compared with axons (Nathanson, 1976). Subsequent work by a variety of investigators has confirmed this hypothesis and suggested that a number (although not all) of the receptor actions of octopamine appear to be mediated intracellularly by the formation of cyclic AMP (e.g., see Robertson and Juorio, 1976; Evans, 1984 and this volume).

THE FIREFLY LIGHT ORGAN

Physiological studies indicate that the flash of the firefly adult and the intermittant glowing of the firefly larva are under neural control, originating from a pacemaker present in the animal's brain (see reviews by Carlson, 1969; Case and Strause, 1978; Nathanson, 1986). The light organ, located in the ventral part of the terminal two or three or abdominal segments, receives innervation from the last two abdominal ganglia. The pacemaker sends volleys of electrical impulses down the ventral nerve cord and into the distal abdominal ganglia. Recent studies have indicated that, in both the larva and adult, octopamine-containing dorsal unpaired median (DUM) neurons in these ganglia send processes into the light organ, where they terminate on photocytes (in the larva) or near the junction of photocytes and tracheolar accessory cells (in the adult) (Christensen and Carlson, 1981; 1982; Christensen et al, 1983). Octopamine has been shown to be present in the light organ (Robertson and Carlson, 1976; Copeland and Robertson, 1982; Christensen et al, 1983), and octopamine applied exogenously to either normal or denervated lanterns in larvae and adults causes a bright spontaneous glowing which lasts for several minutes (Carlson, 1968; 1972). Recently, a high affinity reuptake system for octopamine has been observed in larval light organs (Carlson and Evans, 1986), and octopamine release from the larval light organs has been detected under conditions associated with light production (Carlson and Jalenak, 1986).

CYCLIC AMP AS A MEDIATOR OF OCTOPAMINE ACTION
IN THE FIREFLY LIGHT ORGAN

In 1979, Nathanson identified an extremely active octopamine-sensitive adenylate cyclase in the light organs of both larval and adults fireflies (Nathanson 1979; Nathanson and Hunnicutt, 1979a). The degree of stimulation of this enzyme by octopamine is as much as 100-fold (compared to the 2 or 3-fold stimulation seen in insect nerve cord, muscle or brain), making the firefly light organ one of the richest sources of hormone-activated adenylate cyclase in the animal kingdom. The enzyme is enriched in the photocyte layer of the light organ (Nathanson, unpublished observations), and shows a pharmacological profile of hormone activation which is quite similar to that for exogenous compounds in causing light emission. For example, the rank order potency of phenylethylamines in stimulating both the enzyme and light emission is synephrine > octopamine > epinephrine > norepinephrine > tyramine > phenylethanolamine > phenylethylamine > dopamine (Nathanson and Hunnicutt, 1979b).

The correlation between enzyme activation and light emission extends, also, to synthetic octopamine analogs. For example, N-demethyl-formamidine, which is 3-5 times more potent than octopamine in stimulating light organ adenylate cyclase (Nathanson and Hunnicutt, 1981), is similarly more potent than octopamine in stimulating light emission (idem; Hollingworth and Murdock, 1980). Likewise, 2,6-diethylphenyliminoimidazolidine, which is about 20 times more potent than octopamine in activating light organ adenylate cyclase, is about 15-20 times more potent than octopamine in stimulating light emission (Nathanson, 1985a,b). This latter compound is the most potent octopamine agonist yet identified for the firefly light organ. These and related data indicate that there is a strong association between the ability of agents to stimulate cyclic AMP production in the lantern and their ability to elicit light emission. However, these experiments do not rule out the possibility that cyclic AMP production might be an epiphenomenon unrelated to the action of these agents at some non-cyclic AMP-associated octopamine receptor.

To investigate this latter possibility, we have undertaken additional studies using cholera toxin, a

substance which is able to activate adenylate cyclase
independently of receptor stimulation (Nathanson, 1985c).
Cholera toxin causes a delayed but irreversible inhibition
of the GTPase activity of the G-regulatory protein
involved in adenylate cyclase activation. Normally, as a
result of receptor activation, GTP binds to this protein,
thereby allowing the protein, in turn, to bind to and
activate adenylate cyclase. When GTP is hydrolyzed, the
G-protein no longer binds to adenylate cyclase, and cyclic
AMP synthesis is turned off. However, when GTPase activity is inhibited by cholera toxin, GTP remains intact and
the G-protein keeps the adenylate cyclase in a persistently active state with continued synthesis of cyclic AMP.

We have observed that, when cholera toxin is injected
into living fireflies which are not spontaneously flashing, light organ adenylate cyclase becomes irreversibly
activated after about eight hours. Coincident with this,
light organs begin to emit light spontaneously, gradually
increasing in intensity until the light output equals that
seen during the peak of a spontaneous flash. This level
of light emission is then maintained for many hours, until
substrates for light production are exhausted. If cyclic
AMP levels are measured during the period of maximal light
emission, tissue content is elevated about 8-fold over the
basal levels seen in uninjected resting fireflies.

From what is known of cholera toxin action in other
tissues, it would be predicted that the rate at which
adenylate cyclase is irreversibly activated would be
related to the rate at which GTP is bound to the
G-protein. In the absence of receptor stimulation, this
binding should occur slowly over many minutes, whereas in
the presence of receptor stimulation, the rate of GTP
binding should be greatly increased. We have actually,
indirectly, observed this acceleration of GTP binding by
stimulating cholera toxin-pretreated fireflies to emit a
spontaneous flash about 4-6 hours after being treated with
the toxin, at a time when they are not yet glowing. By
flashing, the fireflies presumably release endogenous
octopamine which activates light organ octopamine receptors. (Fireflies usually show little or no spontaneous
flashing in the laboratory, but can be stimulated to flash
by pinching an antenna or leg.) We have observed that,
with each flash, light illumination suddenly increases.
However, instead of returning to the original level of

illumination, the light output remains partially elevated. With each succeeding flash, there is an additional increment of irreversible increase in illumination. The effect of flashing is like a rachet, increasing the level of light output by steps, until the lantern is brilliantly and permanently aglow. Presumably, with each flash, the fireflies are binding more and more GTP to the G-regulatory proteins. Once bound, the GTP can not be hydrolyzed (because of the effect of the toxin) and more and more adenylate cyclase is placed into an active state. This type of activation is in agreement with second messenger theory and further serves to support the hypothesis that octopamine-stimulated cyclic AMP formation is an integral part of the chain of events which results in the initiation of light production. This hypothesis is also consistent with prior observations of Oertel and Case (1976) showing that theophylline, an inhibitor of cyclic AMP breakdown, can cause spontaneous glowing of larval lanterns. We have confirmed these results of Oertel and Case under conditions which block the calcium-mobilizing effects of theophylline, and have shown, further, that phosphodiesterase inhibitors like theophylline can act synergistically to increase the potency of exogenous octopamine in stimulating light emission from isolated lanterns.

Recent studies by Nathanson and Carlson (1987) have provided additional evidence for the involvement of cyclic AMP production in the neural control of light emission. In these experiments, a larval lantern and its associated lantern nerve were isolated, and the nerve was electically stimulated to cause glowing of the lantern. The lantern was then immediately frozen and the amount of cyclic AMP present was determined. Among a large number of lanterns stimulated at different intensities, there was a highly significant correlation between the degree of light produced and the content of cyclic AMP present in the lantern. These results, together with the data described previously, strongly support a role for octopamine and cyclic AMP in mediating the neural control of light emission in the firefly.

ACKNOWLEDGEMENTS

I wish to thank Edward J. Hunnicutt and Christopher Owen for technical assistance. This work was supported, in part, by USDA 8600090 and by grants from the McKnight Foundation and the Katharine Daniels Research Fund.

REFERENCES

Carlson A.D. (1968) Effect of drugs on luminescence in larval fireflies. J. exp. Biol. 49, 195-199.

Carlson A.D. (1972) A comparison of transmitter and synephrine on luminescence induction in the firefly larva. J. exp. Biol. 57, 737-743.

Carlson A.D. (1969) Neural control of firefly lumines-cence. Advances in Insect Physiology 6, 51-96.

Carlson A.D. and Evans P.D. (1986) Inactivation of octopamine in larval firefly light organs by a high affinity uptake mechanism. J. exp. Biol. 122, 369-385.

Carlson A.D. and Jalenak M. (1986) Release of octopamine from the photomotor neurons of the larval firefly lantern. J. exp. Biol. 122, 453-457.

Case J.F. and Strause L.G. (1978) Neurally controlled luminescent systems. In, Bioluminescence in Action (Herring, P.J., ed) pp. 331-366, Academic Press, New York.

Christensen T.A. and Carlson A.D. (1981) Symmetrically organized dorsal unpaired median DUM neurones and flash control in the male firefly, Photuris versicolor. J. exp. Biol. 93, 133-147.

Christensen T.A. and Carlson A.D. (1982) The neurophysi-ology of larval firefly luminescence: direct activation through four bifurcating DUM neurons. J. comp. Physiol. 148, 503-514.

Christensen T.A., Sherman T.G., McCaman R.E. and Carlson A.D. (1983). Presence of octopamine in firefly photomotor neurons. Neuroscience 9, 183-189.

Copeland J. and Robertson H.A. (1982) Octopamine as the transmitter at the firefly lantern: Presence of an octopamine-sensitive and dopamine-sensitive adenylate cyclase. Comp. Biochem. Physiol. 72C, 125-127.

Evans P.D. (1984) The role of cyclic nucleotides and calcium in the mediation of the modulatory effects of octopamine on locust skeletal muscles. J. Physiol. (Lond.) 348, 325-340.

Hollingworth R.M. and Murdock L.L. (1980) Formamidine pesticides: octopamine-like action in a firefly. Science 208, 74–76.

Jurio A.W. and Molinoff P.B. (1974) The normal occurrence of octopamine in neural tissues of the octopus and other cephalopods. J. Neurochem. 22, 271–280.

Lingle C.J., Marder E. and Nathanson J.A. (1982) The role of cyclic nucleotides in invertebrates. In, Handbook of Experimental Pharmaoclogy, Vol 58/II (Kebabian, J.W. and Nathanson, J.A., eds) pp. 787–845, Springer-Verlag, New York.

Molinoff P.B. and Axelrod J. (1972) Distribution and turnover of octopamine in tissues. J. Neurochem. 19, 157–163.

Nathanson J.A. (1976) Octopamine-sensitive adenylate cyclase and its possible relationship to the octopamine receptor. In, Trace Amines and the Brain (Usdin, E. and Sandler, M., eds) pp. 161–190, Marcel Dekker, New York.

Nathanson J.A (1979) Octopamine receptors, adenosine 3',5'-monophosphate, and neural control of firefly flashing. Science 203, 65–68.

Nathanson J.A. (1985a) Characterization of octopamine-sensitive adenylate cyclase: elucidation of a class of potent and selective octopamine-2 receptor agonists with toxic effects in insects. Proc. Natl. Acad. Sci. USA 82, 599–603.

Nathanson J.A. (1985b) Phenyliminoimidazolidines: characterization of a class of potent agonists of octopamine-sensitive adenylate cyclase and their use in understanding the pharmacology of octopamine receptors. Mol. Pharmacol. 28, 254–268.

Nathanson J.A. (1985c) Cholera toxin, cyclic AMP, and the firefly flash. J. Cyclic Nucleotdie and Protein Phosphor. Res. 10(2), 157–166.

Nathanson J.A. (1986) Neurochemical regulation of light emission from photocytes. In, Insect Neurochemistry and Neurophysiology 1986 (Borkevec, A. and Gelman, D. eds) pp. 263–266, Humana Press, Clifton, N.J.

Nathanson J.A. and Carlson A.D. (1987) Induction of cyclic AMP by electrical stimulation of lantern nerves in the larval firefly lantern, submitted for publicatoin.

Nathanson J.A. and Greengard P. (1973) Octopamine-sensitive adenylate cyclase: evidence for a biological role of octopamine in nervous tissue. Science 180, 308–331.

Nathanson J.A. and Greengard P. (1974) Serotonin-sensitive adenylate cyclase in neural tissue and its similarity of the serotonin receptor: a possible site of action of lysergic acid diethylamide. Proc. Natl. Acad. Sci. USA 71, 797-801.

Nathanson J.A. and Hunnicutt E.J. (1979a) Neural control of light emission in Photuris larvae: identification of octopamine-sensitive adenylate cyclase. J. exp. Zool. 208, 255-262.

Nathanson J.A. and Hunnicutt E.J. (1979b) Octopamine-sensitive adenylate cyclase: properties and pharmacological characterization. Soc. Neurosci. Abstr. 5, 346.

Nathanson J.A. and Hunnicutt E.J. (1981) N-demethylchlordimeform: a potent partial agonist of octopamine-sensitive adenylate cyclase. Mol. Pharmacol. 20, 68-75.

Oertel D. and Case J.F. (1976) Neural excitation of the larval firefly photocyte: slow depolarization possibly mediated by a cyclic nucleotide. J. exp. Biol. 65, 213-227.

Robertson H.A. and Juorio A.V. (1976) Octopamine and some related noncatechol amines in invertebrate nervous systems. Intl. Rev. Neurobiol. 19, 173-224.

Robertson H.A. and Steele J.E. (1972) Activation of insect nerve cord phosphorylase by octopamine and adenosine 3',5'-monophosphate. J. Neurochem. 19, 1603-1605.

Robertson H.A. and Steele J.E. (1973) Effect of monophenolic amines on glycogen metabolism in the nerve cord of the American cockroach, Periplaneta americana. Insect Biochem. 3, 53-59.

Walker R.J., Ramage A.G. and Woodruff G.N. (1972) The presence of octopamine in brain of Helix aspersa and its action on specific snail neurones. Experientia (Basel) 28, 1173-1174.

THE REGULATION OF INSECT VISCERAL MUSCLE BY OCTOPAMINE

Ian Orchard and Angela B. Lange

Dept. Zoology, University of Toronto,

Toronto, Ontario, Canada, M5S 1A1

INTRODUCTION

Insect visceral muscles are striated, yet they display properties similar to smooth muscle of vertebrates, with contractions that are slow and rhythmic and often co-ordinated to form peristaltic waves (Davey, 1964; Miller, 1975; Lange et al., 1984). Whilst these contractions may continue spontaneously when the muscles are isolated from the central nervous system, they can be modified by hormonal and/or nervous input. Insect visceral muscles have therefore provided useful preparations for studying the pharmacological activities of a variety of putative neuroactive chemicals. These studies suggest some interesting facets of control, in that many visceral muscles have been shown to be extremely sensitive to both peptides and to biogenic amines (see Cook and Holman, 1979; Lange and Orchard, 1984a,b). It seems possible, therefore, that insect visceral muscles may provide useful model systems for the examination of both aminergic and peptidergic regulation. Unfortunately, there is a general lack of information regarding identified neurons which innervate insect visceral muscle, and until such studies are done, and the neurons shown to contain these amines or peptides, there will remain doubt as to the physiological significance of these pharmacological observations.

With this in mind we set out to define the neural control of an insect viseral muscle with a view to

41

establishing a model system for the examination of peptidergic and aminergic regulatory mechanisms. We identified neurons containing the pentapeptide proctolin (H-Arg-Tyr-Leu-Pro-Thr-OH), and neurons containing the monophenolic amine octopamine (1-(p-hydroxyphenyl) -2-aminoethanol) innervating the visceral musculature of the oviducts of the locust, Locusta migratoria (Lange and Orchard, 1984b; Orchard and Lange, 1985; Lange et al., 1986). The present paper will review the information pertaining to the octopaminergic modulation of this insect visceral muscle.

NEURAL ORGANISATION

The lateral and common oviducts of Locusta are innervated from the penultimate abdominal ganglion. Each sternal nerve bifurcates soon after leaving the ganglion, sending one branch, the oviducal nerve, towards the oviducts (see Fig. 1A). Cobalt backfilling of the oviducal nerves reveals the presence of 3 ventral, bilaterally paired motoneurons (OVN1-3) and 2 dorsal, unpaired median neurons (DUMOV1 and 2) (Fig. 1B) within the ganglion. The OVN1-3 will not be dealt with here, but are the neurons believed to contain the neuropeptide proctolin (Lange et al., 1986). DUMOV 1 and 2 have large cell bodies (50-70 µm diameter) and possess both ipsilateral and contralateral axons emerging from the ganglion and projecting to the oviducts.

PRESENCE AND RELEASE OF OCTOPAMINE

The electrophysiological and morphological properties of DUMOV 1 and 2 (Lange and Orchard, 1984b) are similar to those of an identified octopaminergic neuron (DUMETi) found in the metathoracic ganglion of locusts (Evans and O'Shea, 1978). We therefore examined the octopaminergic nature of the DUMOV neurons using the sensitive radioenzymatic assay of Molinoff et al., (1969) as modified by Orchard and Lange (1985). Table 1 illustrates the octopamine content of various structures associated with the oviducts. As can be seen, octopamine is present in the DUMOV cell bodies (assay performed on microdissected somata), oviducal nerves, and innervated regions of the oviducts. Assuming the cell

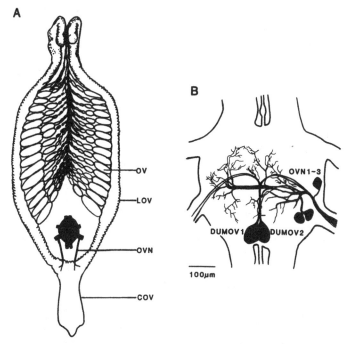

Figure 1A. The female reproductive tract of <u>Locusta</u> <u>migratoria</u>. B. Cell bodies and main arborisations of neurons with axons passing to the oviducts (based upon camera lucida drawings of cobalt backfillings). COV, common oviduct; DUMOV1,2, dorsal unpaired median neurons innervating the oviducts; LOV, lateral oviduct; OV, ovaries; OVN, oviducal nerve; OVN1-3, oviducal neurons 1-3. Redrawn from Lange <u>et</u> <u>al</u>. (1984); Lange and Orchard (1984b).

Table 1. Octopamine content of structures associated with the oviducts of Locusta migratoria.

Structure	pmol	nmol.g^{-1}wet wt
DUMOV1 (per soma)	0.35 ± 0.14	1310 ± 522
DUMOV2 (per soma)	0.33 ± 0.09	1231 ± 335
Oviducal nerve (per cm)	1.62 ± 0.30	---
Oviducal muscle (innervated region)	9.3 ± 2.1	13.5 ± 4.5

Data from Orchard and Lange (1985)

bodies to be spheres of approx. 80 μm diameter with a tissue density of 1 g.cm^{-3}, then the concentration of octopamine within the soma is 1.27 x 10^{-3}M. This is comparable to other estimates for identified octopaminergic neurons (range 8.9 x 10^{-4}M – 2.8 x 10^{-3}M) (see Evans and O'Shea, 1978; Christensen et al., 1983).

Incubation of oviducts of Locusta in high potassium saline (100 mM) results in a calcium–dependent release of octopamine (Orchard and Lange, 1987a). Approximately 19% of the total content of octopamine may be released by a 5 min incubation in high potassium saline. Using more physiological stimuli (Orchard and Lange, 1987a) we have been able to demonstrate a calcium–dependent, neurally–evoked release of octopamine from oviducts using axonal stimulation or intrasomatic stimulation. More recently (Orchard and Lange, 1987b) we have shown that the oviducts of the cockroach, Periplaneta americana, also contain considerable amounts of octopamine (2.6 ± 0.3 pmol/oviduct; 31.0 ± 0.3 pmol/mg protein), with 60 mM high potassium saline capable of releasing about 14% of the total store of octopamine. These results would suggest a generalized control of insect oviducts by octopamine.

PHYSIOLOGICAL EFFECTS

The oviducts of locusts are myogenically active as judged by the persistance of rhythmic contractions when isolated from the central nervous system. Electrical stimulation of the motor axons in the oviducal nerves results in contraction of the lower lateral and upper common oviducts. Excitatory junction potentials (EJPs) evoked by nerve stimulation have rise times of 35–75 msec and decay times of 150–300 msec and are confined to the region of the oviduct lying between the upper common oviduct and the insertion of the ovarioles (Orchard and Lange, 1986a). The amplitude of evoked EJPs increases stepwise with a maximum of 3 steps, illustrating polyneuronal innervation from three motor units. Summation and facilitation are evident.

The presence of octopamine within DUMOV 1 and 2 and its release from the oviducts following electrical stimulation indicate that octopamine plays a regulatory role on the oviducts. The physiological effects of octopamine were revealed following exogenous application of octopamine to oviducts attached to a force transducer. Octopamine reversibly reduces the amplitude of neurally–evoked contractions, lowers basal tension and inhibits myogenic contractions (Lange and Orchard, 1984a; Orchard and Lange, 1985). In addition, octopamine results in a hyperpolarisation of membrane potential (5 mV) and a reduction in the amplitude of EJPs (Orchard and Lange, 1986a). The effects of octopamine upon the amplitude of neurally–evoked contractions is dose-dependent (Fig. 2) with a threshold between 5×10^{-10} and 7×10^{-9} and half–maximal at about 5×10^{-7} M. The order of potency of a variety of amines is octopamine = synephrine > metanephrine > tyramine > dopamine (which has no effect at 5×10^{-6} M). Nor-epinephrine and serotonin have the opposite effect to octopamine, increasing both the amplitude of neurally–evoked contractions and frequency of myogenic contractions.

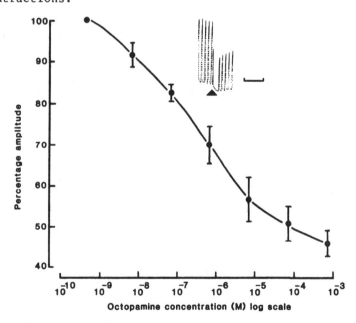

Figure 2. Dose–response curve for the effect of octopamine on the amplitude of neurally-evoked contractions of lateral oviducts. Inset shows an example of the reduction in amplitude of neurally-evoked contractions induced by 10^{-6} M octopamine. Time scale, 1 min. Redrawn from Orchard and Lange (1985); Lange and Orchard (1984a).

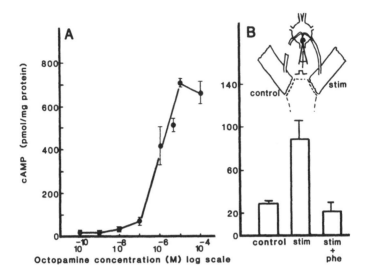

Figure 3. A) Dose response curve for the effect of octopamine (in the presence of 5 x 10^{-4} M IBMX) on cyclic AMP content of lateral oviducts. B) One minute stimulation of a DUMOV cell body elevates cylcic AMP content of lateral oviducts. The effect is blocked in the presence of 5 x 10^{-6} M phentolamine (phe). Redrawn from Lange and Orchard (1986).

MODE OF ACTION

There has been increasing evidence for the presence of octopamine-sensitive adenylate cyclase in invertebrate tissue. The elevated levels of cyclic AMP induced by octopamine are believed to act as the cellular mediators for the physiological effects (see Lange and Orchard, 1986). To examine for a similar phenomenon in insect visceral muscle we have performed a series of experiments involving cyclic AMP in locust oviducts. Octopamine, in the presence of the phosphodiesterase inhibitor 3-isobutyl-1-methylxanthine (IBMX), elevates the cyclic AMP content of the lateral oviducts of Locusta in a dose-dependent manner (Lange and Orchard, 1986; Fig. 3A). A two-fold elevation in cyclic AMP is induced by 10^{-8} M octopamine and half-maximal stimulation by 7 x 10^{-7}M. Maximal stimulation (35 fold) occurs with 10^{-5}M octopamine.

In addition intrasomatic stimulation of one DUMOV results in an accumulation of cyclic AMP in the lateral oviducts in the presence of IBMX (Fig. 3B). The specificity of the response of the lateral oviducts to a variety of amines reveals the same order of potency as for the physiological effect described earlier. In view of the ability of both octopamine and DUMOV stimulation to elevate cyclic AMP levels in the lateral oviducts we examined the possibility that cyclic AMP may mediate the physiological effects of octopamine. Elevation of cyclic AMP levels in the lateral oviducts by means of IBMX, the diterpene forskolin, or dibutyryl cyclic AMP, mimics the major physiological effects of octopamine (Lange and Orchard, 1986). Thus the effects of octopamine upon locust oviducts, like other insect systems (see Orchard and Lange, 1987c) appear to be mediated via cyclic AMP.

PHARMACOLOGY

The pharmacological properties of octopamine receptors mediating an elevaion in cyclic AMP in the lateral oviducts of locusts has recently been examined (Orchard and Lange, 1986b). The α-adrenergic agonists clonidine and naphazoline are weak agonists of these receptors, with tolazoline showing no agonistic activity at 10^{-5}M. The formamidine, demethylchlordimeform is a potent though partial agonist (Fig. 4), with no additive effect at maximal concentrations of octopamine. A number of aminergic antagonists are capable of blocking octopamine-mediated elevations in cyclic AMP in a competitive manner. Phentolamine is a potent antagonist with an IC_{50} of 6.5 x 10^{-8}M when tested against 10^{-6} M octopamine. The IC_{50} values indicate a rank order of potency of phentolamine > gramine > metoclopramide > mianserin > cycloheptadine > yohimbine. The rank order against demethylchlordimeform is the same for selected antagonists. These octopamine receptors, therefore, share many properties with the octopamine-2 receptor described by Evans (1981), revealing greater sensitivity to metoclopramide than to chlorpromazine or yohimbine, and apparently being linked to an adenylate cyclase.

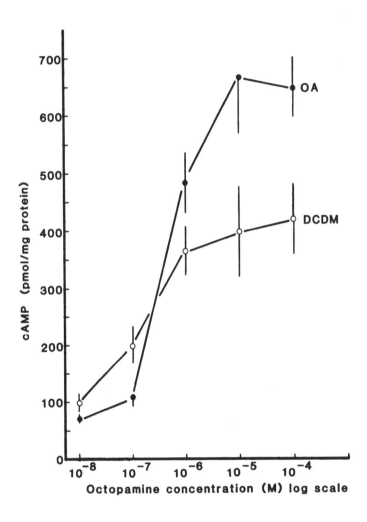

Figure 4. Dose-response curves for the effects of
octopamine (OA) and demethylchlordimeform (DCDM) on cyclic
AMP content of lateral oviducts. All values obtained in
the presence of 5×10^{-4} M IBMX. Redrawn from Orchard
and Lange (1986b).

CONCLUSIONS

The visceral musculature of the oviducts of the locust, Locusta migratoria, receive innervation from identified octopaminergic neurons. These neurons appear to be of a similar class to other identified insect octopaminergic neurons, ie; they appear to be dorsal, unpaired median neurons. Octopamine is released from these DUMOV neurons and induces physiological effects upon locust oviducts. Contraction properties are altered in such a way as to relax the muscle and make it less sensitive to excitatory motor input. The mechanism for at least some of these effects appear to be via the second messenger cyclic AMP. The receptors appear to be coupled to adenylate cyclase, and pharmacologically may be classified as octopamine-2 receptors (Evans, 1981). Of particular importance is the fact that these receptors do not conform to any one recognised vertebrate receptor. Thus, while being sensitive to α-antagonists, the octopamine receptor is also antagonised by vertebrate serotonergic and histaminergic antagonists such as cyproheptadine, gramine and mianserin (Evans, 1981; Orchard and Lange, 1986b). Furthermore, in contrast to octopamine receptors, vertebrate α-receptors are not typically coupled to adenylate cyclase.

The body of work now available upon the locust oviducts reveals that they are an ideal insect visceral muscle for studying the actions of octopamine, and present a model system for future studies upon the interactions of aminergic and peptidergic regulators.

This work was supported by the NSERC of Canada.

REFERENCES

Christensen T.A., Sherman T.G., McCaman R.E. and Carlson A.D. (1983) Presence of octopamine in firefly photomotor neurons. Neuroscience 9, 183-189.
Cook B.J. and Holman G.M. (1979) The pharmacology of insect visceral muscle. Comp. Biochem. Physiol. 64C, 183-190.
Davey K.G. (1964) The control of visceral muscles in

insects. In, Advances in Insect Physiology, Vol. 2
(Beament J.W.L., Treherne J.E. and Wigglesworth V.B.,
eds.) pp. 219-245,Academic Press, London.
Evans P.D. and M. O'Shea (1978) The identification of an
octopaminergic neurone and the modulation of a
myogenic rhythm in the locust. J. Exp. Biol. 73,
235-260.
Evans P.D. (1981) Multiple receptor types for octopamine
in the locust. J. Physiol. 318, 99-122.
Lange A.B. and Orchard I. (1984a) Some pharmacological
properties of neuromuscular transmission in the
oviduct of the locust, Locusta migratoria. Arch.
Insect Biochem. Physiol. 1, 231-241.
Lange A.B. and Orchard I. (1984b) Dorsal unpaired median
neurons, and ventral bilaterally paired neurons,
project to a visceral muscle in an insect.
J. Neurobiol. 15, 441-453.
Lange A.B., Orchard I. and Loughton B.G. (1984)
Spontaneous and neurally evoked contractions of
visceral muscles in the oviduct of Locusta migratoria.
Arch. Insect Biochem. Physiol. 1, 179-190.
Lange A.B. and Orchard I. (1986) Identified octopaminergic
neurons modulate contractions of locust visceral
muscle via adenosine 3',5'-monophosphate (cyclic
AMP). Brain Res. 363, 340-349.
Lange A.B., Orchard I. and Adams M.E. (1986) Peptidergic
innervation of insect reproductive tissue: The
association of proctolin with oviduct visceral
musculature. J. Comp. Neurol. 254, 279-286.
Miller T. (1975) Insect visceral muscle. In, Insect
Muscle (Usherwood, P.N.R., ed.) pp 545-606, Academic
Press, New York.
Molinoff P.B., Landsberg L. and Axelrod J. (1969) An
enzymatic assay for octopamine and other
β-hydroxylated phenylethylamines. J. Pharmac. Exp.
Ther. 170, 253-261.
Orchard I. and Lange A.B. (1985) Evidence for
octopaminergic modulation of an insect visceral
muscle. J. Neurobiol. 16, 171-181.
Orchard I. and Lange A.B. (1986a) Neuromuscular
transmission in an insect visceral muscle.
J. Neurobiol. 17, 359-372.
Orchard I. and Lange A.B. (1986b) Pharmacological profile
of octopamine receptors on the lateral oviducts
of the locust, Locusta migratoria. J. Insect Physiol.

32, 741-745.

Orchard I. and Lange A.B. (1987a) The release of octopamine and proctolin from an insect visceral muscle: effects of high-potassium saline and neural stimulation. Brain Res. (in press).

Orchard I. and Lange A.B. (1987b) Cockroach oviducts: The presence and release of octopamine and proctolin. J. Insect Physiol. 4, 265-268.

Orchard I. and Lange A.B. (1987c) Octopamine in insects, with special reference to the control of haemolymph lipid and visceral muscle in locusts. In, Toxicants affecting GABA, octopamine, and other neuroreceptors in invertebrates, ACS Symposium Series (Green M., Hollingworth R. and Hedin P.A., eds.), ACS books, (in press).

MONOAMINES AS TARGETS FOR INSECTICIDE DISCOVERY

K. R. Jennings, D. G. Kuhn, S. Trotto and W. K. Whitney
Agricultural Research Division
American Cyanamid Company
P. O. Box 400
Princeton, NJ 08540

Although many chemical insecticides are available for crop protection, grain storage, homeowner and public health uses, the majority of these materials fall into a few small classes of chemistry when classified by modes of action. When these insecticide classes are examined, it becomes clear that the majority of insecticides in use today exert their toxic action on the insect nervous system.

The insecticides active on the nervous system are generally viewed as relatively fast-acting toxins, protecting crops from feeding damage by a rapid insect kill. One class of insecticide/acaricide that acts on nerve and muscle targets produces a unique set of poisoning symptoms in the pest that disrupts behavior in a characteristic manner. This class of chemistry, the formamidines, is represented by the cotton insecticide chlordimeform and the tickicide amitraz. This class of chemicals produce behavior such as spin down and walk off in mites on treated leaves, a pronounced antifeedant effect in larval Lepidoptera exposed to treated leaves and hyperactivity in moths. These materials also possess ovicidal properties that are thought to result from a direct interference with the neuromuscular events associated with the neonate larva emerging from the egg.

It is believed that the formamidines exert their effects by acting as octopamine agonists in invertebrates (Hollingworth and Murdock, 1980; Nathanson and Hunnicutt, 1981). The symptoms of poisoning are consistent with

53

stimulation of the octopaminergic system and the pharmacological studies conducted on insects demonstrate that the
active, demethyl forms of the formamidines are potent
agonists of both OA-1 and OA-2 receptor types.

Two other major biogenic amine transmitter systems
exist in insects for which there is less information on
function and no known insecticide class that exerts its
effect on these amines. These are dopamine and serotonin
which are localized to the central nervous system and are
known to have pharmacological actions in insects and have
been shown to play a role in some neuroeffector systems such
as the salivary gland (Evans, 1980; Omar et al., 1982).

When defining possible target sites for insecticide
design it is valuable to be able to identify the target site
as a critical function, which when perturbed, will lead to
effective control of the insect pest in the field. One
approach to identifying a critical lesion in insects is the
use of genetic mutants – this strategy has been very
successfully applied in Drosophila both with behavioral
mutants and mutations of biogenic amine metabolism. In
Drosophila, 40 genetic loci have been identified as being
involved in biogenic amine metabolism yet only 11 of these
loci have a clearly identified function (Wright, 1987).
Many of these genes as mutants are lethal and clearly serve a
vital function – often a single mutant gene is pleiotropic,
with disturbances evident in both behavioral and developmental processes. While it is often a priori difficult to
demonstrate this, the precedent of having a known pesticide
that acts at that site can provide an impetus to work in a
given area.

Catecholamines have an important role to play in
sclerotization in addition to other roles as a neurotransmitter and this may represent a plausible target for future
insecticides. Sclerotization, which involves the incorporation and oxidation of catecholamines in the cuticle to cross
link proteins and chitin is important throughout the
development of an insect, but especially so at eclosion and
molting. The benzoylurea insecticides control insects by
perturbing cuticle formation during molting (Mitsui, 1985)
and represent one case example of how a developmentally
targeted insecticide would perform. The benzoylureas,
typified by diflubenzuron, although effective against
certain pests and having a reduced potential for affecting
predator populations due to the poor contact action, have
had limited commercial success. The reason for this is the

perceived slowness of action of these materials by the farmer in contrast to the faster acting nerve poisons such as the pyrethroids, phosphates and carbamates. This may represent a potential limitation of novel biogenic amine targeted insecticides that disrupt cuticular function without also exerting a more immediate pesticidal action by interfering with behavioral processes.

Compared with the other biogenic amines, a great deal more is known about octopamine. Looking at octopaminergic function from the point of insect control - the question can be asked "What aspects of octopaminergic transmission could be important targets for insect control?" If you consider the function of octopamine as a neuromodulator of muscle function, it is not clear that just any disruption of function would be critical to the insect. In the case of acelytcholine function we know that inhibiting breakdown of the transmitter (Eldefrawi, 1985), activating the receptor by agonists (Eldefrawi, 1985) or blocking the receptor with antagonists (Jennings et al., 1987) can lead to death. With octopamine, use of a receptor antagonist may not be a useful strategy. If octopamine is not normally required for muscle function but serves a modulatory function, fine tuning or potentiating normal muscular responses, then blockade of the receptor would not be expected to produce a critical physiological change in the insect.

The formamidine chlordimeform (CDM) is a weak octopamine receptor agonist/antagonist in the parent form. It is the demethyl metabolite (DCDM), which is a potent octopamine agonist activating the receptor, that disrupts behavior and leads to effective pest control. Similarly, a material which would prevent the normal inactivation (uptake) process for this biogenic amine would be expected to cause overactivity of the octopamine synapses and lead to hyperactivity, hypereflexia, muscular incoordination and all the other sequelae of poisoning by formamidines. Other, less specific targets include the various enzymes of the second messenger systems (Evans, 1985).

With this reasoning and the example of the formamidine agonist class we decided to examine a series of different chemical classes for agonist activity at the octopamine receptor.

In view of the pronounced miticidal effects (including the unique behavioral actions observed with the formamidines) it was decided that a series of organic chemicals with known miticidal activity but not representing any known class of insecticide or miticide would be chosen for screening purposes. Screening was carried out against the biogenic amine receptors linked to adenyl cyclase enyzme prepared from the Periplaneta americana nerve cord and pharmacological studies and additivity experiments carried out to identify the correct biogenic amine receptor: DA, 5HT or OA.

In the process of this screening, a compound was identified (1) with agonist properties at the adenylcyclase linked-octopamine receptor. This compound, an amino-oxazoline was also known to have miticidal action, killing Tetranychus uriticae at 100 PPM. When related structures were examined it was found that molecules of this general type were agonists of the octopamine receptor in the cockroach nerve cord having Ka in the range of 10^{-5} - 10^{-6} M. It was also found that non-phenyl structures of this type displayed octopaminergic activity, including cyclohexyl cis and trans forms:(2) Interestingly, the trans form was substantially more active as an octopamine agonist than the cis form. Because of the knowledge that this class of miticide was exerting its effect at the octopaminergic receptor in a manner similar to the formamidine insecticide demethylchlordimeform, it was decided to synthesize the 2-methyl-4-chloro analogue of the aminooxazoline class: 2-methyl-4-chloro phenyl aminooxazoline (3).

This molecule was very active as an octopaminergic agonist, having a Ka=10^{-6}M. In addition, it showed enhanced ovicidal activity against Lepidoptera over the other members of the oxazoline chemical series. In specially designed tests to look for the characteristic symptomology of formamidine poisoning, this chemical displayed a number of effects seen with chlordimeform. These included hyperactivity in the moths Spodoptera eridania, (southern armyworm) and Heliothis virescens ovicidal activity against Heliothis virescens (tobacco budworm), and S. eridania eggs and antifeedant properties at high concentrations against S. eridania (Table 1). Direct toxicity to Aphis fabae (bean aphid) was also observed. In addition to these formamidine-like properties, when injected into firefly adults the aminooxazolines activated the light organ, causing a pronounced, long-lasting glow. This evidence confirms in vivo, the octopamine agonist activity seen in the cockroach (Periplaneta americana) nerve cord receptor preparation (Fig. 2) using the techniques described by Nathanson and Greengard (1973). The aminooxazolines were also potent

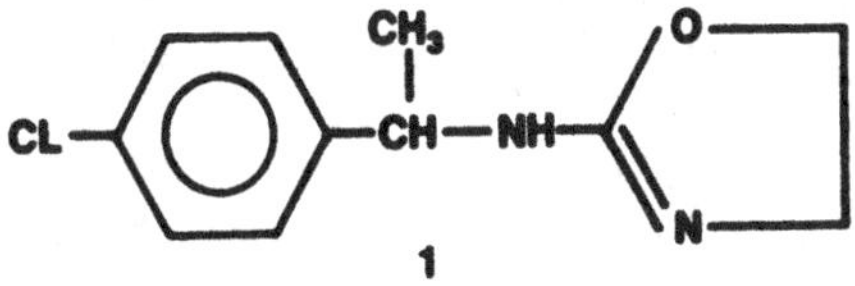

1

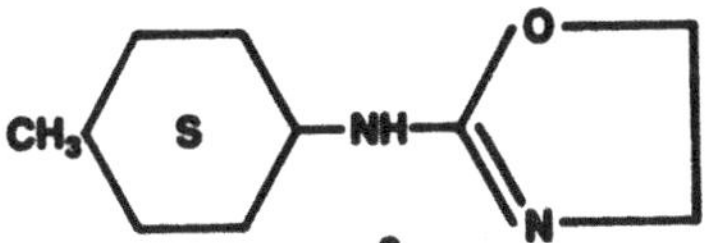

2

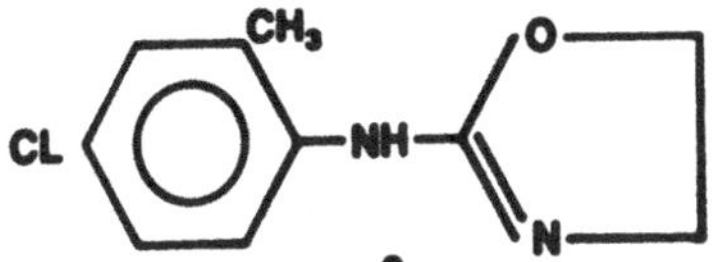

3

Figure 1. Aminooxazoline Structures

agonists on the lepidopteran (S. eridania) octopamine receptor.

Table 1. Effects of Three Octopaminergic Compounds on Insects

Compound	% Reduced Feeding by *S. eridania*				% Unhatched Eggs of *S. eridania*				% Unhatched Eggs of *H. virescens*				% Kill of *A. fabae*		
	1000	100	10	1ppm	1000	100	10	1ppm	1000	100	10	1ppm	300	100	10ppm
3	97	30	1	–	100	100	57	–	81	62	28	–	100	98	90
2(Trans)	25	10	5	–	86	91	14	–	44	49	4	–	0	–	–
CDM	65	20	5	–	100	100	100	84	98	75	50	45	94	48	26

(– = not tested)

The aminooxazoline class of chemistry and in particular the 4-chloro-2-methyl phenyl analog represent a novel class of octopaminergic agonists with pesticidal activity. In structure, these materials have some resemblance to the amidine pesticides such as clenpyrin and the phenylimidazolidines reported by Nathanson (Nathanson, 1985).

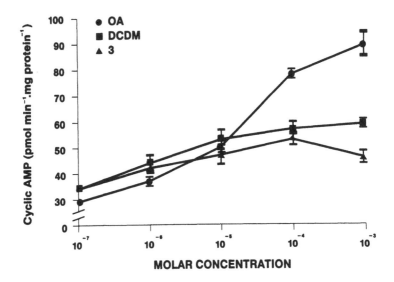

Figure 2.   Dose-response of formamidine, aminooxazoline
            and octopamine on <u>P</u>. <u>americana</u> nerve cord
            homogenate adenylcyclase (mean ± SD).

The octopamine receptor, while unique to invertebrates, has been shown to bear some pharmacological cross-sensitivity to adrenergic drugs (Evans, 1980). In particular, many $\alpha2$-adrenergic agonists and antagonists are known to exert effects at the octopaminergic synapse. This is true of the phenethylimidazolidines (Nathanson, 1985) and indeed the formamidine pesticides are known to exert adrenergic effects. A survey of the patent literature disclosed that aminooxazolines had been shown to possess adrenergic properties and some members of this class of chemistry had reached the stage of clinical trials (Weerasuriya et al., 1984). Indeed, among aminooxazolines identified as α-adrenergic agonists are included representatives of the phenyl aminooxazolines (Bay-c6014) and the cyclohexyl aminooxazolines (Bay-a6781). These aminooxazoline α-agonists demonstrate hypotensive activities comparable to clonidine in rats (Timmermans et al., 1981).

In addition to these aminooxazolines identified for pharmaceutical uses, the class of aryliminooxazolines and oxazolidines have been patented for ectoparasite (acaricidal) activity (Wollweber et al., 1971). Included in this claim for acaracidal activity are compounds of the phenyl-aminooxazoline type including structure (3).

Interestingly, these aminooxazoline ectoparasiticides are reported to be effective against carbamate and organophosphate resistant acarides. In view of our findings that this class of chemistry represents a novel type of octopaminergic agonist, the finding of efficacy against pests resistant to conventional acetylcholinesterase inhibitors is not surprising. The aminooxazolines thus represent another class of octopaminergic agonist with α2-adrenergic properties. It would be of interest to conduct a systematic comparison of the comparative pharmacology of these two receptors, one specific for a monophenolic amine, one for a catecholamine.

Octopamine and octopamine sensitive receptors have been shown to occur in many invertebrate phyla including nematodes, molluscs, annelids and arthropods. In <u>Caenorhabditis elegans</u>, a free living nematode, octopamine has been shown to occur and been implicated in an important behavior, egg-laying (Horvitz et al., 1982).

Octopamine acts to inhibit spontaneous (<u>E. coli</u> supported) or serotonin-stimulated egg-laying in <u>C. elegans</u>. In a series of experiments to investigate the <u>in vivo</u> pharmacology of this octopaminergic response, we examined the effects of other octopamine agonists including the pesticidal formamidines and aminooxazolines on this behavior. These tests were conducted by incorporating serotonin (as the creatinine sulfate) at a concentration of 10 mg/ml into the agar medium for the nematode (Horvitz et al., 1982) with or without test compounds.

In addition to octopamine as a standard, DCDM (as HCL salt) and the aminooxazoline (3) were evaluated as OA agonists. The results (Fig. 3) demonstrate that both the formamidine and aminooxazoline compounds act similarly to octopamine, suppressing serotonin-stimulated egg laying in <u>C. elegans</u>. These experiments provide further support for the direct <u>in vivo</u> octopaminergic agonist activity of the formamidines and aminooxazolines, observed both in insects and nematodes. These findings also suggest some degree of homology in the pharmacology of the octopaminergic response in these two very different organisms. Further experiments examining the octopamine receptor in vitro at the bio- chemical level in C. elegans are required to investigate the degree of homology between this receptor and the more thoroughly studied insect receptor.

EFFECTS OF OCTOPAMINE AGONISTS ON EGG LAYING IN C. *elegans*

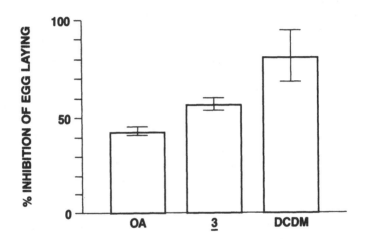

Figure 3.   Inhibition of serotonin induced egg laying in C. elegans.   10 mg/ml 5HT was present in 1 ml volume of agar in all treatments. Octo- pamine, aminooxazoline (3) and DCDM.HCL were present at 5 mg/ml. Readings were taken on 1-3 mature C. elegans 90 min. after treatment and held at room temperature (Horvitz et al., 1982) mean ± SD.

Because of the interest on the part of the agrichemical industry in developing selective insecticides toxic to insects but of low mammalian toxicity, the pharmacological cross-sensitivity of the invertebrate specific transmitter octopamine and a class of receptors for the vertebrate transmitter norepinephrine represents a potential barrier to the development of an octopaminergic insecticide. The formamidine insecticides are known to have effects in mammals that could involve adrenergic mechanisms (HSU et al., 1986).

Taking into account the sensitivity of the octopamine receptor to agonists and antagonists known to be active at the α2-adrenergic receptor, it is of some interest to speculate on how a catecholamine receptor and a receptor for a monophenolic amine could arise in evolution yet apparently share some degree of receptor binding site homology.

If octopamine and it's receptors evolved early enough in phylogeny to be represented in all members of the invertebrate line from nematodes to arthropods, it may be that the close similarity to the α2 pharmacology represents a vestige of a shared ancestral receptor that evolved to respond to divergent catechol and monophenol amine chemical messengers. If this is the case, it might be expected that a true octopaminergic pharmacology, based on a study of the receptor in a number of different phyla (and not relying on borrowed vertebrate adrenergic pharmacology), could lead to the design of agonists and antagonists with some specifity for the monophenolic receptor. It would be a valuable exercise to use the techniques of molecular biology now being applied to the adrenergic receptors to elucidate the structural specificity of the octopaminergic receptor in invertebrates. In view of the similarity in pharmacological responsiveness between adrenergic (expecially α2-adrenergic) receptors and octopaminergic receptors it may be possible to use the mammalian adrenergic sequence information to develop probes to isolate the homologous sequence(s) from readily available gene libraries in <u>Drosophila</u>.

In the Insecticide Discovery process the biogenic amines represent a number of possible critical targets for insecticidal design. These amines are involved in cuticular formation and hardening, developmental processes, sensory and exocrine function as well as modulation of behavior. Most attention from industry has been focused on the monophenolic amine, octopamine. This emphasis on octopamine is due to the precedent of a class of chemical pesticide, the formamidines, known to act at this site and the much greater body of knowledge about octopamine's function in insects compared with the other biogenic amines, dopamine and serotonin.

We have demonstrated here the potential for using the mode of action in helping to guide the synthesis of a novel class of pesticidal octopaminergic agonists, the amino-oxazolines.

The long-term goals for success in this area will require an increased understanding of the roles played by the biogenic amines in insect physiology and the greater use of mode-of-action studies to identify novel classes of chemistry that exert their pesticidal action by perturbing amine function.

ACKNOWLEDGEMENTS

The authors would like to acknowledge the able technical assistance of C. Kukel, R. Borysewicz and D. Wright, Jr. and thank Dr. W. Wood for providing use of C. elegans.

REFERENCES

Eldefrawi A.T. (1985). Acetylcholinesterases and anticholin-esterases. In, Comprehensive Insect Physiology Bio-chemistry and Pharmacology Vol. 12 (Kerkut G.A. and Gilbert L.I., eds) pp 115-130 Pergamon Press, U.K.

Eldefrawi M.E. (1985). Nicotine. In, Comprehensive Insect Physiology Biochemistry and Pharmacology Vol. 12 (Kerkut G.A. and Gilbert L.I., eds) pp 263-272 Pergamon Press, U.K.

Evans P.D. (1980). Biogenic amines in the insect nervous system. Adv. Ins. Physiol. 15, 317-473.

Evans P.D. (1985). Biogenic amines and second messenger systems in insects. In, Approaches to New Leads for Insecticides (von Keyserlingk H.C., Jager A. and von Szczepanski C.H., eds) pp 117-131 Springer-Verlag, FRG.

Hollingworth R.M. and Johnstone E.M. (1985). Pharmacology and toxicology of octopamine receptors in insects. In, Pesticide Chemistry Human Welfare and the Environment Vol. 1 (Miyamoto J. and Kearney P.C., eds) pp 187-192 Pergamon Press, U.K.

Hollingworth R.M. and Murdock L.L. (1980). Formamidine pesticides: octopamine-like actions in a firefly. Science. 208, 74-76.

Horvitz R.H., Chalfie M., Trent C., Sulston J.E. and Evans P.E. (1982). Serotonin and octopamine in the nematode Caenorhabditis elegans. Science 216, 1012-1014.

Hsu W.H., Lu Z.X. and Hembrough F.B. (1986). Effect of amitraz on heart rate and aortic blood pressure in conscious dogs: influence of atropine, prazosin, tolazoline and yohimbine. Toxicol. Appl. Pharmacol. 84, 418-422.

Jennings K.R., Brown D.G., Wright Jr. D.P. and Chalmers A.E. (1987). Methyllycaconitine - a potent natural insecticide active on cholinergic receptors. In, The Search for Novel Pest Control Agents-Receptors for GABA, Octopamine and Other Neuroreceptors. ACS Symposium Series IN PRESS.

Mitsui T. (1985). Chitin synthesis inhibitors: benzoylarylurea insecticides. Jap. Pest. Inform. 47 3-7.

Nathanson J.A. (1985). Phenyliminoimidazolidines: characterization of a class of potent agonists of octopamine-sensitive adenylate cyclase and their use in understanding the pharmacology of octopamine receptors. Mol. Pharmacology 28, 254-268.

Nathanson J.A. and Greengard P. (1973). Octopamine-sensitive adenylatecyclase: Evidence for a biological role of octopamine in nervous tissue. Science 180, 308-310.

Nathanson J.A. and Hunnicutt E.J. (1981). N-Demethylchlordimeform: A potent partial agonist of octopamine-sensitive adenylate cyclase. Mol. Pharmacology 20, 68-75.

Omar D., Murdock L.L. and Hollingworth R.M. (1982). Actions of pharmacological agents on 5-hydroxytryptamine and dopamine in the cockroach nervous system (Periplaneta americana L.). Comp. Biochem. Physiol. 73C, 423-429.

Timmermans P.B.M.W.M., de Jonge A., van Meel J.C.A., P., Slothorst-Grisdijk F., Lam E. and Van Zwieten P.A. (1981). Characterization of α-adrenoceptor populations. Quantitative relationships between cardiovascular effects initiated at central and peripheral α-adrenoceptors. J. Med. Chem. 24, 5 502-507.

Weerasuriya K., Shaw E., Turner P. (1984). Preliminary clinical pharmacological studies of S3341, a new hypotensive agent, and comparison with clonidine in normal males. Europ. J. Clin. Pharmacol. 27, 281-286.

Wollweber H., Hiltmann R. and Stendel W. (1971). 2-(arylimino)oxazolidines. Ger. Offen. 1,963,192.

Wright T.R.F. (1987). The genetics of biogenic amine metabolism, sclerotization and melanization in *Drosophila melanogaster*. Advances in Genetics 25 (in press).

N-ACETYL 5-HYDROXYTRYPTAMINE AND N-ACETYL DOPAMINE

IN THE COCKROACH (Periplaneta americana)

B. Duff Sloley[1] and Roger G.H. Downer[2]

[1]Neuropsychiatric Research Unit, CMR Building, University
of Saskatchewan, Saskatoon, Saskatchewan, Canada S7N 0W0
and
[2]Department of Biology, University of Waterloo,
Waterloo, Ontario, Canada N2L 3G1

N-acetyl 5-hydroxytryptamine (NA-5-HT) and N-acetyl
dopamine appear to be the major metabolites of 5-hydroxy-
tryptamine (5-HT) and dopamine (DA) in insects. N-acetylated amines are produced by a number of insect tissues
including nervous tissue (Murdock and Omar, 1981; Sloley
and Downer, 1984) and malpighian tubules (Sloley, 1985).
In addition DA derivatives including dopamine-3-0-sulphate,
β-alanyl dopamine and NADA are involved in the tanning of
insect cuticle (Morgan et al., 1987; Sloley and Downer,
1987). N-acetyl transferase is the enzyme used for the
metabolism of amines in the adult insect while monoamine
oxidase activity is almost totally absent. However, when
insects ecdyse the large amount of DA eventually required
for cuticular tanning is stored or transported as
dopamine-3-0-sulphate (Sloley and Downer, 1987) and NADA
phosphate (Bodnaryk et al., 1974). These reserves are
rapidly converted to NADA, β-alanyldopamine and possibly
other compounds which are then incorporated into the cuticle. After the final ecdysis the insect loses the ability
to produce or metabolise large amounts of dopamine-3-0-
sulphate (Sloley and Downer, 1987) and appears to rely to a
great extent on N-acetylation for amine inactivation.
However, several anomalies arise when investigating NA-5-HT
or NADA formation by adult insects injected with 5-HT or
DA. These include observations that the N-acetylated
products do not account for all of the precursor injected,
very little of the injected amine is stored as the original

compound within nervous or other tissues and injected NA-5-HT and NADA are also very rapidly removed from the circulation.

In order to resolve these anomalies high performance liquid chromatography with electrochemical detection (HPLC-EC) has been used to simultaneously measure 5-HT, DA, NA-5-HT, NADA, 5-hydroxyindoleacetic acid, dihydroxyphenyl-acetic acid, homovanillic acid and other compounds in single samples of insect tissues. This capability makes HPLC-EC ideal for the resolution of the catabolic pathways of amines and their N-acetylated products in insects.

The current study examines the disposition of 5-HT, DA, NA-5-HT and NADA injected into adult cockroaches. Haemolymph from insects injected with 5-HT and DA is also examined for the presence of sugar and phosphate conjugates of the amines and their N-acetylated metabolites in order to establish the major route of amine inactivation in the adult cockroach.

MATERIALS AND METHODS

Procedures for maintaining insects (Downer, 1972), determining amine and metabolite concentrations (Sloley and Downer, 1984; Sloley et al., 1986), incubating malpighian tubules (Sloley, 1985), injecting insects and collecting haemolymph (Sloley and Downer, 1987) have been described previously.

Haemolymph (5 µl) collected from insects injected 15 minutes previously with 5-HT or DA was incubated at 20°C with 95 µl of 0.2 N acetate buffer with glucosidase at pH 8.5 or purified phosphatase at pH 5.0 for 20 minutes. The incubation was stopped with 100 µl 0.2 N perchloric acid, centrifuged and an aliquot of the supernatant injected onto the HPLC-EC system to determine whether conjugates of amines or their N-acetylated metabolites were present. Glucosidase had some phosphatase activity but the purified phosphatase (Cooper Biomedical, Mississauga, Ontario) contained no glucosidase activity as indicated by the supplier. No sulphatase activity was present in either enzyme preparation as indicated by failure to hydrolyse p-nitrocatechol sulphate.

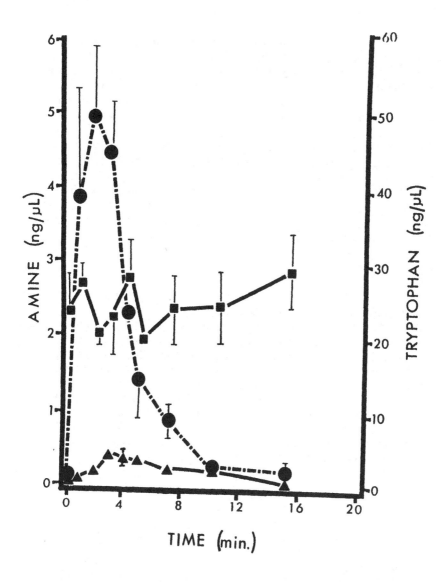

Figure 1. Levels of tryptophan, 5-hydroxytryptamine and
N-acetyl-5-hydroxytryptamine in cockroach haemolymph at
various times after the injection of 2 µg
5-hydroxytryptamine. Tryptophan: —■——■—— ;
5-hydroxytryptamine: –•——•— ;
N-acetyl-5-hydroxytryptamine ——▲——▲—— .
Values are X ± S.E.M. based on six determinations.

RESULTS

5-Hydroxytryptamine injected into the cockroach is rapidly removed from the haemolymph. This is associated with a transient increase in NA-5-HT concentrations (Figure 1). Injected DA is also rapidly removed and again a transient elevation of the N-acetylated product can be demonstrated (Sloley and Downer, 1987). Crude homogenates and organ incubations of nervous tissue (Sloley and Downer, 1984) and malpighian tubules (Table 1) are capable of producing N-acetylated amines when provided with substrate. When 5-HT is provided in incubations of malpighian tubules NA-5-HT is produced. However, the production of NA-5-HT does not account for all of the 5-HT metabolised. In addition, malpighian tubules metabolise NA-5-HT to an unknown product (Table 2). The sum of the rate of production of NA-5-HT and the rate of its removal accounts for most of the 5-HT metabolised.

Incubation of haemolymph from 5-HT injected insects with glucosidase or phosphatase (Table 3) indicates that most of the 5-HT is converted to a NA-5-HT-sugar conjugate. Incubation with purified phosphatase releases some NA-5-HT but this is only about 15% of that released by an equivalent activity of glucosidase. 5-HT levels in enzyme incubations are not significantly different from unincubated controls. Similar experiments using haemolymph from DA-injected insects produced equivocal results due to the presence of endogenous levels of conjugated DA and NADA and the ability of haemolymph to release NADA from conjugates in incubations in the absence of added enzymes.

DISCUSSION

The production of NA-5-HT and NADA from 5-HT and DA by adult cockroach tissues without a concomitant production of 5-hydroxyindoleacetic acid or dihydroxyphenylacetic acid indicates that N-acetylation is the major route of amine catabolism in these animals. This agrees with the work of Evans and Fox (1975a,b) and Murdock and Omar (1981). The production of N-acetylated products in vivo is very rapid. Large amounts of injected amines (2 and 5 μg/insect) can be completely metabolised within 20 minutes indicating a remarkable capacity for conversion of amines. Malpighian tubules are capable of N-acetylating 5-HT as well as

TABLE 1

Effect of incubation of malpighian tubules with dopamine and 5-hydroxytryptamine on levels of tryptophan, N-acetyl dopamine and N-acetyl 5-hydroxytryptamine

Treatment	n	TP	5-HT	NA-5-HT	NADA
			ng/mg protein		
Control (no amine)	5	77.02±16.40	N.D.	N.D.	N.D.
Dopamine (100 ng) (2.1 µM)	5	67.29±15.67	N.D.	N.D.	20.26±3.45
5-HT Creatine SO$_4$ (100 ng) (1.0 µM)	5	103.00±27.37	off scale*	6.56±1.51	N.D.

Values are $\overline{X}$ ± S.E.M. based on n determinations. Incubations were conducted at 20°C for 20 min in 250 µl of Hoyle's (1953) saline. *The presence of 5-HT added to the incubation medium is detected but the level exceeds accurate quantitation in this system. No DOPAC, HVA or 5-HIAA was produced by any sample.

TABLE 2

Levels of indoleamine metabolites in cockroach malpighian tubules after incubation with 20 ng 5-hydroxytryptamine or N-acetyl 5-hydroxytryptamine

Treatment	5-HT remaining ng/sample*	NA-5-HT remaining ng/sample*	5-HT used ng/mg protein/min	NA-5-HT used ng/mg protein/min	NA-5-HT produced ng/mg protein/min
No 5-HT 20 min	N.D.	N.D.	N.D.	N.D.	N.D.
20 ng 5-HT 0 min	20.0	N.D.	N.D.	N.D.	N.D.
20 ng 5-HT 20 min	11.76±0.96	1.76±0.30	2.12±0.14	-	0.45±0.07
20 ng NA-5-HT 0 min	N.D.	20.0	N.D.	N.D.	N.D.
20 ng NA-5-HT 20 min	N.D.	15.13±0.61	-	1.40±0.17	-

Values are $\overline{X}$ ± S.E.M. based on 5 determinations. *Each sample represents the malpighian tubules from one cockroach. NA-5-HT used (1.40 ± 0.17) and NA-5-HT produced (0.45 ± 0.07) sum to 1.85 ± 0.24 ng/mg protein minute which is 87% ± 11% of the 5-HT used (2.12 ± 0.14 ng/mg protein/minute).

TABLE 3

Effect of incubation with glucosidase or phosphatase on
5-hydroxytryptamine and N-acetyl 5-hydroxytryptamine
concentrations in haemolymph of cockroaches injected with
5-hydroxytryptamine 15 minutes prior to haemolymph
collection

Treatment	5-hydroxytryptamine ng/μl	N-acetyl 5-hydroxytryptamine ng/μl
0 μg 5-HT/insect no incubation	< 30 pg/μl	< 100 pg/μl
5 μg 5-HT/insect 15 min	2.13 ± 0.82	0.57 ± 0.16
0 μg 5-HT/insect 15 min followed by 20 min incubation with 1 unit glucosidase	< 30 pg/μl	< 100 pg/μl
5 μg 5-HT/insect 15 min followed by 20 min incubation without glucosidase	2.17 ± 1.18	0.35 ± 0.09
5 μg 5-HT/insect 15 min followed by 20 min incubation with 1 unit glucosidase	3.69 ± 0.78	14.65 ± 2.67*
5 μg 5-HT/insect 15 min followed by 20 min incubation with 1 unit phosphatase	2.93 ± 1.25	2.28 ± 0.30*

Values are the mean ± the standard error of the mean based
on 5 determinations. *Significantly different from controls injected with 5-HT, $p < 0.01$. Incubations stopped at
zero time were no different from incubations without enzymes in their respective buffers indicating no degradation
of amines by haemolymph.

converting NA-5-HT to other compounds. Incubation of
haemolymph from 5-HT injected insects also supports the
contention that most of the amines are converted to sugar
conjugates of the N-acetylated compounds. This is in
agreement with Trimmer (1985). However, the presence of
other conjugates such as phosphates cannot be ruled out
because of the inability to obtain sufficiently pure
enzymes with which to perform incubation studies. Studies
using DA as the amine of interest are difficult as endogenous DA and NADA conjugates exist in cockroach haemolymph.
These may be reserves which can be used for cuticle repair
in the advent of damage to the exoskeleton. The ability of
the haemolymph to release conjugated DA and NADA without
the addition of glucosidase or phosphatase further comp-
licates these experiments.

Although it appears that 5-HT forms a NA-5-HT-sugar
conjugate, the exact nature of the sugar involved remains
to be determined. It would be interesting to know if
trehalose which is the major circulating sugar in insects
is the source of the conjugating material.

ACKNOWLEDGEMENTS

This work was supported by a Natural Sciences and
Engineering Research Council of Canada grant to R.G.H.D.
and a Saskatchewan Health Post-doctoral Fellowship to
B.D.S. The authors would like to thank Dr. A.A. Boulton
for his enthusiastic support of the research.

REFERENCES

Bodnaryk R.P., Brunet P.C.V. and Koeppe J.K. (1974) On the
    metabolism of N-acetyl dopamine in Periplaneta
    americana. J. Insect Physiol. 20, 911-923.
Downer R.G.H. (1972) Interspecificity of lipid regulating
    factors from insect corpus cardiacum. Can. J. Zool. 50,
    63-65.
Evans P.H. and Fox P.M. (1975a) Comparison of various bio-
    genic amines as substrates of N-acetyl transferase from
    Apis mellifera CNS. Comp. Biochem. Physiol. 51C,
    131-141.

Evans P.H. and Fox P.M. (1975b) Enzymatic N-acetylation of indolealkylamines by brain homogenates of the honey bee, Apis mellifera. J. Insect Physiol. 21, 343-353.

Hoyle G. (1953) Potassium ions and insect nerve muscle. J. Exp. Biol. 30, 121-135.

Morgan T.D., Hopkins T.L., Kramer K.J., Roseland C.R., Czapla T.H., Tomer K.B. and Crow F.W. (1987) N-β-alanylnorepinephrine: Biosynthesis in insect cuticle and possible role in sclerotization. Insect Biochem. 17, 255-263.

Murdock L.L. and Omar D. (1981) N-acetyl dopamine in insect nervous tissue. Insect Biochem. 11, 161-166.

Sloley B.D. and Downer R.G.H. (1984) Distribution of 5-hydroxytryptamine and indolealkylamine metabolites in the American cockroach, Periplaneta americana L. Comp. Biochem. Physiol. 79C, 281-286.

Sloley B.D. (1985) Metabolism of 5-hydroxytryptamine and dopamine in the American cockroach, Periplaneta americana. Ph.D. thesis, University of Waterloo, Waterloo, Canada.

Sloley B.D., Downer R.G.H. and Gillott C. (1986) Levels of tryptophan, 5-hydroxytryptamine and dopamine in some tissues of the cockroach, Periplaneta americana. Can. J. Zool. 64, 2669-2673.

Sloley B.D. and Downer R.G.H. (1987) Dopamine, N-acetyl dopamine and dopamine-3-0-sulphate in tissues of newly ecdysed and fully tanned adult cockroaches (Periplaneta americana). Insect Biochem. 17, 591-596.

Trimmer B.A. (1985) The inactivation of exogenous serotonin in the blowfly, Calliphora. Insect Biochem. 15, 435-442.

OCTOPAMINERGIC MODULATION OF LOCUST FLIGHT MUSCLE

MATTHEW D. WHIM

and

PETER D. EVANS.

A.F.R.C. Unit of Insect Neurophysiology and Pharmacology,
Dept. of Zoology, University of Cambridge, Downing Street,
Cambridge, CB2 3EJ, U.K.

Octopamine has been suggested to be a regulator of
various aspects of insect flight. Its levels rise rapidly
during the first few minutes of flight (Goosey and Candy,
1980) which may be related to its ability to regulate
carbohydrate and lipid metabolism (Candy, 1978). In
addition it may also function as a modulator of
neuromuscular transmission in flight muscle (Klaassen,
Kammer and Fitch, 1986) and of the central pattern
generator for flight motor patterns (Sombati and Hoyle,
1984). However in the latter experiments it is not clear
if octopamine has a specific direct action on the central
pattern generator neurones or a more general effect due to
an enhanced efficacy of sensory transmission producing an
increase in arousal.

To date, however, the source of any endogenous
octopamine mediating any of the above effects has not been
definitely determined. It is not clear if the actions on
flight muscle described above are brought about by the
increased levels of octopamine in the haemolymph, perhaps
produced by the release of octopamine from octopamine
containing neurohaemal organs, such as the corpora
cardiaca and median neurohaemal organs (see Evans, 1985).
Alternatively the raised haemolymph levels of octopamine
could be a secondary consequence of the spill over into
the haemolymph of octopamine released from the terminals
of modulatory octopaminergic neurones innervating specific
skeletal muscles. The neurones of one such group of
octopaminergic neurones, the dorsal unpaired median (or

DUM) cells (Hoyle et al., 1974) have been shown to be directly responsible for many of the physiological actions of octopamine in insects (see Evans, 1980, 1985 and this volume). Anatomical evidence has been presented for the innervation of the dorsal longitudinal flight muscle (DLM) of the locust by a DUM cell, termined DUMDL (Hoyle, Colquhoun and Williams, 1980). The present paper will compare the physiological and pharmacological effects of stimulating DUMDL and of applying exogenous octopamine to the DLM of the locust, Schistocera gregaria, and provide evidence for the octopaminergic nature of this neurone.

Octopamine produces a marked modulation of twitch tension parameters in the DLM of the locust. Fig. 1 shows the effect of introducing a 30s pulse of 10^{-6}M DL-octopamine into the muscle superfusate whilst the metathoracic DLM is being stimulated to contract at a frequency of 1Hz. The muscle is stimulated by a pair of hook electrodes on metathoraic nerve 1 after it has fused with mesothoraic nerve 6. The metathoracic DLM is innervated by five motorneurones. One in the metathoracic ganglion which exists via nerve 1 and four in the mesothoracic ganglion that exit via nerve 6. In these experiments the proximal end of the muscle was dissected clear of its attachment to its mesothoracic homologue and attached to a force transducer which measured tension almost isometrically. Octopamine increases the amplitude of twitch tension by about 18% and the effect gradually declines to resting values over the next 5 min. The rate of contraction of twitch tension also increases (by about 6%) and there is a marked increase (53%) in the rate of relaxation of twitch tension. The latter effect is largely responsible for the decrease in twitch duration (not shown) from 102 msec to 84 msec. The above effects are dose-dependent with thresholds between 10^{-8} and 10^{-9}M DL-octopamine with maximal effects occurring in the range of 10^{-6}M - 10^{-5}M. The responses are all blocked by phentolamine, an alpha adrenergic antagonist that blocks octopamine receptors in other preparations (Evans, 1981) but not by DL-propranolol. The responses are stereospecific for the D(-)-isomer of octopamine which is the naturally occurring form in the locust (Goosey and Candy, 1980). In general the pharmacological profile of the octopaminergic responses of the locust DLM is very similar to that observed for octopamine receptors in other preparations (Evans, 1981).

DUMDL can be stimulated by a pair of hook electrodes

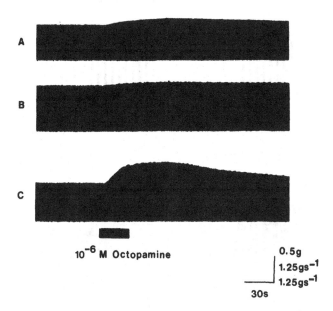

Fig. 1   The effect of a 30s pulse of 10^{-6}M DL-octopamine
(black bar) on the twitch tension generated in the DLM in
response to stimulation of the motorneurones at a
frequency of 1Hz. A, the response of twitch tension,
(B,C) the effects on contraction and relaxation rates
respectively. The parameters returned to baseline values
after a 10 min period of superfusion in saline.

on the contralateral nerve 1 of the locust metathoracic
ganglion. When stimulated it produces effects that
closely mimic the effects of superfusion of octopamine
over the preparation. The effects depend on the frequency
of stimulation of DUMDL and are again blocked by
phentolamine but not by DL-propranolol.
    The effects of stimulating DUMDL and of applying
exogenous octopamine both depend on the frequency of
stimulation of the DLM motorneurones. Fig. 2 shows the
effect of 10^{-5}M DL-octopamine on tension generated in the
DLM by firing the motorneurones for 10s periods at
different frequencies. It can be seen that octopamine
increases the amplitude of the individual tension

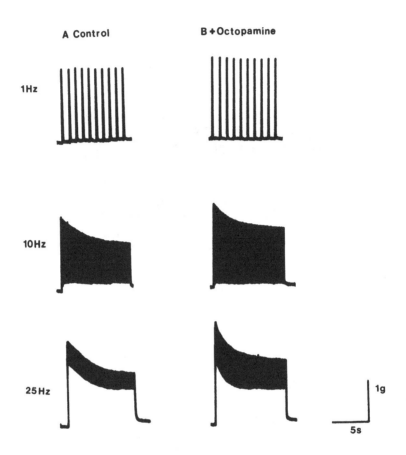

Fig.2 The potentiating effect of octopamine upon neurally evoked tension in the DLM depends upon the frequency of motorneurone stimulation. The figure shows the effects of stimulating the motorneurones for 10s periods at frequencies of 1, 10 and 25Hz in the absence (A) and presence (B) of 10^{-5}M DL-octopamine.

transients at all frequencies tested. This results from a combination of two effects. First, a decrease in the amount of tension maintained between individual contractions and second, an increase in the maximal levels of tension generated by each contraction. Thus octopamine serves to decrease the degree of fusion between the individual twitches and to increase the force generated by each contraction. Stimulation of DUMDL at 5Hz for 10s produces a qualitatively similar effect.

In the work presented above all the motorneurones were stimulated together. However, in subsequent experiments where tension was recorded from the five individual motor units of this muscle (Neville, 1963) no significant differences in the form of the octopaminergic modulation were observed.

The present study shows that the elevations in the levels of octopamine in insect haemolymph during the first few minutes of flight (Goosey and Candy, 1980) to values of 2×10^{-7} M would be sufficient to affect the contractile properties of the DLM. However, it seems likely that any modulation of DLM contractile properties by circulating octopamine will clearly be secondary to the effects produced by the octopamine released locally in the flight muscles due to the activation of DUMDL. Indeed the former increase in haemolymph levels may well be a secondary spill over effect resulting from the activation of DUMDL and other octopaminergic neurones innervating flight muscles during the first few minutes of flight. The results of the present study provide physiological and pharmacological evidence for the octopaminergic nature of DUMDL.

More recent experiments indicate that as in the extensor-tibiae muscle of the locust (Evans, 1984) the effects of octopamine on DLM are mediated by increases in cyclic AMP levels. Further, it appears that both the physiological responsiveness and cyclic AMP responses of the locust DLM to octopamine change during the first few days of adult life. The significance of these observations remains unclear at present but may well be related to the many other maturation properties that have been observed in this muscle during the same time period.

In behavioural terms the release of octopamine onto the DLM during the first few minutes of insect flight will have several important consequences. First, it will stimulate carbohydrate metabolism which is the main source of energy for the first few minutes of locust flight (Candy, 1978). Second, it will increase the amount of force generated by the flight muscles which could be important when the insect initially takes off. Third, it will reduce the amount of overlap in the duration of twitches in antagonistic muscles when the locust body temperature is low during the initial minutes of flight. This could prevent a considerable loss of energy in the period before the locust reaches its normal internal body flight temperature (Neville and Weis Fogh, 1963). In conclusion, the results of the present study suggest that the peripheral release of octopamine from DUMDL during the first few minutes of locust flight is likely to be an important modulatory factor influencing the kinetics of contraction of the DLM's and may represent an energy saving adaptation.

Candy, D.J. (1978) The regulation of locust flight muscle metabolism by octopamine and other compounds. Insect Biochem. 8, 177–181.

Evans, P.D. (1980) Biogenic amines in the insect nervous system. Adv. Insect Physiol. 15, 317–473.

Evans, P.D. (1981) Multiple receptor types for octopamine in the locust J. Physiol. 318, 99–122.

Evans, P.D. (1984) A modulatory octopaminergic neurone increases cyclic nucleotide levels in locust skeletal muscle. J. Physiol. (Lond.). 348, 307–324.

Evans, P.D. (1985) Octopamine. In Comprehensive Insect Biochemistry, Physiology and Pharmacology. Eds. G.A. Kerkut and L. Gilbert pp499–530, Pergamon Press, Oxford.

Goosey, M.W. and Candy, D.J. (1980) The D-octopamine content of the haemolymph of the locust, Schistocerca americana gregaria and its elevation during flight. Insect Biochem. 10, 393–397.

Hoyle, G., Colquhoun, W. and Williams, M. (1980) Fine structure of an octopaminergic neuron and its terminals. J. Neurobiol. 11, 103–126.

Hoyle, G., Dagan, D., Moberly, B. and Colquhoun, W. (1974) Dorsal unpaired median insect neurons make neurosecretory endings on skeletal muscle. J exp Zool 187, 159–165.

Klassen, L.W., Kammer, A.E. and Fitch, G.K. (1986) Effects of octopamine on miniature excitatory junctional potentials from developing and adult moth muscle. J. Neurobiol. 17, 291–302.

Neville, A.C. (1963) Motor unit distribution of the dorsal longitudinal flight muscles in locusts. J. exp. Biol. 40, 123–136.

Neville, A.C. and Weis-Fogh, T. (1963) The effect of temperature on locust flight muscle. J. exp. Biol. 40, 111–121.

Sombati, S. and Hoyle, G. (1984) Generation of specific behaviours in a locust by local release into neuropile of the natural neuromodulator octopamine. J. Neurobiol. 15, 481–506.

Neurobiology: Vertebrates

COMPARISON OF THE BINDING OF [3H]TRYPTAMINE IN RAT BRAIN WITH THAT OF TWO LIGANDS FOR MONOAMINE OXIDASE, [3H]MPTP AND [3H]PARGYLINE

David C. Perry[1], Linda J. Grimm[1], Andrea M. Martino-Barrows[2], Margaret L. Jones[2] and Kenneth J. Kellar[2]. Departments of Pharmacology, George Washington University School of Medicine[1] and Georgetown University School of Medicine and Dentistry[2], Washington, D.C.

[3H]Tryptamine labels high affinity (2-3 nM) binding sites in rat brain and other tissues with a unique pharmacological profile (Cascio and Kellar, 1983). Other than tryptamine itself, compounds exhibiting high affinity for this site include β-carbolines (including harmaline and related drugs) and phenethylamines (Cascio and Kellar, 1982, 1983; Bruning and Rommelspacher, 1984; Wood et al, 1984) and kynuramines and related compounds (Charlton et al., 1985). 5-Hydroxytryptamine (serotonin) has only modest affinity for these sites. In vitro autoradiography studies revealed that [3H]tryptamine binding sites are widely distributed throughout the brain, with highest levels in forebrain and limbic areas, including cortex, caudate putamen, hippocampus, olfactory tubercle, nucleus accumbens, amygdala and interpeduncular nucleus (Perry, 1986; Altar et al., 1986; Kaulen et al., 1986; McCormack et al., 1986). Autoradiography also revealed very dense binding of [3H]tryptamine in the choroid plexus. Structure-activity studies indicated that the binding in this structure was pharmacologically distinct from that in the rest of the brain. For instance, kynuramine binds with 11 nM potency to brain sites (labeled "T-1"), whereas it showed little or no affinity to the choroid sites (labeled "T-2") (Perry, 1986).

Tryptamine is known to be a substrate for oxidation by MAO; administration of MAO inhibitors (MAO-I) in vivo greatly increases brain levels of tryptamine (Philips and Boulton, 1979). Also, many of the compounds that exhibit high affinity vs. [³H]tryptamine binding are either MAO substrates (i.e. beta-carbolines, phenethylamine, kynuramine) or inhibitors (i.e. harmaline). As membrane-bound enzymes such as MAO can serve as high affinity binding sites, it is possible that the [³H]tryptamine binding sites in brain membrane preparations may not be neurotransmitter receptors but in fact brain MAO (eg. Kienzl et al., 1985). Several findings would seem to argue against this suggestion. For instance, most behavioral effects of tryptamine are only seen in the presence of MAO-I, (for review see Jones, 1982). Also, the standard binding assay for [³H]tryptamine (Cascio & Kellar, 1983) is performed in the presence of 10 µM pargyline. Finally, two recent investigations reported that chronic administration of MAO-I to increase brain levels of tryptamine caused a decrease in the number of brain [³H]tryptamine binding sites (Cascio and Kellar, 1986; Wood et al., 1985), and that chronic reserpine given to deplete potential storage pools of tryptamine caused an increase in binding (Cascio and Kellar, 1986). These responses are characteristic of the up- and down-regulation seen with chronic blockade or stimulation of neurotransmitter receptors. Nevertheless, one could explain the results obtained with chronic MAO-I as a result of direct inhibition of [³H]tryptamine binding to MAO (Greenshaw and Dewhurst, 1987). Thus we have attempted to further clarify the question of the identity of [³H]tryptamine binding sites in rat brain by comparison with two ligands for MAO, [³H]MPTP (1-methyl-4-phenyl-1,2,5,6-tetrahydropyridine) and [³H]pargyline (Rainbow et al., 1985; Javitch et al., 1984; Reznikoff et al., 1985).

METHODS

Male Sprague-Dawley rats were employed in all assays (150-300 g). Homogenate binding for [³H]tryptamine was done by the method of Cascio and Kellar (1983). Briefly, tissue was suspended in 50 mM TrisHCl buffer (pH 7.5 at 0°C) containing 5.6 mM ascorbic acid and homogenized with a Brinkmann Polytron. Homogenates were centrifuged at 58,000 x g for 10 min at 4°C, and pellets were resuspended and recentrifuged. The final pellets were rehomogenized in buffer plus 10 μM pargyline HCl and incubated at 37°C for 5 min to inhibit MAO (this step was omitted in some experiments, as noted). Binding of [^{3}H]tryptamine (2-4 nM; 25-40 Ci/mmol, NEN-Dupont) was with 5-10 mg tissue in a final volume of 1 ml; all points were done in triplicate. Pargyline (10 μM) was included in most assays except where noted. Non-specific binding was assayed by addition of 10 μM tryptamine to some tubes. Incubations were for 60 min on ice; mixtures were then filtered over Whatman GF/B filters followed by 3x4 ml rinses with cold buffer. Binding of [^{3}H]MPTP (2-5 nM; 55.5 Ci/mmol, Amersham) and [^{3}H]pargyline (2-5 nM; 24.9 Ci/mmol, NEN-Dupont) was done using essentially the same methods, with the following differences. Tissue for these two assays was not preincubated at 37°C with pargyline, and the incubation medium did not include pargyline. [^{3}H]MPTP incubation as for 90 min on ice (Javitch et al., 1985; Parsons & Rainbow, 1984) and [^{3}H]pargyline incubations for 30 min at room temperature (Rainbow et al., 1985; Reznikoff et al., 1985). Non-specific binding for [^{3}H]MPTP was defined by 2 μM MPTP, and for [^{3}H]pargyline by 10 μM pargyline.

For subcellular distribution, tissue was prepared as described before (Cascio and Kellar, 1983). Briefly, whole brain minus cerebellum and medulla/pons was homogenized with a glass-teflon homogenizer in 0.32 M sucrose plus 10 mM Tris (pH 7.5 at 0°C). Following a 10 min spin at 900 x g to get rid of broken cells and nuclei, the supernatant was centrifuged 20 min at 13,500 x g, the pellet resuspended in 0.32 M sucrose, and a fraction removed and set aside as "total homogenate" (S1). The remaining suspension was layered on top of 0.8 M and 1.2 M layers of sucrose, and centrifuged for 2 hours at 55,000 x g in a swinging bucket rotor. Tissue at the 0.32 M/0.8 M interface was removed as the myelin fraction, tissue at the 0.8 M/1.2 M interface was removed as the synaptosomal fraction, and the pellet was removed as the mitochondrial fraction. These three fractions plus the total homogenate (S1) fraction were resuspended in 50 mM Tris buffer, and thereafter treated as fresh homogenates using the standard tissue preparation. Binding analysis was done using a non-linear curve fitting program (McPherson, 1983).

RESULTS

Results of the subcellular distribution are seen in Table 1, expressed as percent of binding in the total homogenate. [3H]Tryptamine binding was significantly higher in the synaptosomal fraction compared to the myelin and mitochondrial fractions. A previous experiment found a larger ratio of synaptosomal to mitochondrial binding for this ligand (Cascio and Kellar, 1982). [3H]MPTP binding was essentially evenly distributed between the synaptosomal and mitochondrial fractions. In contrast especially to [3H]-tryptamine, [3H]pargyline binding was largely concentrated in the mitochondrial fraction. The ratio of mitochondrial to synaptosomal binding was 2.2 for [3H]pargyline, 0.7 for [3H]tryptamine and 0.9 for [3H]MPTP.

[3H]Pargyline is ideal for labeling MAO sites in comparative experiments such as autoradiography and sub-cellular distribution, because it binds irreversibly to the flavin portion of the enzyme (see McCauley, 1976). This irreversible binding, however, precludes its use for quantitative competitive binding studies which are analyzed using the assumption of reversibility. Thus a reversible labeled ligand for MAO was needed. [3H]Harmaline has been

Table 1. Subcellular Distribution in Rat Brain Membranes

fraction	[3H]Tryptamine % total	[3H]Pargyline % total	[3H]MPTP % total
Total Homogenate	[100]	[100]	[100]
Synaptosomal	115.1 ± 6.7	88.8 ± 19.0	38.3 ±7.0
Myelin	22.0 ± 5.2	9.5 ± 1.2	14.0 ±1.7
Mitochondrial	81.6 ±11.4	194.9 ± 10.1	34.0 ±4.8

Binding was measured using 2.2 nM [3H]tryptamine, 3.6 nM [3H]pargyline and 3.3 nM [3H]MPTP, and is expressed as the percent of the specific binding obtained in the total homogenate (S1 fraction, minus nuclei). Values are means from four animals ± SEM.

used in this capacity (Nelson et al., 1979), but this compound was not readily available. MPTP is a neurotoxin that causes Parkinsonian symptoms; an important step in its toxicity is thought to be its high-affinity binding to MAO-B and its subsequent conversion to MPP^+ in astrocytes (Javitch et al., 1985). High affinity reversible binding of [^{3}H]MPTP in rat brain appears to be to MAO, especially to MAO-B (Javitch et al., 1984, 1985; Parsons and Rainbow, 1984; Rainbow et al., 1985; Reznikoff et al., 1985). We thus chose this ligand to reversibly label MAO for comparison with [^{3}H]tryptamine binding in rat brain.

Saturation binding of [^{3}H]MPTP to rat cortex yielded a biphasic Scatchard curve; the high affinity component had a K_d of 29 nM (B_{max} 54 pmol/g tissue) and the low affinity K_d was 360 nM (B_{max} 249 pmol/g tissue). Javitch et al. (1989) also found biphasic binding using whole brain membranes, with the high affinity K_d = 28 nM. They suggested that the lower affinity component (K_d = 2600 nM) may represent saturable binding to the filters, and noted that this component could be excluded by defining non-specific binding with 300 nM MPTP. The higher affinity and lower capacity of the second binding component of [^{3}H]MPTP seen in our experiments (using a blank of 2 μM MPTP) raised some questions as to the identity of this site.

In subsequent experiments, competition potencies for several drugs were determined in parallel assays with [^{3}H]MPTP and [^{3}H]tryptamine; the K_I values are presented in Table 2, along with those for inhibition of [^{3}H]harmaline binding to MAO-A in rat brain (Nelson et al., 1979). Tryptamine has only micromolar potency vs. [^{3}H]MPTP and [^{3}H]harmaline; conversely, MPTP is 8-fold less potent vs. [^{3}H]tryptamine than [^{3}H]MPTP. Harmaline has similar high affinities vs. [^{3}H]tryptamine and [^{3}H]harmaline; however, these two ligands are distinguished most clearly by disparate potencies of tryptamine and clorgyline.

The binding assay for [^{3}H]tryptamine normally includes 10 μM pargyline to protect the radioligand. In order to more clearly determine the effect of MAO-I in this assay, varying amounts of pargyline, clorgyline, (-)deprenyl and (+)deprenyl were added to the assay either in the presence or absence of the standard 10 μM pargyline; results are summarized in Table 3. At low concentrations (down to 10 nM) there was little effect on [^{3}H]tryptamine binding.

TABLE 2. Inhibition of [^{3}H]Tryptamine, [^{3}H]MPTP and [^{3}H]-
Harmaline Binding

Competitor	[^{3}H]Tryptamine	[^{3}H]MPTP	[^{3}H]Harmaline*
Tryptamine	3.5 nM	2440 nM	2450 nM
MPTP	286 nM	34 nM	------
Harmaline	9.1 nM	355 nM	7.2 nM
Pargyline	⁻1000 μM	406 nM	2790 nM
Clorgyline	⁻10 μM	112 nM	7.6 nM
(-)Deprenyl	>100 μM	264 nM	13.2 μM
(+)Deprenyl	>100 μM	588 nM	------

Competition binding was performed with from 6-10
concentrations of each drug, in triplicate, using 2-3 nM
[^{3}H]tryptamine and 3-4 nM [^{3}H]MPTP.   Values are means from
two separate determinations, using $K_I = IC_{50}/(1+F/K_d)$.
*[^{3}H]Harmaline values taken from Nelson et al., 1979

However, at higher concentrations, the specific binding
increased when compared to that with no drug added.
Maximal stimulation of binding ranged from 65-105% above
control. This effect occurred at concentrations ranging
from 1 μM (clorgyline) to 100 μM ((-)deprenyl). Note that
the maximal stimulation with pargyline occurred at 30 μM,
close to the normally employed concentration of 10 μM
(binding at 10 μM was 150% of control). Maximal stimulation of binding in the presence of 10 μM pargyline was
similar to that obtained in its absence, but occurred at
lower concentration of MAO-I.

TABLE 3. Effect of MAO-I on [^{3}H]Tryptamine Binding

| MAO-I | Maximal Stimulation of Binding | | IC$_{50}$, μM |
	% control	concentration, μM	
Pargyline	170	30	~ 1000
Clorgyline	205	1	~ 10
(-)Deprenyl	195	100	------
(+)Deprenyl	165	10	------
Clorgyline +10 μM pargyline	176	0.1	~ 6
(-)Deprenyl +10 μM pargyline	184	30	------
(+)Deprenyl +10 μM pargyline	157	10	------

Competition curves were run with from 5-8 concentrations of each drug using 2.3 nM [^{3}H]tryptamine in rat cortex, either with or without the addition of 10 μM pargyline to the incubation. The maximal <u>stimulation</u> of binding (compared to control with no drug added) is shown in the first column, and the concentration yielding that stimulation is shown in the second column. The final column is the approximate concentration yielding 50% <u>inhibition</u> of binding (neither isomer of deprenyl showed any appreciable inhibition of [^{3}H]tryptamine binding).

DISCUSSION

High affinity binding sites for [^{3}H]tryptamine in rat brain may represent receptors for this trace amine (or possibly for some other endogenous compound such as tetra-hydro-beta-carboline, phenethylamine or kynuramine). Knowledge of such receptors will aid in understanding the various pharmacological effects of tryptamine (see Jones, 1982). However, the possibility that this radioligand is

merely labeling membrane-bound MAO needs to be seriously
considered (Greenshaw and Dewhurst, 1987). Certainly
tryptamine is a good substrate for MAO oxidation, with
approximately equivalent preference for MAO-A or MAO-B in
vitro (Suzuki et al., 1982; Tipton et al., 1982). It can
also act to inhibit oxidation of other substrates in vitro,
such as serotonin by MAO-A (Nelson et al., 1979).

In vivo tryptamine is also apparently a substrate for
both forms, as administration of either the selective MAO-A
inhibitor clorgyline or the selective MAO-B inhibitor
(-)deprenyl results in increased brain levels of tryptamine
(Phillips and Boulton, 1979). However, Wood et al. (1985)
found that chronic administration of clorgyline, but not
deprenyl, led to down-regulation of tryptamine binding
sites. This finding suggests that only tryptamine associated with MAO-A in brain is functional (that is, coupled to
an active receptor system). MAO-B in the brain, which is
localized primarily in astrocytes and serotonin neurons, is
thought to primarily serve a function of scavenging exogenous and improperly located amines, whereas MAO-A, localized mostly in neurons, primarily acts to deactivate neurotransmitter amines (Denney and Denney, 1985). Thus only
neuronal tryptamine appears to be associated with receptors. It should be noted that Wood et al. (1985) included
an important control experiment: they administered a large
dose (11 mg/kg) of clorgyline acutely and measured [^{3}H]-
tryptamine binding; they found no changes. If clorgyline
decreases [^{3}H]tryptamine binding by directly inhibiting its
ability to bind to MAO, this effect should have been seen
acutely as well as chronically. Thus [^{3}H]tryptamine binding
sites behave just as neurotransmitter receptors: they are
up- and down-regulated by treatments that lower or raise
levels of the endogenous ligand, and only changes in
neuronal pools of the endogenous ligand lead to receptor
regulation.

MAO is known to be located in mitochondrial membranes;
a neurotransmitter receptor, on the other hand, would be
likely to be concentrated at nerve endings, and thus in a
synaptosomal fraction. Previous analysis of the subcellular distribution of [^{3}H]tryptamine binding found a
synaptosomal/mitochondrial ratio of almost 10; in the
current study, the ratio was considerably lower, but it
still favored binding to the synaptosomal fraction. In
contrast, the binding of [^{3}H]MPTP and especially [^{3}H]pargy-

line was clearly greater in the mitochondrial fraction. Regardless of the specific numbers and fractions, if [^{3}H]tryptamine was labeling MAO, it should exhibit the same subcellar patterns as [^{3}H]pargyline, which it does not.

The presence of 10 μM pargyline in the binding assay argues against the identity of [^{3}H]tryptamine as MAO; nevertheless, one can imagine that pargyline inhibition of MAO (which is irreversible and non-competitive) may not actually block the substrate recognition site. For MAO-A a close correlation between the K_I for inhibition of MAO enzyme activity and the K_I for inhibition of the binding of the competitive inhibitor [^{3}H]harmaline <u>has</u> been demonstrated (Nelson et al., 1979). Nevertheless, tryptamine may also be a substrate for MAO-B, and pargyline has some MAO-B inhibitory activity, so the question still needs answering. Kienzl et al. (1985) found that using 10^{-3}M tryptamine (instead of 10^{-5} M) to define non-specific binding doubled the B_{max}, while lowering the affinity and increasing n_H. If they additionally left pargyline from the assay, the B_{max} increased to almost 10 times the original value, while the apparent K_D dropped to 100 nM and the Hill slope increased to over 2. This suggests that pargyline is blocking a low affinity binding site for [^{3}H]tryptamine, which could well represent MAO. Conversely, the MAO inhibitors pargyline and clorgyline were shown here to inhibit binding to the regular high affinity [^{3}H]tryptamine site, but with very low potency (10-1000 μM).

Such binding-site "cross-talk" is common when one utilizes high concentrations of inhibitors. What is important is to compare the potencies of a series of drugs at inhibiting binding to the sites in question. The K_I values in Table 2 readily demonstrate the fundamental differences in pharmacological specificity between the high-affinity binding sites labeled by [^{3}H]tryptamine, [^{3}H]MPTP, and [^{3}H]harmaline. Tryptamine is almost 700 times less potent at inhibiting binding to the latter two ligands than at its own site. Differences of more than an order of magnitude are apparent in the potencies of every one of these compounds when comparing [^{3}H]tryptamine to the two other ligands. Because [^{3}H]harmaline and [^{3}H]MPTP appear to label MAO-A and MAO-B respectively, these data argue strongly for an independent identity for the high affinity binding site of [^{3}H]tryptamine.

REFERENCES

Altar C.A., Wasley A.M. and Martin L.L.   (1986) Autoradio-
    graphic localization and pharmacology of unique
    [^{3}H]tryptamine binding sites in rat brain. <u>Neurosci.</u>
    <u>17</u>, 263-273.
Bruning G. and Rommelspacher H.   (1984) High affinity
    [^{3}H]tryptamine binding sites in various organs of the
    rat. <u>Life Sci.</u> <u>34</u>. 1441-1446.
Cascio C.S. and Kellar K.J.   (1982) Tetrahydro-beta-
    carbolines: affinities for tryptamine and serotonergic
    binding sites. <u>Neuropharmacol.</u> <u>21</u>, 1219-1221.
Cascio C.S. and Kellar K.J.   (1983) Characterization of
    [^{3}H]tryptamine binding sites in brain. <u>Eur. J.</u>
    <u>Pharmacol.</u> <u>95</u>, 31-39.
Cascio C.S. and Kellar K.J.   (1986) Effects of pargyline,
    reserpine, and neurotoxin lesions on [^{3}H]tryptamine
    binding sites in rat brain. <u>Eur. J. Pharmacol.</u> <u>120</u>,
    101-105.
Charlton K.G., Johnson T.D., Maurice R.W. and Clarke D.E.
    (1985) Kynuramine: high affinity for [^{3}H]tryptamine
    binding sites. <u>Eur. J. Pharmacol.</u> <u>106</u>, 661-664.
Denny R.M. and Denny C.B.   (1985) An update on the identity
    crisis of monoamine oxidase: new and old evidence for
    the independence of MAO A and B. <u>Pharmac. Ther.</u> <u>30</u>,
    227-259.
Fuller R.W. and Roush B.W.   (1972) Substrate-selective and
    tissue-selective inhibition of monoamine oxidase.
    <u>Arch. Int. Pharmacodyn. Ther.</u> <u>198</u>, 270-276.
Greenshaw A.J. and Dewhurst W.G. (1987) Tryptamine
    receptors: fact, myth, or misunderstanding? <u>Brain Res.</u>
    <u>Bull.</u> <u>18</u>, 253-256.
Javitch J.A., Uhl G.R. and Snyder, S.H.   (1984) Parkinson-
    ism-inducing neurotoxin, <u>N</u>-methyl-4-phenyl-1,2,3,6-
    tetrahydropyridine: characterization and localization
    of receptor binding sites in rat and human brain.
    <u>Proc. Natl. Acad. Sci. USA</u> <u>81</u>, 4591-4595.
Javitch J.A., D'Amato R.J., Strittmatter S.M. and Snyder
    S.H. (1985) Parkinsonism-inducing neurotoxin, <u>N</u>-
    methyl-4-phenyl-1,2,3,6-tetrahydropyridine: uptake of
    the metabolite <u>N</u>-methyl-4-phenylpyridine by dopamine
    neurons explains selective toxicity. <u>Proc. Natl.</u>
    <u>Acad. Sci. USA</u> <u>82</u>, 2173-2177.

Jones R.S.G. (1982) Tryptamine: a neuromodulator or neurotransmitter in mammalian brain? Prog. Neurobiol. 19,
117-139.

Kaulen P., Bruning G., Rommelspacher H. and Baumgarten,
H.G. (1986) Characterization and quantitative autoradiography of [^{3}H]tryptamine binding sites in rat
brain. Brain Res. 366, 72-88.

Kienzl E., Riederer P., Jellinger K. and Noller N. (1985)
A physicochemical approach to characterize [^{3}H]trypt-
amine-binding-sites in human brain. In, Neuropsychopharmacology of the Trace Amines (Boulton, A.A.,
Maitre, L., Bieck, P.R. and Riederer, P., eds) pp.469-
485, Humana Press, New Jersey.

Knoll J. and Magyar K. (1972) Some puzzling
pharmacological effects of monoamine oxidase inhibitors. Adv. Biochem. Psychopharmacol. 5, 393-408.

McCauley R. (1976) 7[^{14}C]Pargyline binding to
mitochondrial outer membranes. Biochem. Pharmacol.
25, 2214-2216.

McCormack J.K., Beitz A.J. and Larson A.A. (1986) Autoradiographic localization of tryptamine binding sites
in the rat and dog central nervous system. J.
Neurosci. 6, 94-101.

McPherson G.A. (1983) Analysis of radioligand binding
experiments on a microcomputing system. Trends
Pharmacol. Sci. 4, 369-370.

Nelson D.L., Herbert A., Petillot Y., Pichat L., Glowinski
J. and Hamon M. (1979) [^{3}H]Harmaline as a specific
ligand of MAO-A. J. Neurochem. 32, 1817-1827.

Parsons B. and Rainbow T.C. (1984) High-affinity binding
sites for [^{3}H]MPTP may correspond to monoamine
oxidase. Eur. J. Pharmacol. 102, 375-377.

Perry D.C. (1986) [^{3}H]Tryptamine autoradiography in rat
brain and choroid plexus reveals two distinct sites.
J. Pharmacol. Exp. Ther. 236, 548-559.

Phillips S.R. and Boulton A.A. (1979) The effect of monoamine oxidase inhibitors on some arylalkylamines in
rat striatum. J. Neurochem. 33, 159-167.

Rainbow T.C., Parsons B., Wieczorek C.M. and Manaker S.
(1985) Localization in rat brain of binding sites for
parkinsonian toxin MPTP: similarities with [^{3}H]pargy-
line binding to monoamine oxidase.

Reznikoff G., Manaker S., Parsons B., Rhodes C.H. and
Rainbow T.C. (1985) Similar distribution of monoamine
oxidase (MAO) and parkinsonian toxin (MPTP) binding
sites in human brain. Neurol. 35, 1415-1419.

Suzuki O., Katsumata Y. and Oya, M. (1982) Substrate specificity of type A and type B monoamine oxidase. In, <u>Monoamine Oxidase: Basic and Clinical Frontiers</u> (Kamijo, K., Usdin, E. and Nagatau, T. eds.), pp. 74-86, Excerpta Medica, Princeton, New Jersey.

Tipton K.F., Fowler C.J. and Houslay M.D. (1982) Specificities of the two forms of monoamine oxidase. In, <u>Monoamine Oxidase: Basic and Clinical Frontiers</u> (Kamijo, K., Usdin, E. and Nagatau, T. eds.), pp. 87-99, Excerpta Medica, Princeton, New Jersey.

Wood P.L., Martin L.L. and Altar C.A. (1985) [^{3}H]Tryptamine receptors in rat brain. In, <u>Neuropsychopharmacology of the Trace Amines</u> (Boulton, A.A., Maitre, L., Bieck, P.R. and Reiderer, P. eds.) pp. 101-114, Humana Press, New Jersey.

Wood P.L., Pilapil C., LaFaille F., Nair N.P.V. and Glennon R.A. (1984) Unique [^{3}H]tryptamine binding sites in rat brain: distribution and pharmacology. <u>Arch. Int. Pharmacodyn.</u> <u>268</u>, 194-201.

TRYPTAMINE TURNOVER: EFFECTS OF DRUGS

Louis L. Martin[1], Glen B. Baker[2] and Paul L. Wood[1]
[1]Neuroscience Research Dept., Pharmaceuticals Division
CIBA-GEIGY Corporation, Summit, New Jersey, 07060, U.S.A.
and [2]Department of Psychiatry, University of Alberta
Edmonton, Alberta, Canada, T6G 2B7

INTRODUCTION

Tryptamine (T) is a neuroactive compound which is found in trace amounts throughout the mouse CNS with the highest levels occurring in the caudate nucleus (Juorio and Durden, 1984). In addition, subcellular fractionation studies have revealed that T is primarily localized in the synaptosomal fraction and, thus, may be neuronal in origin (Boulton and Baker, 1975). Furthermore, Juorio and Green-shaw (1986) have recently demonstrated by unilateral elec-trolytic lesions of the rat substantia nigra that T ap-pears to exist in neurons of the nigrostriatal pathway. Autoradiographic studies of the distribution of T binding sites in rat brain have similarly demonstrated that the striatum also contains moderately high concentrations of T receptors (Altar et al., 1986). Thus, evidence for the presence of both pre- and post-synaptic components of a tryptaminergic system indicate that T may play a neuro-transmitter role in the striatum.

The accumulation of T following pargyline administra-tion has been used as a measure of the rate of T synthe-sis, or turnover, in the brains of rats and mice (Durden and Philips, 1980; Juorio, 1982). Since the striatum may be an important site of action for T in the mammalian CNS, we have evaluated the effects of several psychoactive compounds on rat stratal T turnover. These compounds included tetrahydro-β-carboline (THBC) and norharmane

95

which have been found to have high affinity for T receptors in vitro (Cascio and Kellar, 1982), three serotonin
(5-HT) antagonists, metergoline, cyproheptadine and methysergide which have been shown to block the effects of T in
vivo (reviewed by Jones (1982)), two classical neuroleptics, haloperidol and chlorpromazine, which have previously been shown to increase T turnover (Juorio, 1982) and
two atypical neuroleptics, clozapine and thioridazine.
The accumulation of T following pargyline administration
was used as a measure of T turnover.

METHODS

Male adult Sprague-Dawley rats, (Tac:N(SD)fBR; Tac-
onic Farms, Germantown, NY; 200-225 g body weight) were
used in all studies. Drugs were administered intraperitoneally. Pargyline HCl (Sigma, St. Louis, MO) was dissolved in saline and administered in a volume of 1 ml/kg
body weight. Haloperidol (McNeil, Fort Washington, PA),
thioridazine HCl, clozapine, methysergide maleate (Sandoz,
East Hanover, NJ), chlorpromazine HCl (Smith, Kline &
French, Philadelphia, PA), cyproheptadine HCl (Merck,
Sharp & Dohme, West Point, PA), metergoline (Farmitalia,
Milan, Italy), norharmane HCl (Sigma, St. Louis, MO) and
tetrahydro-β-carboline (THBC; synthesized by Dr. Charles
Huebner at CIBA-GEIGY) were dissolved in 1% methocel 65 HG
(Fluka, Ronkonkoma, NY) and administered in a volume of 2
ml/kg body weight.

Rats were sacrificed by focused microwaves using a
Thermex Thermatron Metabostat Controller Model 4094 (Ger-
ling Moore, Santa Clara, CA). The striata were removed
and homogenized in 2 ml of 1 N HCl containing 10 pmol of
internal standard ($\alpha,\alpha,\beta,\beta$-tryptamine ($[^2H_4]$T; MSD Stable
Isotopes, St. Louis, MO). The homogenates were centri-
fuged at 35,000 X g for 30 min at 4° C. The supernatants
were decanted into tubes containing 300 ul of 5 N NaOH,
250 ul of 0.5 M sodium phosphate buffer (pH 7.6; containing 1 g of sodium azide per liter) and 10 ul of indi-
cator (100 mg of bromophenol red and 100 mg of bromophenol
blue in 0.4 N NaOH). The pH was adjusted to approximately
7.5 with 2.5 N NaOH using a color change from yellow to
purple as the end point. Standard tubes containing 10
pmol each of T and $[^2H_4]$T were also prepared. The tube
contents were then extracted into 2.5 ml of a 2.5% solu-

tion of diethylhexylphosphate in chloroform. After aspirating the upper layer, the organic layer was extracted with 1 ml of 0.5 N HCl. The upper layer was then transferred to a clean tube containing 150 ul of acetic anhydride. Solid sodium bicarbonate was slowly added with vortexing until bubbling ceased and the solution was saturated. The mixture was then extracted with 2 ml of ethyl acetate. The upper ethyl acetate layer was transferred to a clean tube, dried under nitrogen flow and reacted with 50 ul of ethyl acetate and 100 ul of pentafluoroproprionic anhydride at 70° C for 60 min. At the end of the incubation, the reaction mixture was extracted with 250 ul of cyclohexane and 250 ul of a saturated borax solution and centrifuged. The upper organic layer was transferred to a clean tube and evaporated to dryness. The residues were finally redissolved in 20 ul of ethyl acetate.

T analysis was performed on a Hewlett-Packard 5987 GC-MS (Sunnyvale, CA). Splitless injections of 4 ul aliquots of the samples were made on a 6 ft glass column packed with 3% SP2250 on 100/120 Supelcoport (Supelco, Bellefonte, PA). Helium was used as the carrier (15 ml/min) and methane as the reagent gas (0.5 to 1.0 Torr). The gas chromatographic conditions were: injection port temperature, 250° C and column temperature, 220° C. The mass spectrometric conditions were: interface temperature, 275° C; ion source temperature, 110° C; and electron energy, 230 eV. The retention time for the T derivative was approximately 1.8 min (Fig. I).

Single ion monitoring analyses were carried out in the negative chemical ionization mode, monitoring the ion currents at m/z values of 290 and 310 for the pentafluoroproprionyl (PFP) derivative of acetylated, authentic T and 293 and 313 for the corresponding internal standard derivative. After assessing the identity of endogenous T by coelution with the internal standard and its response ratio at m/z 290 and 310, routine assays were performed by focusing only on m/z 290 and 293, corresponding to T-PFP and $[^2H_4]$T-PFP, the PFP derivatives of acetylated T and internal standard, respectively. The absolute amount of T in each sample was obtained by comparing the ratio of the integrated peak areas of T-PFP (m/z 290) and $[^2H_4]$T-PFP (m/z 293) for the sample with that of the standard.

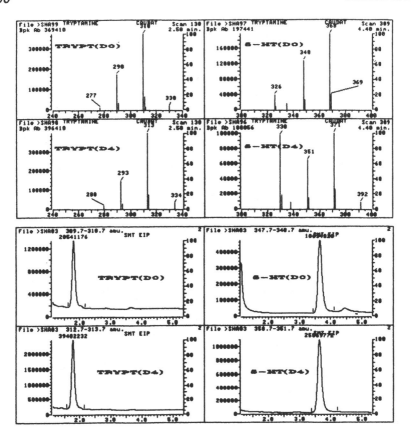

Figure I. Selected ion current profiles and mass spectra of T-PFP, [²H₄]T-PFP, 5-HT-PFP and [²H₄]5-HT-PFP from a sample of rat striatum.

RESULTS

Figure I shows the selected ion current profiles and mass spectra of T-PFP (m/z 290, 310) and [²H₄]T-PFP (m/z 293, 313) for a sample of rat striatum. Also shown are selected ion current profiles and mass spectra for 5-HT-PFP and [²H₄]5-HT-PFP which eluted later from the same sample. Peaks for T-PFP and [²H₄]T-PFP were free from apparent interferences as assessed by the identity ratio of m/z 290/310 which was identical for both standards and biological extracts.

Striatal T levels in saline-treated rats were found to be 67 ± 22 fmol/mg protein (N = 5). This value agrees well with a value of 0.6 ng/g tissue (approximately 40 fmol/mg protein assuming protein constitutes 10% of wet tissue weight) obtained by Juorio and Greenshaw (1986). T levels rose rapidly to 2.75 ± 0.41 pmol/mg protein 1 h following administration of pargyline (75 mg/kg) (Fig. II). Linear regression analysis revealed that T accumulation was linear through the first hour following pargyline administration and that the turnover rate was 2.6 pmol/mg protein/h (fractional rate constant, K = 39 min^{-1}). This result agrees well with a value of 0.24 nmol/g tissue/h (again assuming that protein constitutes 10% of wet tissue weight) obtained by Durden and Philips (1980). Since the amount of T in the striatum increased to approximately 10 times control levels in the first 15 min following pargy-

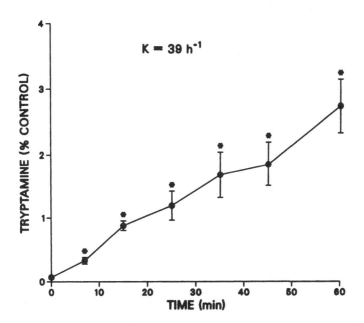

Figure II. Time course of the effects of pargyline (75 mg/kg) on rat striatal T levels. Rats were injected with pargyline at various time intervals prior to sacrifice. Values represent the mean ± SEM of five determinations. * p < .05 versus vehicle, Dunnett's multiple range test on log-transformed data.

line administration, controls were not routinely run and T turnover was estimated by striatal T levels determined 15 min following pargyline injection. Thus, in turnover studies, a test compound (or vehicle) was administered 45 min before pargyline (75 mg/kg) which was in turn injected 15 min prior to sacrifice.

THBC reduced T turnover by 33 and 36% following pargyline administration at doses of 10 and 30 mg/kg, respectively (Fig. III). Lower doses had no effect. Norharmane (30 mg/kg), on the other hand, was without effect on T turnover even at a dose of 30 mg/kg (Table I).

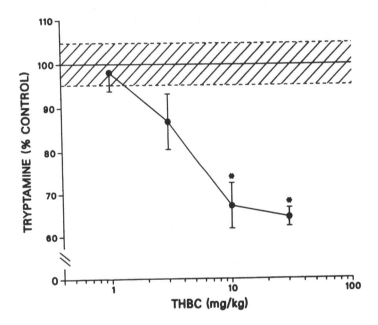

Figure III. Effects of THBC on striatal T levels 15 min after pargyline administration. Rats were injected with THBC or vehicle (1% methocel) 45 min before administration of pargyline (75 mg/kg) which was given to all animals 15 min prior to sacrifice. Values represent the mean ± SEM of seven or eight determinations. The shaded area represents the mean ± SEM of the vehicle-treated group (.714 ± .034 pmol/mg protein; N = 21).
* p < .05 versus vehicle, Dunnett's multiple range test.

TABLE I

Effects of Drugs on Rat Striatal Tryptamine Levels 15 min
   Following Pargyline Administration (75 mg/kg i.p.)

Treatment (mg/kg i.p.)	N	Tryptamine ± SEM (fmol/mg protein)	% of Control
Vehicle	6	300 ± 22	
Norharmane (30)	7	239 ± 33	80
Vehicle	5	330 ± 31	
Methysergide (30)	6	551 ± 42 *	167
Vehicle	7	444 ± 35	
Cyproheptadine (30)	6	658 ± 54 *	134
Metergoline (30)	5	490 ± 34	110
Chlorpromazine (30)	6	568 ± 42 *	128
Vehicle	7	529 ± 30	
Haloperidol (0.6)	6	654 ± 36 *	123
Clozapine (10)	8	634 ± 29 *	120
" (30)	6	752 ± 52 *	142
Vehicle	7	387 ± 54	
Haloperidol (0.2)	6	706 ± 99 *	183
Thioridazine (30)	7	607 ± 65 *	156

Rats were injected with a test compound or vehicle
(1% methocel) 45 min before administration of pargyline
(75 mg/kg) which was given to all animals 15 min prior to
sacrifice.   * $p < .05$ versus vehicle, Dunnett's multiple
range test.

The 5-HT antagonists, methysergide and cyproheptadine, increased T turnover by 67 and 34%, respectively, at
doses of 30 mg/kg (Table I). The 5-HT antagonist, metergoline (30 mg/kg), on the other hand, had no effect on T
turnover. All four neuroleptics, chlorpromazine (30
mg/kg), haloperidol (0.2, 0.6 mg/kg), clozapine (10, 30
mg/kg) and thioridazine (30 mg/kg), increased T turnover.

DISCUSSION

THBC is formed in vivo from T (Hsu, 1985) and exerts many behavioral effects in rats (Rommelspacher et al., 1977). The present results demonstrate that THBC decreased rat striatal T turnover and, therefore, suggest that THBC may function as a central T receptor agonist. Thus, central T receptor agonism may be one mechanism by which THBC produces some of its behavioral actions. Furthermore, THBC may be an endogenous transmitter/neuromodulator at central T receptors. With the exception of amphetamine (Juorio, 1982), no other compound has been shown to reduce T turnover. Thus, THBC may also prove to be an important tool in the study of central tryptaminergic receptor function.

These results contrast with those of Hicks and Langer (1983) who have reported that THBC selectively blocks T-induced vasoconstriction of the isolated perfused rat tail artery. This would suggest that central and peripheral T receptors may be pharmacologically dissimilar and may be distinguished by THBC which acts as a T receptor agonist centrally and as an antagonist peripherally.

Norharmane, which is a THBC analog and also has high affinity for central T receptors (Wood et al., 1984), does not reduce T turnover. This suggests that T (and norharmane) may may bind to multiple T receptors (i.e. those which modulate T turnover and those which do not).

Several 5-HT antagonists (cyproheptadine, methysergide, metergoline and others) and neuroleptic agents (chlorpromazine and trifluoperazine) have been reported to block T-mediated effects (reviewed by Jones (1982)). In addition, Juorio (1982) has demonstrated that several neuroleptics, including haloperidol and chlorpromazine, increase mouse striatal T turnover. In the present study, the 5-HT antagonists, cyproheptadine and methysergide, and the neuroleptics, haloperidol, chlorpromazine, thioridazine and clozapine, all increased T turnover, indicating that they may be either antagonizing T receptors or affecting afferents to tryptaminergic neurons. Only metergoline was without effect on T turnover. Thus, the results of the present study generally agree with previous reports that some 5-HT antagonists block the actions of T in vivo (Vaupel and Martin, 1976) and that neuroleptics increase T turnover (Juorio, 1982). Interestingly, 5-HT antagonists and neuroleptics are essentially inactive in displacing T receptor binding (Cascio and Kellar, 1983; Wood et al., 1984). This argues further for either a difference between T binding sites and T receptors which modulate T turnover or for a transsynaptic action of these agents on tryptaminergic neurons.

REFERENCES

Altar, C.A., Wasley, A.M. and Martin, L.L. (1986) Autoradiographic localization and pharmacology of unique [^{3}H]tryptamine binding sites in rat brain. Neuroscience 17, 263-273.

Boulton, A.A. and Baker, G.B. (1975) The subcellular distribution of β-phenylethylamine, p-tyramine and tryptamine in rat brain. J. Neurochem. 25, 477-481.

Cascio, C.S. and Kellar, K.J. (1982) Tetrahydro-β-carbolines: affinities for tryptamine and serotonergic binding sites. Neuropharmacology 21, 1219-1221.

Cascio, C.S. and Kellar, K.J. (1983) Characterization of [^{3}H]tryptamine binding sites in brain. Eur. J. Pharmacol. 95, 31-39.

Durden, D.A. and Philips, S.R. (1980) Kinetic measurements of the turnover rates of phenethylamine and tryptamine in vivo in the rat brain. J. Neurochem. 34, 1725-1732.

Hicks, P.E. and Langer, S.Z. (1983) Antagonism by tetrahydro-β-carboline of the vasoconstrictor responses to tryptamine in rat tail arteries. Eur. J. Pharmacol. 96, 145-149.

Hsu, L.L. (1985) In vivo formation of 1,2,3,4-tetrahydro-β-carboline from [^{14}C]-tryptamine in the brain. IRCS Med. Sci. 13, 1054-1055.

Jones, R.S.G. (1982) Tryptamine: a neuromodulator or neurotransmitter in mammalian brain? Prog. Neurobiol. 19, 117-132.

Juorio, A.V. (1982) The effects of some antipsychotic drugs, D-amphetamine, and reserpine on the concentration and rate of accumulation of tryptamine and 5-hydroxytryptamine in the mouse striatum. Can. J. Physiol. Pharmacol. 60, 376-380.

Juorio, A.V. and Durden, D.A. (1984) The distribution and turnover of tryptamine in the brain and spinal cord. Neurochem. Res. 9, 1283-1293.

Juorio, A.V. and Greenshaw, A.J. (1986) Tryptamine depletion in the rat striatum following electrolytic lesions of the substantia nigra. Brain Res. 371, 385-389.

Rommelspacher, H., Kauffmann, H., Cohnitz, C.H. and Coper, H. (1977) Pharmacological properties of tetrahydronor-harmane (tryptoline). Naunyn-Schmiedeberg's Arch. Pharmacol. 298, 83-91.

Vaupel, D.B. and Martin, W.R. (1976) Actions of methoxa-mine and tryptamine and their interactions with cypro-heptadine and phenoxybenzamine on cat spinal cord seg-mental reflexes. J. Pharmacol. Exp. Ther. 196, 87-96.

Wood, P.L., Pilapil, C., La Faille, F., Nair, N.P.V. and Glennon, R.A. (1984) Unique [^{3}H]tryptamine binding sites in rat brain: distribution and pharmacology. Arch. Int. Pharmacodyn. Ther. 268, 194-200.

TRACE AMINE BINDING SITES IDENTICAL TO RECEPTORS?

E. Kienzl[1], P. Riederer[2]

[1]Ludwig Boltzmann Inst. Clin. Neurobiology, Lainz-Hospital, Vienna, Austria and
[2]Clin. Neurochemistry, Dept. Psychiat.,
Univ. Würzburg, Würzburg, FRG

INTRODUCTION

One factor deeply influenced research to establish tryptamine (T) as a neurotransmitter or neuromodulator in the central nervous system (CNS): the use of "binding studies" as an experimental method and the assumption that the physiological T-receptor site should be carried by a protein distinct of other binding systems.

When studying a putative receptor system, it is necessary to evaluate the biochemical properties of the system in relation to behaviour that might be mediated by this respective receptor system. Biochemically, ligand binding should be reversible, of high affinity and of finite binding capacity. Furthermore, the interaction of a ligand with its receptor is the first step in several events which finally leads to a biological response. However, little is known about the mechanisms which account for the coupling between ligand binding and the primary response that elicits the biological signal.

T is present in the CNS, is synthesized there and its pharmacology and biochemistry including metabolism, release process etc are known. T is deaminated by MAO intracellularly. Simple diffusion from the synaptic cleft into neurons or glia is assumed for that amine or alternatively, it might enter neurons via uptake pumps.

Experimental studies indicate a clear differentiation of effects mediated by T and serotonin (5-HT). According to recent studies T has strong pharmacological effects on CNS function (Jones 1982; Wood et al. 1984; Cascio and Kellar 1986). 1) Iontophoretic application of T modifies the firing rate of rat cerebral cortical neurons in response to 5-HT probably by modifying the catecholaminergic neurotransmission process. 2) Iontophoretic application of T produces depression of the firing rate on cortical neurons that appeared to be independent of 5-HT systems and 3) T and 5-HT produce opposite effects on rat body temperature when injected into the preoptic area of the hypothalamus.

Binding studies which favor T as a potential transmitter are carried out at the temperature of $0°$ C and reveal a low capacity in the f-molar range. These kinetic properties are complicated in binding experiments under physiological conditions at $37°$ C. The difference may be due to different structural configurations of the reaction product between receptor and ligand, or to the existence of more than one T binding site with different affinities (Kienzl et al. 1985). Some experimental studies focus these important physiological aspects.

MATERIAL AND METHODS

Human brain homogenates of the medial frontal cortex (case 270/86; female; age 54 years; carcinoma; no history of any treatment with psychotropic drugs) were studied after

Figure 1

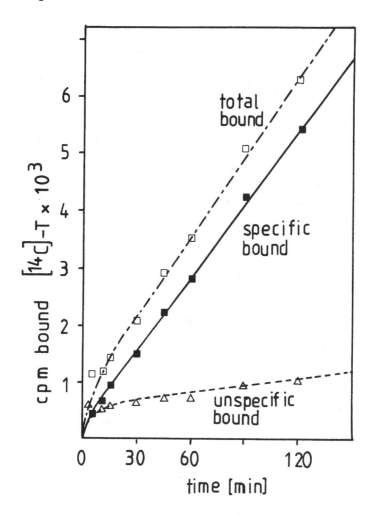

Time course of ⟨¹⁴C⟩-T binding characteristics (1,002 nM) at 37° C to a crude membrane fraction of human frontal cortex of control tissue.

Homogenate: 0,3 mg protein/ml

dissection 5 hours post mortem. Neuropathologic examination using routine histological methods did not show any abnormalities. Brain samples were stored frozen at -80° C. Tissue preparations and radio assay techniques have been described recently (Kienzl et al. 1985) or were performed according the method of Kellar and Cascio (1983).

Dried plant leaves were disrupted by sonification in aqua dest and T binding studies have been performed in homogenates according to Kellar and Cascio (1983). The acid phosphatase activity of DIONAEA MUSCIPULA was determined by an assay kit of Boehringer, Mannheim, FRG, fifteen to thirty minutes after stimulation of digesting mechanisms.

RESULTS AND DISCUSSION

$<^3H>$-T binding in brain homogenate according to Kellar and Cascio (1983) is carried out in 50 mM Tris-HCl buffer with addition of pargyline (inhibition of monoamine oxidase, MAO) and ascorbic acid to prevent oxidation. At 37° C it is predicted that $<^3H>$-T dissociates with a $t_{1/2}$ of less than 1 min resulting in much lower affinities than that obtained at 0° C (Cascio and Kellar 1983). Current studies reveal another high capacity system with lower K_d-values in the p-molar range, which is abolished to a large extent in the presence of MAO-inhibitors (Kienzl et al. 1985). These findings could be demonstrated with $<^3H>$-T as well as with $<^{14}C>$-T. As prolonged incubation of mitochondrial membrane fragments with T promotes the formation of hydrogen peroxide by the MAO-FADH$_2$ complex additional experiments with $<^{14}C>$ labeled were performed to prevent tritium-labeled hydrogen exchange as well as lipid peroxidation with $<^3H>_2$ O$_2$. Fig. 1 shows that this theoretical possibility is not the reason for the long-term kinetic of T-binding at 37° C. Therefore the anomalous behaviour in time, i.e. the longterm

kinetic properties of this process suggest
time-dependent biochemical processes, e.g.
degradation of T and binding of aldehydes
rather (see discussion) than the possibly
simultaneous presence of multiple receptor
membrane proteins. For example, it may well be
that T affects other aminergic binding sites.
Activation experiments of 5-HT-bindg sites with
T might be a step to a more functional
explanation of some still contradictory results
of current experimental developments.

5-HT has a relatively high affinity for
the T-binding site: K_1-value: 280 nM (Kellar
and Cascio 1983). T obviously is a possible
factor in regulating or modulating the 5-HT
transmitter-receptor interaction (Weiner and
Wesemann 1982). Membranes were incubated for 10
min at 37° C mit 7 nM $<^3H>$-5-HT in the presence
of T, 5-methoxy-T and tyramine respectively
prior to the binding assay according the
filtration procedure of Snyder and Bennett
(1976). The data given by Weiner and Wesemann
(1982) demonstrate that under these
experimental conditions all substances are
potent activators of 5-HT binding.
Interestingly they are all substrates of MAO,
which catalyzes the oxidation of the amine via
the corresponding aldehydes. Although T under
certain conditions may act as a MAO-inhibitor,
it was concluded to be unlikely that T
sensitizes 5-HT binding by MAO-inhibiton
(Weiner and Wesemann 1982) because 1) the
effect mediated by T exceeds by far that
observed in the presence of pargyline 2) the
effect of the MAO inhibitors pargyline and
clorgyline is not additive to T induced
activation and 3) the tuberculostatic agent
isoniazid decreases T-induced activation of 5-
HT binding though it has no inhibitory action
on MAO.

The results of these experiments suggest at
least that one site of T-binding present under
physiological assay conditions exhibits

catalytic activities in the serotonergic transmission. However, the underlying mechanism of the long-term kinetic properties of T-binding cannot be explained by these studies and might be caused by slowly synthetized degradation products. Biogenic aldehydes produced from biogenic amines by MAO <indole-3-acetaldehyde (IAL) derived from T> could have the ability to interfere in several physiological processes (Tipton et al. 1977).

Some investigators reported binding studies of IAL to neuronal membranes (Alivisatos and Tabaroff 1973; Nilsson and Tottmar 1985, 1987). Biogenic aldehydes were incorporated into rat membranes in vitro and in vivo. In vitro incorporation was increased by acetaldehyde and inhibitors of aldehyde reductase and aldehyde dehydrogenase. In vivo such effects were caused by ethanol and barbiturates. The absorbing material was shown to be membrane-bound, it could be precipitated with perchloric acid (PCA) and it showed properties of phospholipid or proteolipid products. The binding reaction was non-enzymatic (Nilsson and Tottmar 1985). As T is assumed not to be stored in vesicles it is highly accessible to MAO-degradation, and its rate constant of oxidation is 300 times higher than that of 5-HT indicating a considerable formation of IAL in spite of low concentrations of that trace amine (Nilsson and Tottmar 1985). Thus, the binding data for T at 37° C might reveal the reaction between the biogenic aldehyde and membrane components. If MAO is inhibited such a binding can't detected. It is tempting to speculate that the reaction of IAL with phospholipids influences membrane properties and by such effects may affect 5-HT binding characteristics as well as physiological effects (Sabelli and Giardina 1970). Evidences underlying this assumption is given by recent T-binding studies using butanol extracted myelin (Miyakawa et al. 1982). Brain stems including the hypothalamus, midbrain and medulla oblongata including pons were homogenized in sucrose and

Figure 2

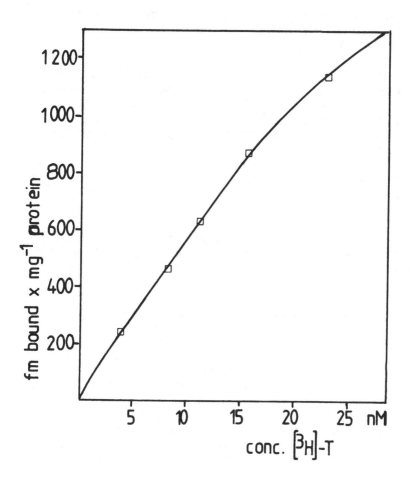

Specific ‹3 3H›-T binding to preparations from dried plant leaves (Genista germanica L.)

B_{Max}: 7797 fm/mg protein, K_D: 0,128 µM
Temperature: 0° C
Incubation time: 1 hour

extracted with 10 vol. of water-saturated
butanol for 2 hours at room temperature.
Incubations were performed at room temperature
with several concentrations of an aqueous
(^{14}C)-T solution.

The binding curve was resolved into
saturable and non saturable binding components
and evidence has been presented that the
specific binding capacity for (^{14}C)-T consisted
of proteolipid components. Displacement studies
indicated that among various compounds (eg.
indoleamine analogues and other
neurotransmitters) only T and 5-methoxy-T
inhibited T-binding.

On the basis of these findings it is
suggested that lipids and proteolipids can act
directly as binding sites themselves (Loh and
Law 1980).

TRYPTAMINE BINDING IN PLANT TISSUE

T-binding sites also exist in plant tissue.
The binding studies were performed according the
method of Kellar and Cascio (1983) to dried
Indigo-plants. Specific T-binding at 0° C
revealed a maximum binding capacity in the p-
molar range (Fig. 2). Table 1 shows that these
binding sites are dependent from various assay
conditions and that they are specific with
regard to Scatchard and Hill analysis. In
addition Table 2 indicates selectivity of T-
binding among tissues of various plant sources.
What could contribute to a physiological
relevance of biogenic amine-binding sites in
plant tissue? Plants contain low molecular
weight substances, active at very low
concentrations in regulating growth and
development. These compounds are capable of
evoking responses in regions distinct from sites
of synthesis. These substances are not expected
to function as co-factors of enzymes and they
are active without metabolic conversion. In

Table 1 *Specific <³H>-tryptamine binding to homogenates of plant leaves*

> GENISTA GERMANICA
> B_{Max} pmol/mg prot.   6,69
> K_D nM   39,86
> Hill N   1,6
>
> CASSIA ACUTIFOLIA DEL.
> B_{Max} pmol/mg prot.   7,92
> K_D nM   147,59
> Hill N   2,2

<³H>-T binding (31,4 nM) in Genista germanica L depends from the variation of the assay conditions and from storage conditions and state of ripening. For example at 20° C and an incubation time of 45 min. 2,83 pmol/mg prot. <³H>-T are bound, while this value is slightly higher (3,25 pmol/mg prot.) at 0° C and 120 min. of incubation. Indole acetic acid (10^{-7} M) inhibits T-binding to 43,7 % of controls. 5-hydroxyindole acetic acid (10^{-6} M) blocks T-binding < 20,9 nM <³H>-T > to 17 % . However, this inhibitory effect varies in other plant tissues (unpubl. observations).

Table 2 *Specific <³H>-tryptamine binding (31.4 nM) in various plant tissues*

PLANT TISSUE	<³H>-T bound (pmol/mg protein)
PHASEOLUS VULGARIS L. ssp. vulgaris	0.23
CASSIA ACUTIFOLIA DEL.	0.70
MYRISTICA FRAGRANS HOUTT.	2.16
GENISTA GERMANICA L.	2.83
SOLANUM CYCOPERSICUM	0.12

plants there might be simple systems to show the
response of a tissue preparation to a given
concentration of a "plant-hormone" which might
be adequate to some change in "receptor
properties". Also structure – activity
relationships can be studied in such simple
models. Only a certain charge distribution in
the energetically optimum geometric conformation
allows a substance to act as growth-hormone or
auxin. Its activity depends upon a
conformational change in the receptor induced
binding of the auxin molecule (Venis 1985).

Plant hormones also elicit specific
qualitative changes in RNA and protein synthesis
(Higgins et al. 1976). Studies confirmed that
auxins produce rapid and specific changes in
gene transcription. It became evident that the
lag period preceding auxin-induced growth was
in the range of only 8-15 minutes and was even
shorter under certain experimental conditions.
It is suggested that the rapid auxin-induced
changes in plasma membrane structures seem to be
caused by interaction with membrane-bound
receptors. In contrast, long lasting auxin-
mediated growth response and long-term effects
on enzyme synthesis may reflect selective
transcription-dependend changes probably
requiring intervention of soluble receptors of
nuclear and/or cytoplasmic origin.

Such mechanisms might explain the response
of the secretory cells of carnivorous DIONEA
MUSCIPULA. This plant captures insects which
serve mainly as exogenous nitrogen source and
supply of mineral nutruients (Heslop-Harrison
1978)

Preliminary studies give evidence for a
stimulation of the acid phosphatase secretion by
feeding the plant with flies or after
administration of a solution of harmaline (µ-
molar to molar range). It is suggested that this
T-analogue exerts this stimulatory activity via
T- or tetrahydro-ß-carboline binding sites or

Figure 3

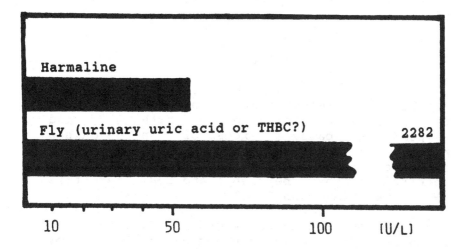

acid phosphatase activity in the digestion medium of dionea muscipula (pH 3,5)

One fly was applied to one Venus's-flytrap. After touching the trigger hairs located at the inner leaf surface the two sides of the leaf move quickly together closing the trap. It is suggested that urinary constituents are excreted by the anxious fly thus stimulating the production and secretion of digesting medium including acid phosphatase. 15 to 30 minutes after capturing the fly or after administration of substances like glycine, glutamate, valine, leucine, isoleucine, trpytophan, tyrosine, phenylalanine, harmaline (each in a range of 1 µmol to 1 mmol) and optimum secretion of digesting medium was expected. Acid phosphatase activity was measured with an assay kit obtained from Boehringer, Mannheim, FRG; p-nitrophenyl-phosphate was used as substrate. Only harmaline led to a significant increase in the activity of acid phosphatase.

receptors (Fig. 3). Therefore, aminergic binding sites or receptors seem to be essential to establish certain physiological functions in plant tissues.

CONCLUSION

Gaddum (1953) was the first who attempted to link T to receptor function. He identified receptors which responded to both 5-HT and tryptamine because an excess of T on the guinea pig ileum abolished the response to both substances. In fact, there seems to be an interrelationship between T and 5-HT receptors (Weiner and Wesemann 1982).

Current studies raise the possibility that phospho- or proteolipids might act as T-acceptor sites thus modulating membrane properties and receptor activities (Miyakawa et al. 1982). Agreeing with these experiments oxidative deamination products of T (conversion of T via MAO to IAL) bind with long-term kinetic properties to brain tissue and thus could offer new possibilities for further studies on physiological and pathological functions (Riederer et al. 1985; Nilsson and Tottmar 1987). However, such regulative principles in neural systems await future investigations.

ACKNOWLEDGMENT

We thank Mrs J. Philipp for the careful preparation of the manuscript.

REFERENCES

Alivisatos S.G.A. and Tabaroff B. (1973) Formation and metabolism of "biogenic" aldehydes. In, <u>Chemical Modulation of Brain Function</u> (Sabelli H., ed) pp 41-46, Raven Press, New York.

Cascio C.S. and Kellar K.J. (1986) Effects of pargyline, reserpine and neurotoxin lesions on <³H>-tryptamine binding sites in rat brain. Eur. J. Pharmacol. 120, 101-105.

Gaddum J.H. (1953) Tryptamine Receptors. J. Physiol. 119, 363-368.

Heslop-Harrison Y. (1978) Carnivorous Plants. Scientific American, February, 104-115.

Higgins T.J.V., Zwar J.A. and Jacobsen J.V. (1976) Gibberellic acid enhances the level of translatable mRNA for alpha-amylase in barley aleurone layers. Nature (London) 260, 166-169.

Jones R.S.G. (1982) A comparison of the responses of cortical neurones to iontophoretically applied tryptamine and 5-hydroxytryptamine in the rat. Neuropharmacology 21, 209-214.

Kellar K.J. and Cascio C.S. (1983) Characterization of <³H>-tryptamine binding sites in brain. Eur. J. Pharmacol. 95, 31-39.

Kienzl E., Riederer P., Jellinger K. and Noller H. (1985) A physicochemical approach to characterize <³H>-tryptamine binding sites in human brain. In, Neuropsychopharmacology of the Trace Amines (Boulton A.A., Bieck P.R., Maitre L. and Riederer P., eds) pp 469-485, Humana Press, New Jersey.

Loh H.H. and Law P.Y. (1980) The role of membrane lipids in receptor mechanisms. Annu. Rev. Pharmacol. Toxicol. 20, 201-234.

Miyakawa A., Kim S., Kameyama T. and Ishitani R. (1982) Butanol extracts from myelin fragments: Some characteristics of tryptamine binding. Japan. J. Pharmacol. 32, 1051-1057.

Nilsson G.E. and Tottmar O. (1985) Biogenic aldehydes in brain: Characteristics of a reaction between rat brain tissue and indole-3-acetaldehyd. J. Neurochem. 45 (3), 744-751.

Nilsson G.E. and Tottmar O. (1987) Biogenic aldehydes in brain: On their preparation and reactions with rat brain tissue. J.

Neurochem. <u>48</u> <u>(5),</u> 1566-1572.

Riederer, P., Kienzl E., Jellinger K. and Noller H. (1985) Characterization of tryptamine binding sites in human brain tissue. In, <u>Biological Psychiatry 1985</u> (Shagass Ch., Josiassen R.C., Bridger W.H., Weiss K.J., Stoff D., Simpson G.M., eds) pp 338-340, Elsevier, New York, Amsterdam, London.

Sabelli H.C. and Giardina W.J. (1970) Tryptaldehydes (indoleacetaldehydes) in serotonergic sleep of newly hatched chicks. <u>Arzneimforsch. (Drug Res.)</u> <u>20,</u> 74-80.

Snyder S.H., Bennett J.P. (1976) Serotonin and lytergic acid diethylamide binding in rat brain membranes: Relationship to postsynaptic serotonin receptors.<u>Mol. Pharmacol.</u> <u>12,</u> 373-389.

Tipton K.F., Hoslay M.D. and Turner A.J. (1977) Metabolism of aldehydes in brain. <u>Essays Neurochem. Pharmacol.</u> <u>1,</u> 103-138.

Venis M. (1985) <u>Hormone binding sites in plants.</u> Longman, New York, London.

Weiner N. and Wesemann W. (1982) In vitro studies on the modification of 5-HT receptor sensitivity in the CNS of the rat. In, <u>Biologische Psychiatrie</u> (Beckmann H., ed) pp 247-252, Thieme, Stuttgart, New York.

Wood P.L., Pilapil C. Lafaille F., Nair P.V. and Glennon R.A. (1984) Unique <³H>- tryptamine binding sites in rat brain: distribution and pharmacology. <u>Arch. Int. Pharmacodyn.</u> <u>268,</u> 194-201.

HIGH AFFINITY BINDING OF P-TYRAMINE: A PROCESS

IN SEARCH OF A FUNCTION

Andrea Vaccari

Department of Neurosciences
Chair of Medical Toxicology
University of Cagliari
Cagliari, Italy

INTRODUCTION

The presence of trace amines in the CNS of different animal species, their interactions with monoaminergic path ways, and their intrinsic pharmacological activities have jus tified recent efforts to ascertain whether they may play a neurotransmitter role. The interaction with specific binding sites is a major requisite for a neurotransmitter, though "binding site" does not mean "receptor", i.e. a physiologi cally-relevant entity (Laduron,1984).Additionally, the candi date transmitter must be taken up in the synapse with an energy-dependent process, and must also be released from storage organelles. Para-tyramine (pTA) is actively taken up by nerve tissue preparations such as brain slices (Dyck, 1978), synaptosomes (Ungar et al.,1977) and synaptic vesi cles (Lentzen and Philippu,1977), from where it also under goes both spontaneous and stimulus-evoked release (Lentzen and Philippu,1977; Dyck,1984). Furthermore, pTA displays electrophysiologic (Jones,1984), behavioral (Stoof et al.,1976) and neuroendocrine (Becù-Villalobos et al.,1985) activities. Thus the existence of specific pTA receptors in "tyraminer gic" neurons has been postulated, though the absence of hi stochemically-evident pTA-like immunoreactivity in the rat CNS (Osborne,1985) contrasts with this hypothesis. Recently, high affinity and specific binding sites for pTA in the rat CNS have been identified and characterized (Vaccari,1986).

In the present article two questions will be particularly addressed: (1) What significance, if any, can be attached to pTA binding sites? and (2) Is pTA binding functionally rela ted to the transport process of PTA and dopamine (DA)?

BRAIN AND TISSUE DISTRIBUTION OF pTA SITES

The binding sites of ^{3}H-pTA were highly concentrated in pTA- and DA-rich regions such as the c.striatum (Table 1). This distribution is consistent with the postulated involve ment of pTA in DA-pathways (see Juorio,1982). No binding was detected in heart, kidney and liver membranes, thus gi ving a purely metabolic- rather than functional significance to the high levels of endogenous pTA in those organs (Phi lips,1984). The binding capacity was similar in rat, mouse and calf striatal membranes; the affinity values varied some what in different animal species (Table 1).The purity of la beled pTA used greatly influenced specifically bound radioac tivity. As a matter of fact, when ^{3}H-pTA was sent us in dry ice by the dealer, the B_{max} values were approx. 3-fold as hi gher as those obtained in the absence of storage precautions. Therefore, the B_{max} values from the most recent experiments may be greater than those previously published (Vaccari,1986).

PHARMACOLOGIC CHARACTERIZATION OF pTA BINDING SITES

Displacement Experiments

Involvement of pTA in the DA-system was also suggested by competition experiments where among several agents tested, reserpine and DA were the most potent displacers, with K_i-values of approx. 10 nM (Table 2, see also Vaccari,1986). In contrast, DA-related agonists such as apomorphine and per golide, the DA-precursors tyrosine and DOPA, and the pTA metabolite octopamine were weak or poor displacers of pTA binding, and DA-antagonists known to label D_2 and D_1 recep tors required approx. μM concentrations to inhibit by 50% the specific binding of pTA. Most interestingly, all DA-reup

Table 1

Distribution of pTA binding sites

Tissue	B_{max}	K_D
RAT		
C.Striatum	7240 + 420	14.4 + 1.0
Hypothalamus	670 + 88	12.9 + 1.4
C.Cortex	275 + 52	12.4 + 1.3
Pons-medulla	112 + 14	10.8 + 1.8
Cerebellum	61 + 8	13.8 + 2.1
MOUSE		
C.Striatum	7650 + 490	36.9 + 2.8
CALF		
C.Striatum	6450 + 513	24.3 + 1.9

The C.Striatum included caudate-putamen, n.accumbens and olfactory tubercles. N=4. B_{max} = fmol/mg protein; K_D=nM For further details see Vaccari (1986).

take inhibitors tested, such as nomifensine, benztropine, D-amphetamine, methamphetamine, methylphenidate, and also NA were fairly potent displacers of pTA, with K_i-values ranging from 90 to 150 nM (Table 2). In addition, the potencies of various drugs in competing for pTA binding to rat striatal membranes correlated closely (r=0.96) with their potencies at inhibiting DA-uptake into striatal synaptosomes, and not at all with the displacement of DA-binding to rat or calf striatal membranes (Table 2). Citalopram, a 5-HT uptake inhibitor, and imipramine, a blocker of NA uptake, did not displace pTA binding; in contrast, pTA can inhibit the uptake of DA, 5-HT and NA into synaptic vesicles (Matthaei et al., 1976). Thus the suggestion that 5-HT and NA are transported by the carrier for DA and pTA was here not supported. In summary, these results indicate a functional role of pTA binding sites in the transport system of DA.

Table 2

Pharmacologic characterization of pTA binding sites

Agent	^3H-pTA binding K_i	^3H-DA uptake K_i	^3H-DA binding K_i
Reserpine	5.5	9.6	> 10000
DA	13	66	17
pTA	16	380	> 10000
Tetrabenazine	43	--	--
Benztropine	90	72	30000
L-NA	92	180	200
Nomifensine	95	51	--
Methylphenidate	144	160	> 10000
D-Amphetamine	153	82	> 10000
Haloperidol	556	--	920
Spiperone	1075	--	1400
(+)Butaclamol	3189	--	80
Chlorpromazine	3857	3400	900
Apomorphine	6825	--	8.6
Imipramine	13724	16000	22000
Citalopram	36700	28000	--
Sulpiride	> 50000	> 100000	43000
Chlorgyline	> 100000	--	--
Deprenyl	> 100000	--	--

The K_i-values for inhibition of pTA binding are from Vacca ri (1986); those for DA-uptake into rat striatal synaptosomes are from Javitch et al. (1983) and Richelson and Pfenning (1984); those for DA-binding to rat or calf striatal membra nes are from Burt et al. (1976) and List and Seeman (1982).

Solubilization Experiments

Displacement-based assays had thus indicated that pTA binding sites, while being correlated with the transport pro cess of Da, do not share characteristics of a recognized DA receptor. In order to further demonstrating that DA receptors and pTA binding sites are different molecular entities, rat and calf striatal membranes were incubated 1 hour over ice in the presence of the zwitterionic detergent CHAPS (0.5%), or digitonin (1%) according to procedures which allow solubi lization of ^{3}H-DA and ^{3}H-spiperone receptors (Davis et al., 1982; Kilkpatrick et al.,1985). The supernatants obtained shared binding characteristics of D_2- or ^{3}H-DA receptors, however they did not bind pTA (Table 3), a finding that highlights striking differences in the physico-chemical compo sition of DA and pTA binding sites.

SYNAPTIC LOCALIZATION OF pTA BINDING SITES

The receptor source for present experiments was a cru de membrane preparation, thus an heterogeneous mixture of nuclear, mitochondrial, pre- and postsynaptic membrane frag ments, and varying amounts of intact or broken storage ve sicles trapped down during the centrifugation. This implied a multiple localization of the specifically bound radioactivity (Laduron,1984) and it is, of course, relevant to the purposes of precisely localizing pTA binding sites. No consistent bin ding of pTA could be detected in two mitochondrial pellets ob tained in subcellular fractionation experiments, while both ve sicular bands and synaptosomal membranes bound pTA (unpu blished results). In addition, a target role of MAO proteins could be excluded inasmuch as: (1) the binding reaction of pTA is usually run in the presence of 10 µM pargyline, an irreversible MAOI, and (2) the addition of 100 µM clorgyline (a MAO-A inhibitor) or deprenyl (MAO-B inhibitor) did not affect the binding reaction of pTA. We have, thus, restric ted our attention to pre- or postsynaptic synaptosomal membra nes, and to synaptic vesicles as to the putative "containers" of pTA binding sites.

Table 3

Solubilization of pTA and DA binding sites

Ligand	Native membranes		CHAPS 0.5%		Digitonin 1%	
	Rat	Calf	Rat	Calf	Rat	Calf
^{3}H-pTA						
B_{max}	7.58	5.42	O	O	O	O
K_D	16.9	21.3				
^{3}H-SPIP.						
B_{max}	0.16	0.24	0.13	0.11	0.74	0.09
K_D	0.05	0.21	0.78	2.38	1.21	2.68
^{3}H-DA						
B_{max}	0.48	0.12	0.19	0.04	0.34	0.09
K_D	1.22	1.39	2.13	2.90	2.90	3.60

B_{max} = pmol/mg protein; K_D = nM; N=2, triplicates. Binding conditions were as follows: ^{3}H-pTA 4 nM ($\pm$ 10µM pTA), 10 min./37°C or overnight/4°C. ^{3}H-Spiperone 0.5 nM ($\pm$ 30 µM sulpiride), 2 hours/4°C. ^{3}H-DA 0.8 nM ($\pm$ 1 µM DA), overnight/4°C. Filtration in 0.3% polyethylenimine-soaked GF/B filters.

Effects of Nigro-Striatal Lesions or Reserpine

A presynaptic location of pTA binding sites was inferred from the marked (by 79%) decrease in the density of pTA sites in the lesioned striatum of rats 8 days following surgical nigro-striatal hemitransection (Vaccari, 1986). Similarly, surgical or chemical degeneration of DA-nerve terminals has been shown to strongly decrease DA- and pTA uptakes into striatal slices and synaptosomes, and ^{3}H-nomifensine binding to rat striatal membranes, this antidepressant being a marker for the DA-uptake pump (Dyck, 1984; Dubocovich and

Zahniser,1985). Consistent with these findings, lesioning experiments proved that the number of ^{3}H-spiperone-labeled, D_2 postsynaptic sites was increased (supersensitivity)(Hall et al.,1983), and administration of reserpine to rats severely (by 86%) decreased the number of pTA sites (Vaccari, 1986). Reserpine is known to deplete catecholamine (CA) stores since it inhibits their reuptake into a variety of secretory vesicles (Stitzel,1977), a consequence of its covalent interaction with the CA transporter, rather than with the proton-translocating ATPase (Bashford et al.,1976). Additionally, reserpine enhances the efflux of Ca from vesicular to cytosolic synaptosomal storage sites (Takimoto et al.,1983). A possible explanation for the impaired binding of pTA in DA-depleted striatal membranes is that pTA cannot bind to the DA-carrier inasmuch as it is covalently occupied by reserpine. ^{3}H-Reserpine, in fact, once bound is only displaced by the addition of large excess (in the µM or mM range) of monoamines (Near and Mahler,1983; Deupree and Weaver, 1984; Near,1986). Bearing in mind such considerations, it seems evident that pTA binds to the DA-transporter. Reserpine is, indeed, a specific probe for the CA-transporter protein in synaptic vesicles and adreno-medullary chromaffin granules (Near and Mahler,1983; Deupree and Weaver,1984; Scherman and Henry,1984; Parti et al.,1987). In our experiments (Table 2) reserpine in vitro has proven to be the most powerful displacer of pTA binding among the chemicals used. Tetrabenazine, an additional putative marker for the DA-carrier (Scherman,1986; Near,1986) also actively inhibited specifically bound pTA (Table 2).

Effects of Ca^{2+}-Channel Antagonists on pTA sites

That pTA might interact with the carrier molecule for DA can be also indirectly inferred by the pTA-provoked, reserpine-sensitive release of DA and other CA from various storage organelles (see Philippu,1976) and from synaptosomes (Raiteri et al.,1977). This effect would imply that DA utilizes the membrane carrier in the inside-outside direction (Sweadner,1985). Furthermore it is generally accepted that

the release of DA from storage vesicles is related to an im
pulse-linked influx of Ca^{2+}-ions, thus it is calcium-depen
dent (see Philippu,1976; Vizi,1978). Consistent with the lat
ter statement, Ca^{2+}-channel antagonists have been shown to
directly affect neurotransmitter uptake and release in the
brain (Brown et al.,1986; Miller,1987).

    In our experiments (Table 4), the dihydropyridines ni
fedipine and nimodipine were fairly potent inhibitors of pTA
binding, with K_i-values in the nM range. Nicardipine, the
piperazine flunarizine and the papaverine verapamil acted
at μM concentrations; the benzothiazepine diltiazem did not
displace pTA binding. SKF 525A (proadifen) and $LaCl_3$,
two calcium-uptake blockers, also inhibited the binding of
pTA at fairly low concentrations (Table 4). The alkaloid ve
ratridine, a Na^+-channel agonist which stimulates Ca^{2+}-depen
dent release of vesicular CA, and the release of pTA from
striatal slices (Dyck,1984) displaced pTA binding at concen
trations in the range of those of dihydropyridines (Table 4).

Table 4

Effects of Ca^{2+}-channel antagonists and Calcium-
agents on pTA binding to rat striatal membranes

Drug	$K_i(nM)$	n
Nifedipine	223 $\pm$ 31	3
Nimodipine	445 $\pm$ 150	3
Flunarizine	1205	2
Nicardipine	3913	2
(±)Verapamil	8234	2
Diltiazem	> 50000	2
SKF 525A	2225	2
$LaCl_3$	519	2
Veratridine	650 $\pm$ 160	3

Mean values $\pm$ S.E.M. from n experiments in triplicate.
Drugs were added at the start of binding reaction; ^{3}H-pTA
4 nM.

The differential activity of Ca^{2+}-antagonists would be consi_
stent with the existence of multiple Ca^{2+}-channels in neuro_
nal membranes (Miller,1987). If dihydropyridine receptors in
rat brain are an integral subunit of the Ca^{2+}-channel, then
it seems likely that the carrier-associated pTA binding sites
are structurally linked with the Ca^{2+}-channel.

FACTORS INFLUENCING THE BINDING OF pTA

The binding of pTA depended upon the presence of
Na^+-ions in the incubation medium, inasmuch as it was aboli_
shed in the absence of NaCl (Vaccari,1986). Furthermore,
the rate of association of pTA was low at 4°C, while the equi_
librium was attained by 10 min. at 37°C. After 30 min. of
incubation at 37°C the specifically bound pTA had almost com_
pletely disappeared, what might suggest that the putative
pTA-carrier complex is instable at physiologic temperatures,
a consequence of the need of rendering the carrier molecules
rapidly available for further binding of monoamines. Na^+-
and temperature-dependence are features of those transport
processes where the monoamine carrier is activated by the
electrochemical proton gradient built up by a proton pump
ATPase (Toll et al.,1977). However, they are also requisites
for changes in the affinity state of some membrane receptors
for monoamines such as those which are somehow coupled to
adenylate cyclase, the D_2-receptor included (see Watanabe
et al.,1985). Furthermore, Na^+-ions regulate the membrane
binding of ^3H-nomifensine (Dubocovich and Zahniser,1985),
^3H-mazindol (Javitch et al.,1983), ^3H-cocaine (Kennedy and
Hanbauer,1983), and ^3H-GBR 12783 (Bonnet et al.,1986),
four potent inhibitors and putative markers of the transport
system of DA, nomifensine being also an active displacer of
pTA binding (Table 2). In addition, the binding sites of
^3H-reserpine have been shown to be located at the monoamine
carrier, which can exist in both active and inactive forms,
depending upon microenvironmental conditions (Scherman and
Henry,1984). The binding of pTA was dependent also on the
ATPase integrity. It was, in fact, inhibited by ouabain, an
ATPase inhibitor known to affect synaptosomal uptake of mo_
noamines (Toll et al.,1977); by dinitrophenol, an uncoupler

of oxydative phosphorylation known to block the Na^+-depen
dent uptake of pTA by striatal slices (Dyck,1984); by nige
ricin, which specifically affects the transmembrane pH gra
dient;. by saponin, which dissipates the potential gradient,
and by N-ethylmaleimide, which interacts with -SH groups
in the ATPase protein (Vaccari,1986). In our experiments,
the addition of ATP and/or Mg^{2+}-ions to the incubation me
dium did not affect the binding of pTA; in contrast, the up
take of pTA and DA into synaptic vesicles of the pig caudate
nucleus was enhanced by $ATP-Mg^{2+}$ (Matthaei et al.,1976;
Lentzen and Philippu,1977). Furthermore the presence of
$ATP-Mg^{2+}$ in the medium was necessary for reserpine-provo
ked inhibition of pTA uptake (Lentzen and Philippu,1977),
whereas reserpine did potently displace pTA binding to stria
tal membranes in the absence of $ATP-Mg^{2+}$ (Table 2).

These findings do probably reflect the smaller energe
tic and ionic requirements in the association of a ligand
(pTA) with a carrier molecule, compared to those implied in
the translocation of the ligand through synaptosomal or vesi
cular membranes.

CONCLUSIONS

The specific and high affinity binding sites of ^{3}H-pTA
here described do not share pharmacologic and chemico-phy
sical characteristics of a postsynaptic DA-receptor. In DA-
innervated regions such as the c.striatum, they appear to
be localized at both terminal and vesicular membranes, proba
bly as a component of the carrier molecule for DA-transport.
The putative pTA sites--carrier entity is functionally linked
with dihydropyridine Ca^{2+}-channels and the H^+-translocating
ATPase. In conclusion, pTA binding sites might be conside
red as receptors which trigger their own transport through
neuronal membranes. This conclusion fits well with the hypo
thesis that transmitter uptake is, collectively, receptor inter
nalization at presynaptic terminals or elsewhere (La Bella,
1985).

REFERENCES

Bashford C.L., Casey R., Radda,G.K. and Ritchie G.A. (1976) Energy-coupling in adrenal chromaffin granules. Neuroscience 1, 399-412.

Becù-Villalobos D., Lacau De Mengido I.M. and Libertùn C. (1985) p-Tyramine, a natural amine, inhibits prolactin re lease in vivo. Endocrinology 116, 2044-2048.

Bonnet J.J., Protais P., Chagraoui A. and Costentin J. (1986) High-affinity of ^{3}H-GBR 12783 binding to a speci fic site associated with the neuronal dopamine uptake com plex in the central nervous system. Eur.J.Pharmacol. 126, 211-222.

Brown N.L., Sirugue O. and Worcel M. (1986) The effects of some slow channel blocking drugs on high affinity se rotonin uptake by rat brain synaptosomes. Eur.J.Pharma col. 123, 161-165.

Burt D.R., Creese I. and Snyder S.H. (1976) Properties of ^{3}H-haloperidol and ^{3}H-dopamine binding associated with dopamine receptors in calf brain membranes. Molec.Phar macol. 12, 800-812.

Davis A., Madras B.K. and Seeman P. (1982) Solubilized re ceptors for ^{3}H-dopamine (D_3 binding sites) from canine brain. Biochem.Pharmacol. 31, 1183-1187.

Deupree J.D. and Weaver J.P. (1984) Identification and cha racterization of the catecholamine transporter in bovine chromaffin granules using ^{3}H-reserpine. J.Biol.Chem. 259, 10907-10912.

Dubocovich M.L. and Zahniser N.R. (1985) Binding characte ristics of the dopamine uptake inhibitor ^{3}H-nomifensine to striatal membranes. Biochem.Pharmacol. 34, 1137-1144.

Dyck L.E. (1978) Uptake and release of meta-tyramine, para-tyramine and dopamine in rat striatal slices. Neurochem. Res. 3, 775-791.

Dyck L.E. (1984) Neuronal transport of trace amines: an over view. In, Neurobiology of the Trace Amines (Boulton,A.A., Baker,G.B., Dewhurst,A.G., and Sandler,M., eds) pp 185-204, Humana Press, N.J.

Hall M.D., Jenner P., Kelly E. and Marsden C.D. (1983) Differential anatomical location of ^{3}H-N,n-propylnorapomorphine and ^{3}H-spiperone binding sites in the striatum and substantia nigra of the rat. Br.J.Pharmacol. 79, 599-610.

Javitch J.A., Blaustein R.O. and Snyder S.H. (1983) ^{3}H-Mazindol binding associated with neuronal dopamine uptake sites in corpus striatum membranes. Eur.J.Pharmacol. 90, 461-462.

Jones R.S.G. (1984) Electrophysiological studies on the possible role of trace amines in synaptic function. In, Neurobio - logy of the Trace Amines (Boulton,A.A., Baker,G.B., Dewhurst,A.G., and Sandler,M. , eds) pp 205-223, Humana Press, N.J.

Juorio A.V. (1982) A possible role for tyramines in brain function and some mental disorders. Gen.Pharmacol. 13, 181-183.

Kennedy L.T. and Hanbauer I. (1983) Sodium-sensitive cocaine binding to rat striatal membranes: possible relationship to dopamine uptake sites. J.Neurochem. 41, 172-178.

Kilkpatrick G.I., Jenner P. and Marsden C.D. (1985) Properties of rat striatal D-2 dopamine receptors solubilized with the zwitterionic detergent CHAPS. J.Pharm.Pharmacol. 37, 320-328.

La Bella F.S. (1985) Neurotransmitter uptake and receptor-ligand internalization- are they two distinct processes? Trends Pharmacol.Sci. 6, 319-322.

Laduron P.M. (1984) Criteria for receptor sites in binding studies. Biochem.Pharmacol. 33, 833-839.

Lentzen H. and Philippu A. (1977) Uptake of tyramine into synaptic vesicles of the caudate nucleus. Naunyn-Schm. Arch.Pharmacol. 300, 25-30.

List S.J. and Seeman P. (1982) ^{3}H-Dopamine labeling of D_3 dopaminergic sites in human, rat, and calf brain. J.Neurochem. 39, 1363-1373.

Matthaei H., Lentzen H. and Philippu A. (1976) Competition of some biogenic amines for uptake into synaptic vesicles of the striatum. Naunyn-Schm.Arch.Pharmacol. 293, 89-96.

Miller R.J. (1987) Multiple calcium channels and neuronal function. Science 235, 46-52.

Near J.A. and Mahler H.R. (1983) Reserpine labels the cate cholamine transporter in synaptic vesicles from bovine cau date nucleus. F.E.B.S. 158, 31-35.

Near J.A. (1986) ^{3}H-Dihydrotetrabenazine binding to bovine striatal synaptic vesicles. Molec.Pharmacol.30, 252-257.

Osborne N.N. (1985) Tryptamine, tyramine and histamine im munoreactivity in specific neurones of the gastropod CNS. In, Neuropsychopharmacology of the Trace Amines (Boul ton,A.A., Maitre,L., Bieck,P.R., and Riederer,P. eds) pp 115-124, Humana Press, N.J.

Parti R., Ozkan E.D., Harnadek G.J. and Njus D. (1987) Inhibition of norepinephrine transport and reserpine bin ding by reserpine derivatives. J.Neurochem. 48, 949-953.

Philippu A. (1976) Transport in intraneuronal storage vesi cles. In, The Mechanism of Neuronal and Extraneuronal Transport of Catecholamines (Paton,D.M., ed) pp 215-246, Raven Press, N.Y.

Philips S.R. (1984) Analysis of trace amines: endogenous le vels and the effects of various drugs on tissue concentra tions in the rat. In, Neurobiology of the Trace Amines (Boulton,A.A., Baker,G.B., Dewhurst,A.G., and Sandler, M., eds) pp 127-143, Humana Press, N.J.

Raiteri M., Del Carmine R., Bertollini A. and Levi G. (1977) Effect of sympathomimetic amines on the synaptosomal tran sport of noradrenaline, dopamine and 5-hydroxytryptamine. Eur.J.Pharmacol. 41, 133-143.

Richelson E. and Pfenning M. (1984) Blockade by antidepres sants and related compounds of biogenic amine uptake into rat brain synaptosomes: most antidepressants selectively block norepinephrine uptake. Eur.J.Pharmacol. 104, 277-286.

Scherman D. and Henry J.P. (1984) Reserpine binding to bovine chromaffin granule membranes. Characterization and comparison with dihydrotetrabenazine binding. Molec. Pharmacol. 25, 113-122.

Scherman D. (1986) Dihydrotetrabenazine binding and mono amine uptake in mouse brain regions. J.Neurochem. 47, 331-339.

Stitzel R.E. (1977) The biological fate of reserpine. Pharma col.Rev.28, 179-205.

Stoof J.C., Liem A.L. and Mulder A.H. (1976) Release and receptor stimulating properties of p-tyramine in rat brain. Arch.int.Pharmacodyn. 220, 62-71.

Sweadner K.J. (1985) Ouabain-evoked norepinephrine relea se from intact rat sympathetic neurons: evidence for, car rier-mediated release. J.Neurosci. 5, 2397-2406.

Takimoto G.S., Stittworth J.D. and Stephens J.K. (1983) ^{3}H-Dopamine depletion from osmotically defined storage sites: effects of reserpine, 53 mM KCl, and d-amphetami ne. J.Neurochem. 41, 119-127.

Toll L., Gundersen C.D. and Howard B.D. (1977) Energy utilization in the uptake of catecholamines by synaptic vesicles and adrenal chromaffin granules. Brain Res. 136, 59-66.

Ungar F., Mosnaim A.D., Ungar B. and Wolf M.E. (1977) Tyramine-binding by synaptosomes from rat brain: effect of centrally active drugs. Biol.Psychiat. 12, 661-668.

Vaccari A. (1986) High affinity binding of ^{3}H-tyramine in the central nervous system. Br.J.Pharmacol. 89, 15-25.

Vizi E.S. (1978) Na^+-K^+-activated adenosinetriphosphatase as a trigger in transmitter release. Neuroscience 3, 367-384.

Watanabe M., George S.R. and Seeman P. (1985) Dependen ce of dopamine receptor conversion from agonist high- to low-affinity state on temperature and sodium ions. Biochem.Pharmacol. 34, 2459-2463.

TRYPTAMINE, TRYPTOPHAN AND NOCICEPTION

Alice A. Larson
Department of Veterinary Medicine
University of Minnesota
St. Paul, Minnesota
USA 55108

Rats implanted with a permanently indwelling intrathecal (IT) cannula and naive mice were injected IT with tryptamine (TA) or tryptophan to determine their effects on nociception using various nociceptive tests. Tryptamine IT in rats has been found to decrease the nociceptive threshold measured using the tail flick assay. This effect is also elicited by tryptophan and is potentiated by coadministration or pretreatment with methysergide. In mice, TA IT enhanced the number of writhes evoked in the writhing assay but did not alter the latency of response in the hot plate assay. Tryptophan IT in mice also enhanced the number of writhes, but only after pretreatment with methysergide. These data confirm the apparent hypernociceptive effect of TA in the spinal cord using different assays and species and also suggest that enhanced tryptophan in the spinal area causes sufficient accumulation of indoleamines to alter nociception.

INTRODUCTION

Descending serotonergic activity has been associated with analgesia (Messing and Lytle, 1977). The evidence supporting this concept include the findings that intrathecally (IT) administered serotonin causes antinociception in rat (Yaksh and Wilson, 1979), cat and primate

(Yaksh, 1978). We have previously shown that, in contrast, IT injections of tryptamine decrease the latency of the tail flick response of rats to a radiant heat source (Larson, 1983), suggesting a hypernociceptive effect. Serotonin antagonists further potentiate the hypernociceptive effect of tryptamine, especially in areas of the spinal cord containing a high density of serotonergic innervation (Larson, 1985).

The nociceptive threshold of adjuvant-treated rats has been found to be lower than that of control rats (Pircio, et al, 1975; Larson et al., 1985). The concentration of tryptophan, the amino acid precursor of both serotonin and tryptamine in the CNS, is elevated in the spinal cords of rats made arthritic by the injection of an adjuvant (Weil-Fugazza, et al., 1980). In light of the apparent hyperalgesic effect of tryptamine, an increase in the concentration of tryptophan may thus enhance nociception by increasing the synthesis of tryptamine. Tryptamine in the CNS has, for example, been found to be increased due to isolation-induced stress (Christian and Harrison, 1981; Harrison and Christian, 1983). It was therefore of interest to determine whether increasing tryptophan availability, which potentially increases synthesis of tryptamine as well as serotonin, changes the nociceptive threshold, and whether such a change would reflect enhanced synthesis of serotonin leading to antinociception or enhanced synthesis of tryptamine leading to hypernociception.

METHODS

Male Sprague-Dawley albino rats (375-425 g) and male Cox-Swiss mice (17-23 g) were fed ad libidum. Rats were anesthetized with ether during cannulation and then allowed a minimum of 5 days recovery from surgery. Animals were chronically implanted with IT cannulas by the method of Yaksh and Rudy (1976). Cannulas were contructed of various lengths of PE10 tubing and inserted into the spinal subarachnoid space via the cisterna magna. After each cannulation, an injection of saline was made to clear the catheter of accumulated debris.

Mice were injected IT following the method of Hylden and Wilcox (1980). Injection was made through intact skin between two lumbar vertebrae. Mice were injected intracerebroventricularly (ICV) freehand with a 5 ul volume through the skin to a depth of 2.5 mm at a point 2 mm lateral to the midline and 0.5 mm caudal to bregma.

Drug solutions were made fresh daily. Drugs administered IT were injected in rats in a volume of 20 ul saline followed by a 12 ul flush with sterile water. Drugs were administered IT to mice in a volume of 5 ul saline. Tryptamine HCl, serotonin oxalate, l-tryptophan and pargyline were purchased from Sigma (St. Louis, MO). Methysergide bimaleate was a gift from Sandoz Pharmaceuticals (Hanover, NJ) and L-Deprenyl was a gift from Chinoin Pharmaceuticals (Budapest). All doses are expressed in terms of the salts listed above.

Changes in the pain threshold were measured using the tail-flick assay of D'Amour and Smith (1941) using a 10 sec cutoff, the hot plate assay and the writhing assay. Tail-flick times were determined 5 min prior to drug injections and again at the times indicated after drug injection. The average post-drug latency was then expressed as a percent of the group's average pre-drug control latency. Values greater than 100% of control values reflect antinociception while those less than 100% reflect hypernociception. Mice used in the writhing assay were placed in individual transparent containers and the total number of writhes counted over a 5 min period beginning zero or two minutes after the IP injection of 10 ml/kg of 1% acetic acid. In the hot plate assay, mice were placed on a surface maintained at 52^6C and the latency of a licking response was measured.

IT or ICV injection of either saline or water in rats or mice, using volumes equal to those used for drug injections, produced no change in response in any test used in these studies. The determination of statistical significance of an effect produced by a

particular treatment was analyzed by repeated
measures ANOVA (Winer, 1971) or Student's t-
test. Preplanned comparisons across time and
treatments were made using the l.s.d. test
(Steel and Torrie, 1960) with p < 0.05 as the
accepted level of significance.

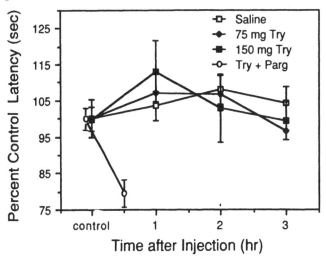

Figure 1. The lack of effect of tryptophan IP
on the tail flick latency of rats is compared
to the enhanced rate of response to 75 mg/kg of
tryptophan administered IP together with 75
mg/kg of pargyline at 30 min post injection.
Tail flick reaction times after 30 min were not
measured due to the intense hyperactivity
syndrome in pargyline-treated rats.

RESULTS

    Administration of tryptophan alone to rats
failed to alter the latency of response in the
tail flick assay (Fig. 1). Injection of 75
mg/kg of tryptophan IP together with 75 mg/kg
of pargyline, however, not only results in the
classic "serotonin syndrome", consisting of
tremor, head shakes and hyperactivity, but also
decreased the tail flick latency 30 min after
injection, as shown in figure 1. Fifteen min
after the SC injection of 5 mg/kg of
methysergide, symptoms of the serotonin
syndrome were attenuated but the rats then

began to vocalize.

Using 5.5 or 8.6 cm IT cannulas, injection of 200 ug of tryptophan in rats produced little or no decrease, respectively, in the tail flick latencies of rats at 30 min, as shown in figure 2. The IT injection of 50 ug of methysergide just 5 min prior to tail flick testing significantly decreased the average tail flick response latency in response to tryptophan.

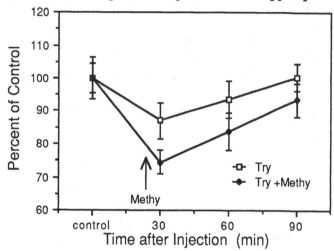

Figure 2. Unmasking of tryptophan's hypernociceptive effect when 200 ug are injected IT into the caudal portion of the rat spinal cord 5 min after an IT injection of 50 ug of methysergide (arrow). All injections were made using 8.6 cm cannulas.

In order to test the effects of TA and tryptophan using the writhing assay in mice, mice were injected with acetic acid IP 3 min prior to the injection of TA or 25 min after the injection of tryptophan. Injection of mice ICV with 5 ug of TA produced a period of profound sedation resembling catatonia which lasted for approximately 30 min. This effect is perhaps a factor in the prolongation of the hot plate latency observed after ICV injection of TA in mice (Table 1). IT injections of TA have not been found to produce such a sedative effect at any dose tested and also failed to

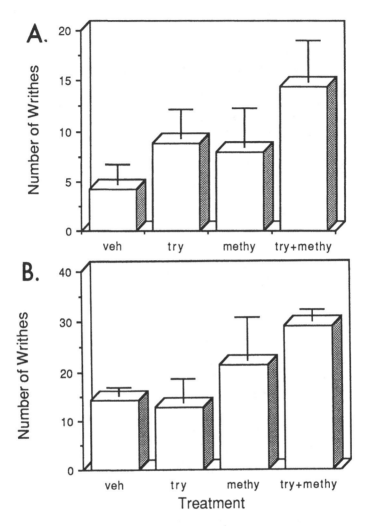

Figure 3. The bar graph in A shows the effect 50 ug of tryptophan injected IT after either saline or 5 mg/kg of methysergide SC on the number of writhes elicted in mice after the IP injection of acetic acid. The bar graph in B shows the effect on writhing of 25 ug of tryptophan injected IT after either saline or 1 mg/kg of methysergide SC. The number of writhes in A were measured for the first 5 min after injection of acetic acid while the number of writhes in B were measured for 5 min beginning 2 min after injection of acetic acid.

alter the hot plate latency in mice. In spite f
the sedative effect of TA administered ICV, no
change was detected in the number of writhes
elicted by IP injected acetic acid compared to
that after injection of saline ICV (Table 1).
In contrast, the number of writhes increased
significantly after injection of 400 ng of TA
IT.

Table 1. Effect of Centrally Administered
Tryptamine (TA) on Antinociception in Mice

Route	Drug	Writhes	Latency (sec)
ICV	Saline	13.2 $\pm$ 1.7 (10)	48.2 $\pm$ 7.2 (10)
ICV	5 ug TA	12.3 $\pm$ 2.2 (8)	69.2 $\pm$ 10.4 (9)
IT	Saline	3.3 $\pm$ 0.7 (10)	35.2 $\pm$ 3.3 (10)
IT	400 ng TA	10.9* $\pm$ 1.8 (10)	37.9 $\pm$ 4.1 (10)

Each value is expressed as as average plus or
minus the standard error. The value in
parentheses is the number of mice used to
generate that value.

As shown in figure 3, injection of either
25 or 50 ug of tryptophan IT did not alter the
response in the writhing assay compared to
control IT injections in mice. However the
response to acetic acid IP was significantly
increased when tryptophan was injected IT to
mice pretreated SC with either 1 or 5 mg/kg of
methysergide.

DISCUSSION

Parenterally admininstered, tryptophan plus
pargyline causes a marked accumulation of
serotonin, as evidenced by the production of
various symptoms of the serotonin syndrome.
Most of the behavioral effects produced by such

treatment are attenuated by a large dose of methysergide, a relatively nonspecific serotonin antagonist. The production of intense vocalization in these animals, however, would suggest that serotonin antagonism also appears to unmask a nociceptive effect.

That the hypernociceptive effect of tryptophan appears to be centrally mediated is supported by the results indicating that IT injection of tryptophan causes a mild hypernociception (Fig. 2). The effects produced by either IT or IP tryptophan occur at comparable time intervals after methysergide and each effect is potentiated by methysergide, suggesting that the synthesis of serotonin from tryptophan attenuates rather than promotes the hypernociceptive effect of tryptophan.

Except for the time course, the potentiation of tryptophan's hypernociceptive effect is identical to the potentiation of TA's hypernociceptive effect which we reported previously (Larson, 1983). The maximal hypernociceptive effect of TA occurred at 5 min after injection IT while that of tryptophan IT is not seen until 30 min. The difference between these two times is consistent with the synthesis of TA after injection of tryptophan (Moir and Eccleston, 1968).

The hypernociceptive effect of TA appears to differ depending on the assay system used to measure nociception. While we have previously shown that TA injected ICV in rats produces a dramatic decrease in the tail flick latency, ICV injections of TA in mice did not alter the number of writhes produced in the writhing assay even though smaller doses of TA injected IT potentiated writhing in mice. The writhing assay depends on a coordinated movement of the whole animal rather than a simple reflex activity produced by a tail flick response. Enhancement of such writhing activity would thus suggest that the hyperalgesic effect of TA and tryptophan is more than a simple potentiation of motor neuronal responsivity.

ACKNOWLEDGEMENTS

This work was supported by NIDA Grant DA04090, DA09190 and NIH Grant NS170147.

REFERENCES

Christian, S.T. and Harrison, R.E. (1981) Stress elevation of peripheral tryptamine tissue content. Neuroscience 7, 869.

D'Amour, F.E. and Smith, D.L. (1941) A Method for determining loss of pain sensation. J. Pharm. exp. Ther. 72, 74-79.

Harrison, R.E.W. and Christian, S.T. (1983) Individual housing stress elevates brain and adrenal tryptamine content, <u>Neurobiology of Trace Amines</u> (Boulton, A.A., Baker, G.B., Dewhurst, W.G., and Sandler, M., eds) pp. 249-255.

Hylden, J.L.K. and Wilcox, G.L. (1980) Intrathecal morphine in mice: a new technique, Eur. J. Pharmacol. 67, 313-316.

Larson, A.A. (1983) Hyperalgesia produced by the intrathecal administration of tryptamine to rats. Brain Res. 265, 109-117.

Larson, A.A. (1985) Distribution of CNS sites sensitive to tryptamine and serotonin in pain processing. In, <u>Neuropsychopharmacology of the Trace Amines</u> (Boulton, A.A., Maitre, L., Bieck, P.R. and Reiderer, P., eds) pp. 241-248.

Larson, A.A., Brown, D.R., El-Atrash, S. and Walser, M.M. (1985) Pain-thresholds in adjuvant-induced inflammation: A Possible model of chronic pain in the mouse. Pharmacol. Biochem. Behav. 24: 49-53.

Messing, R.B. and Lytle, L.G. (1977) Serotonin-containing neurons: their possible role in pain and analgesia. Pain 4: 1-21.

Moir, A.T.B. and D. Eccleston (1968) The effect of precursor loading in the cerebral metabolism of 5-hydroxyindoles. J. Neurochem. 15, 1093-1108.

Pircio, A.W., Fedele, C.T., and Bierwagen, M.E. (1975) A New method for the evaluation of analgesic activity using adjuvant-induced arthritis in the rat. Eurp. J. Pharmacol. 31, 207-215.

Steel, R.G.D. and J.H. Torrie (1960) <u>Principles and Procedures of Statistics</u>, McGraw-Hill, New York.

Weil-Fugazza, J., Godefroy, F., Coudert, D. and

Besson, J.M. (1980) Total and free serum tryptophan levels and brain 5-hydroxytryptamine metabolism in arthritic rats. Pain 9, 319-325.

Winer, B.J. (1971) <u>Statistical Prinicples in Experimental Design</u>, Second Edition, McGraw-Hill, New York.

Yaksh, T.L. (1978) Analgesic actions of intrathecal opiates in cat and primate. Brain Res. 153, 205-210.

Yaksh, T.L. and Rudy, T.A. (1976) Chronic catherization of the spinal subarachnoid space. Physiol. Behav. 17, 1031-1036.

Yaksh, T.L. and Wilson, P.R. (1979) Spinal serotonin terminal system mediates antinociception. J. Pharm. exp. Ther. 208, 446-453.

NEUROBEHAVIORAL ASPECTS OF TRACE AMINE FUNCTIONS:

AFFECTIVE AND MOTORIC RESPONSES.

A. J. Greenshaw,

Dept. of Psychiatry,
University of Alberta,
Edmonton, Alberta
Canada T6G 2B7.

INTRODUCTION

Major problems in the analysis of trace amine function have been (i) the low basal levels of these compounds, hindering their analysis as dependent variables; and (ii) their rapid and extensive oxidative deamination by monoamine oxidase (MAO), necessitating the use of high doses. With respect to the latter problem both the rather massive pharmacological effects of these amines and their problematic pharmaco-kinetic/dynamic profiles have hindered their manipulation as independent variables. Few data are available describing attempts to system-atically assess selective alterations of trace amines at relatively low concentrations in the central nervous system (CNS). The present paper describes some aspects of current approaches to these problems in the context of behavior analysis, particularly in relation to dopamine (DA) function in the CNS. This analysis is restricted to a discussion of beta-phenylethyl-amine (PE). The basic approach outlined here is relevant to the analysis of other trace amines and analogous research is being conducted for tryptamine, p-tyramine and p-octopamine.

THE CURRENT STATUS OF PE IN CNS FUNCTION

It has been established for some time that PE is a biogenic product of the decarboxylation of L-phenylalanine,which is also a major metabolic precursor of catecholamine synthesis (Saavedra, 1974). The structural relationship of PE to sympathomimetic amines and particularly to amphetamine (AMPH, alpha-methylphenylethylamine) corresponds to rather potent effects of PE on uptake and release of amines (Raiteri et al, 1977). Preliminary claims for specific PE binding sites (Hauger et al, 1982) have not been substantiated at this time. The lipophilicity of PE and the fact that its synthesis appears to be governed primarily by L-phenylalanine availability in the presence of the enzyme aromatic L-amino acid decarboxylase (L-AAD) were indicative of a possible widespread distribution in brain and peripheral tissues. Estimations of endogenous PE levels using high resolution mass spectrometry indicate a clearly heterogeneous distribution within brain (Durden et al, 1973). Recent lesion studies (see chapter by Juorio in this book) have now established that PE is to a major extent synthesised in neurons. It has been suggested that it may be co-localised with DA and is significantly associated with the neurons of the nigrostriatal (or striatonigral) systems (Greenshaw et al, 1986). These findings raise important questions concerning the role of PE in brain with particular emphasis on the relationship of PE to DA function in mammalian CNS.

PE AND NIGROSTRIATAL DA FUNCTION: TURNING AFTER 6OHDA LESIONS.

In response to the proposal that specific PE binding sites may exist and the established changes in striatal PE concentrations after 6-hydroxydopamine (6OHDA), the effect of intrastriatal PE on turning has been assessed (Nguyen et al, 1986). Rats with unilateral lesions of the substantia nigra pars compacta exhibit turning towards the lesion side due to to unilateral striatal DA depletion. It is established that direct DA receptor agonists induce turning away from the lesion side by acting on supersensitive DA receptors. Administration of indirect agonists (those which release DA) results in enhanced turning towards the lesion side. For systemic administration of such compounds these observations are well established (see Pycock and Kilpatrick, 1988). Recently the effects of intrastriatal applications of PE on turning have been reassessed (Nguyen et al, 1986). The results of this study clearly demonstrated that administration of PE into the striatum ipsilateral to a nigral lesion results in turning away from the lesioned side. This result was however attributed to the release of residual DA on the lesioned side as AMPH had similar effects which were reversed by the DA synthesis inhibitor alpha-methyl-para-tyrosine. At the present time there is no clear evidence for a direct postsynaptic effect of PE in the striatum, although electrophysiological studies indicate that PE may act postsynaptically to influence the action of catecholamines (Jones and Boulton, 1980). Dopamine is involved in the behavioral effects of pharmacological doses of PE (see Greenshaw et al,1985a).

EFFECTS OF PE ON INTRACRANIAL SELF-STIMULATION

Previous studies have demonstrated that PE has dual effects on rate of responding maintained by electrical stimulation of the lateral hypothalamus (Greenshaw et al, 1985a, 1986). In the presence of a MAO inhibitor this amine induces increases in response rate, in the absence of a MAO inhibitor PE induces a decrease in this measure. These opposite effects have been attributed to the pharmacokinetic/dynamic profile of PE (Greenshaw, 1984). This amine does not affect the duration of reinforcing stimulation under the animals' control (Greenshaw, 1985a, 1986) although it is reported to alter self-stimulation thresholds based on response rate (Stein, 1964) and apparently has abuse potential (Shannon and

De Gregorio, 1982). The analysis of effects of PE
on reinforcement or reward processes is of considerable potential importance as the metabolism
of endogenous PE is altered by exposure to reinforcing hypothalamic stimulation (Greenshaw et
al, 1985a) and a number of antidepressant drugs
(notably MAO inhibitors) profoundly increase
endogenous PE levels in the CNS (Boulton and
Juorio, 1982).

TABLE 1

Levels of PE 30 min following administration of
(i) prodrugs of PE or; (ii) PE in deprenyl
pretreated rats. In each case behavioral effects
were observed as indicated.

Drug Treatment mmol/kg	PE level ng/g	Behavioral change
CEPE 0.10	180 a	* Decrease
PGPE 0.10	1000 b	Increase
PE 0.03	200 c	Increase

a - whole brain level (Baker et al, 1987).
b - whole brain level (Rao et al, 1987).
c - striatal level (Greenshaw et al, 1985a).

*   - based on first 30 min of drug effect
(Stewart et al, 1987).

A recent development in trace amine research
is the use of prodrugs to selectively elevate
levels of these amines (see chapter by Baker et
al, in this book). The effects on self-stimulation of administration of PE in rats pretreated
with the MAO inhibitor deprenyl have been compared to those induced by cyanoethylphenylethyl-

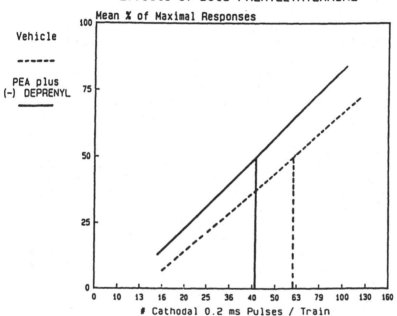

RESPONSE - FREQUENCY FUNCTION
Effects of Beta-PHENYLETHYLAMINE

Figure 1. Effects of beta-phenylethylamine
(4 mg/kg) on response frequency function for
lateral hypothalamic self-stimulation. Plotted
lines are fitted by linear regression.
Beta-phenylethylamine induced a shift to the left
in frequencies that maintained half-maximal response rate. The rats (n=6) were pretreated with
(-) deprenyl (4 mg/kg, 3.5h prior to testing)

amine (CEPE), a prodrug without significant
effects on MAO. In these experiments drug effects
were assessed by measuring response rate as a
function of stimulation frequency, based on the
work of Gallistel and Karras (1984). The procedure was as described by O'Regan et al (1987).
In accord with previous work from this laboratory
(Greenshaw et al, 1985a) PE, after deprenyl pretreatment, induced increases in response rate.
These effects were accompanied by a decrease in
the stimulation frequency which maintained
half-maximal response rate, indicating increased
sensitivity to the reinforcing properties of the
stimulation. Under similar conditions AMPH induced similar effects (Greenshaw and Wishart, 1987).
These observations are illustrated by the
regression plots in Figure 1. In contrast to
these effects, administration of CEPE did not
affect self-stimulation at doses which had
significant effects on spontaneous motor activity
(as described below). At these doses (up to 0.3
mmol/kg) brain PE levels are within the range
(see Table 1) that has been measured after
behaviourally active doses of PE in deprenyl pretreated rats (Greenshaw et al, 1985a) and see
chapter by Baker et al , in this book). These
differential effects of PE and its prodrug are
not easily explained in terms of pharmaco-
kinetic/dynamic factors. Similarly it is unlikely
that the differences are due to an action of
other prodrug metabolites, although this remains
to be determined empirically (Baker et al, in
this book).It is possible that the effect of MAO
inhibition on other amine systems leads to a
qualitative change in PE effects, presumably due
to a PE - amine(s) interaction. This possibility
is discussed below.

EFFECTS OF PE ON SPONTANEOUS LOCOMOTOR ACTIVITY

The effects of PE on locomotor activity have
been measured in both the presence and absence of
MAO inhibitors (Dourish, 1982; Dourish et al,
1983; Turkish et al, 1987). As described above,
stimulant (low dose PE with MAO inhibition) and

depressant (high dose PE) effects have been attributed to pharmacokinetic/dynamic factors. True AMPH-like stimulant effects have apparently never been reported after PE alone because of the rapid metabolism of low doses by MAO and the massive flooding of the system with doses sufficient to achieve sustained levels of this amine. The relationship of stimulant effects of PE to degree of MAO inhibition is clearly demonstrated by a recent study (Turkish et al,1987), as illustrated by the data in Figure 2.

TABLE 2

Effects of prodrugs of PE on locomotor activity (infra-red beam computer-monitored counts as described by Dourish et al, 1983).

Drug Treatment mmol/kg		Activity count
Vehicle		2050 +/- 343
PGPE 0.03		1868 +/- 398
PGPE 0.10		2390 +/- 382
PGPE 0.30	*	5750 +/- 1265
CEPE 0.03		2200 +/- 320
CEPE 0.10	*	750 +/- 211
CEPE 0.30	*	1009 +/- 284

Data are group means +/- SE,(n=6). Activity was measured for 60 min after sc injection. * Denotes different from control P < 0.05 by ANOVA. Results are from Stewart et al (1987).

In an attempt to further characterise this apparent dependence of stimulant effects on MAO inhibition the effects of two prodrugs of PE on locomotor activity were assessed with rats (Stewart et al, 1987). The effects of a range of doses (0.03 - 0.3) of CEPE and of N-propargyl-

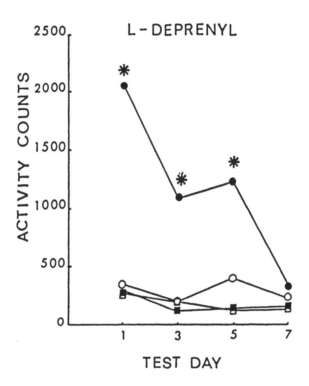

Figure 2. (a) Effects of pretreatment with (-)
deprenyl (5mg/kg) on spontaneous motor activity
after injections of PE (10 mg/kg). Activity
counts include total photobeam counts over a 15
min period immediately following PE or vehicle.
PE or vehicle were then given to animals on each
test day. Data are means, * indicates significantly different from vehicle group. Closed symbols represent (-) deprenyl treatment; open symbols represent vehicle treatment. Circles represent PE treatment; squares represent vehicle
treatment. Adapted from Turkish et al (1987).

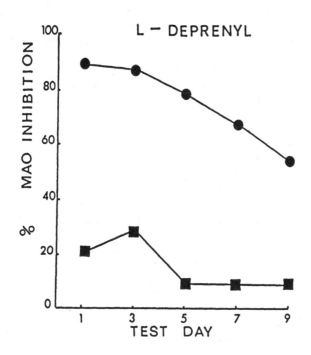

Figure 2. (b) MAO inhibition after treatment with (−) deprenyl expressed as a percentage of vehicle treated controls. Type A inhibition is represented by squares; type B inhibition is represented by circles. See figure 2a for further details.

phenylethylamine (PGPE) respectively were assessed after subcutaneous injection. PGPE is a prodrug of PE which also inhibits MAO (preferentially the type-B form of the enzyme: see chapter by Baker et al, in this book). In these tests CEPE induced a dose-dependent decrease in locomotor activity. Administration of PGPE induced a dose-dependent increase in the activity measures.These effects are illustrated by the data displayed in Table 2.

These opposite effects of increased PE availability in the absence (with CEPE) and presence (with PGPE) of MAO inhibition are unexpected. At the present time, as outlined above, it seems likely that the stimulant / non-stimulant consequence of drug administration in this context is due to an interaction of PE with other amines which are substrates for MAO. This proposal is speculative at the present time; PGPE ,at equimolar doses, results in higher brain levels of PE than does CEPE and, as described by Baker et al in this book, the effects of the unmetabolised prodrugs themselves on uptake and release of catecholamines and 5-hydroxytryptamine must be further investigated. The characteristics of these effects are currently being directly examined in this laboratory in terms of concurrent changes in regional brain amine metabolism, MAO activity and levels of both PE and prodrugs.

CONCLUSION

At the present stage of research in this laboratory it is apparent that PE may be derived largely from neurons in the CNS, possibly DA-containing cells. This amine enhances sensitivity to reinforcing hypothalamic stimulation (in terms of frequency thresholds) in the presence of a MAO inhibitor. Under similar pharmacological conditions PE acts as a psychomotor stimulant. These observations are consistent with the previous proposal that PE may represent an 'endogenous AMPH' (Sandler and Reynolds, 1976). Surprisingly, after administration of the prodrug CEPE (when PE levels were selectively increased) no effects on reinforcement thresholds were observed. At equimolar doses decreases in spontaneous locomotor activity were observed. In contrast to these results, administration of PGPE (a prodrug of PE with inhibitory effects on MAO - preferentially the type-B form of the enzyme) resulted in increased motor activity. These differential effects of PE elevation, in the respective presence and absence of MAO activity, may be interpreted in terms of PE interactions with other MAO substrates. At this time research in this laboratory is directed to an analysis of the hypothesis that a PE - DA interaction is critical for the observed psychomotor stimulant effects. Results with the turning model using unilateral 6OHDA lesions indicate that the level of interaction may not be simply due to a direct post-synaptic effect of PE. The possibility of functional pre- and postsynaptic interactions between PE and DA remains an open question at this time.

A neuromodulatory role of PE and other trace amines in mammalian CNS has been proposed (Boulton, 1979). A comparison of functional effects of prodrugs (with and without MAO-inhibitory activity) in terms of both behavior and neurochemistry, may yield valuable data concerning mechanisms of interaction of this interesting group of amines with other neuroactive substances.

ACKNOWLEDGEMENTS

This research is currently supported by the ALBERTA HERITAGE FOUNDATION FOR MEDICAL RESEARCH and the ALBERTA MENTAL HEALTH ADVISORY COUNCIL, the work described above was also partly funded by SASKATCHEWAN HEALTH and the MRC (CANADA). The author thanks colleagues past and present for their advice and collaborative effort, particularly Professors G.B. Baker, A.A. Boulton, W.G. Dewhurst, A.V. Juorio and T.B. Wishart for their suggestions and discussion of this work.

REFERENCES

Baker G.B. , Coutts R.T. and Rao T.S. (1987) Neuropharmacological and neurochemical properties of N-(cyanoethyl)-2-phenylethylamine, a prodrug of 2-phenylethylamine. Brit. J. Pharmacol. (in the press)

Boulton A.A. (1979) The trace amines: neurohumors (cytosolic, pre- and/or post-synaptic, secondary, indirect?). Behav. Brain Sci. 2, 418.

Boulton A.A. and Juorio A.V. (1982) Brain trace amines. In: Handbook of Neurochemistry, Vol 1, edited by A. Lajtha. New York: Plenum Press pp 189-222.

Dourish C.T. (1982) A pharmacological analysis of the hyperactivity syndrome induced by beta-phenylethylamine in the mouse. Brit J. Pharmacol. 77, 129-139.

Dourish C.T., Greenshaw A.J. and Boulton A.A. (1983) Deuterium substitution enhances the effects of beta-phenylethylamine on spontaneous motor activity in the rat. Pharmacol. Biochem. Behav. 19, 471-475.

Durden D.A., Philips S.R. and Boulton A.A. (1973) Identification and distribution of beta-phenylethylamine in the rat. Can. J. Biochem. 51, 995-1002

Gallistel C.R. and Karras D. (1984) Pimozide and amphetamine have opposing effects on the reward summation function. Pharmacol. Biochem. Behav. 20, 73-77.

Greenshaw A.J. (1984) Beta-phenylethylamine : a functional role at the behavioral level? In: Neurobiology of the Trace Amines. Edited by A.A. Boulton, G.B. Baker, W.G. Dewhurst and M. Sandler. New Jersey:Humana Press pp 351-374.

Greenshaw A.J., Juorio, A.V. and Boulton A.A. (1985a) Behavioral and neurochemical effects of deprenyl and beta-phenylethylamine in Wistar rats. Brain Res. Bull. 15, 183-189.

Greenshaw A.J., Juorio A.V. and Nguyen T.V. (1986) Depletion of striatal beta-phenylethyl-amine following dopamine but not 5-HT denervation. Brain Res. Bull. 17, 477-484.

Greenshaw A.J., Sanger D.J. and Blackman D.E. (1985b) Effects of d-amphetamine and of beta-phenylethylamine on fixed-interval responding maintained by self-regulated electrical hypothalamic stimulation in rats. Pharmacol. Biochem. Behav. 23, 519-523.

Greenshaw A.J. and Wishart T.B. (1987) Effects of (+) amphetamine on response-frequency characteristics of intracranial self-stimulation: a statistical dose response analysis. Prog Neuropsychopharmacol. Biol. Psychiat. (Submitted manuscript).

Hauger R.L., Skolnick P. and Paul S.M. (1982) Specific [^{3}H]-beta-phenylethylamine binding sites in rat brain. Eur. J. Pharmacol. 83, 147-148

Jones R.S.G. and Boulton A.A. (1980) Interactions between p-tyramine, m-tyramine or beta-phenylethylamine and dopamine on single neurons in the cortex and caudate nucleus of the rat. Can. J. Physiol. Pharmacol. 58, 222-227.

Nguyen T.V., O'Regan D., Dewar K.M. and Greenshaw A.J. (1986) Assessment of possible direct post-synaptic effects of d-amphetamine and of beta-phenylethylamine in vivo. Proc. West. Pharmacol. Soc. 29, 271-273.

O'Regan D., Kwok R.P.S., Yu P.H., Bailey B.A., Greenshaw A.J. and Boulton A.A. (1987) Chronic and selective monoamine oxidase inhibition: a behavioral and neurochemical analysis. Psychopharmacol 92, 42-47.

Pycock C.J. and Kilpatrick I.C. (1988) Motor asymmetries and drug effects: behavioral analyses of receptor activation. In: Neuromethods: Psychopharmacology. Edited by A.A. Boulton, G.B. Baker and A.J. Greenshaw. New Jersey: Humana Press (in the press).

Raiteri M., delCarmine R., Bertollini A. and Levi
   G. (1977) Effect of sympathomimetic amines on
   the synaptosomal transport of noradrenaline,
   dopamine and 5-hydroxytryptamine. Eur. J.
   Pharmacol. 41, 133-143.
Rao T.S., Baker G. B. and Coutts R. T. (1987)
   Pharmacokinetic and neurochemical studies on
   N-propargyl-22-phenylethylamine, a prodrug of
   2-phenylethylamine. Naunyn Schmiedeberg's Arch.
   Pharmacol. (in the press)
Saavedra J.M. (1974) Enzymatic isotopic assay for
   and the presence of beta-phenylethylamine in
   brain. J. Neurochem. 22, 211-216.
Sandler M. and Reynolds G.P. (1976) Does beta-
   phenylethylamine cause schizophrenia? Lancet i,
   70-71.
Shannon H.E. and DeGregorio C.M. (1982) Self-
   administration of the endogenous trace amines
   beta-phenylethylamine, N-methylphenylethylamine
   and phenylethanolamine in dogs. J. Pharmacol.
   Exper. Therap. 227, 52-60.
Stein L. (1964) Self-stimulation of the brain
   and the central stimulant action of amphetamine. Fed. Proc. 23, 836-841.
Stewart D.J., Greenshaw A.J., Baker G.B. and
   Coutts R.T. (1987) Prodrugs of 2-phenylethyl-
   amine: comparison of their effects on locomotor
   activity and on concentrations of biogenic
   amines in brain. Soc. Neurosci. Abs. (in the
   press).
Turkish S., Yu P.H. and Greenshaw A.J. (1987)
   Monoamine oxidase B inhibition: a comparison of
   in vivo and ex vivo measures of reversible
   effects. J. Neural Trans. (in the press).

BRAIN TRACE AMINES: MAPPING STUDIES AND EFFECTS OF MESENCEPHALIC LESIONS

Augusto V. Juorio

Neuropsychiatric Research Unit

CMR Building, University of Saskatchewan

Saskatoon, Saskatchewan, Canada S7N OWO

Presence and Philogenetic Distribution of Biogenic Trace Amines

The brain trace amine concentration has been estimated by different research groups and extends over a wide range of values; the general tendency, however, has been towards the confirmation of the lower estimates (see Boulton and Juorio, 1982 for a discussion). The p- and m-isomers of TA, PE and T are present in the mammalian or avian brain; their concentrations, in comparison with that of dopamine or 5-hydroxytryptamine are quite small (Table 1). In contrast, the concentrations of p-TA, p-OA and T in some invertebrate ganglia or nerves are of similar order of that of the classical transmitters (Table 1). The reverse is also true, that is in some neural tissues the concentrations of some of the classical transmitters (noradrenaline, adrenaline or 5-hydroxytryptamine) are within the concentration range as defined for the trace amines (0.1-100 ng g^{-1}) (Table 1) (for a discussion see Reader et al., 1987, this Symposium). The reason for this apparent discrepancy could be found on the fact that the definition of the trace amines is based on a rigid concentration range, it is hoped that with the elucidation of their functional notes a more suitable term will be selected.

TABLE 1

The concentration of some trace amines and classical transmitters in the rat whole brain, fowl whole brain, octopus ganglia, garden snail ganglia, lobster thoracic nerve cord and starfish arm nerve

	Rat ng g^{-1}	Fowl ng g^{-1}	Octopus ng g^{-1}	Snail ng g^{-1}	Lobster ng g^{-1}	Starfish ng g^{-1}
PE	1.8[1]	0.7[7]	3.0[8]	0.6[9]	1.2[13]	4.4[10]
p-TA	2.0[2]	0.5[7]	80[8]	7.5[9]	20[13]	2.2[10]
m-TA	0.3[3]	0.3[7]	0.6[8]	1.1[9]	1.3[13]	<0.8[10]
p-OA	0.9[11]	n.a.	540[8]*	327[9]	290[12]*	163[10]
m-OA	0.2[11]	n.a.	n.a.	1.5[9]	n.a.	<2[10]
DA	600[4]	290[7]	4690[8]	2200[9]	110[12]	5950[10]
NA	490[4]	418[7]	3900[8]	80[15]	140[12]	2133[10]
A	40[16]	46[7]	n.d.[14]	n.d.[15]	n.d.[12]	n.d.[10]
T	0.5[5]	<0.3[7]	<0.6[8]	6.4[9]	1.7[13]	1250[10]
5-HT	400[6]	700[7]	4180[8]	2250[7]	150[12]	<3.0[10]

Values are in ng g^{-1} of fresh tissue. The superscripts indicate the source of reference. [1]Durden et al. (1973); [2]Philips et al. (1974a); [3]Philips et al. (1974); [4]Bertler and Rosengren (1959); [5]Philips et al. (1974b); [6]Bogdanski et al. (1956); [7]Juorio (1976); [8]Juorio and Philips (1976); [9]Juorio and Kazakoff (1984); [10]Juorio and Robertson (1977); [11]Juorio (obtained by the radioenzymatic dansylation technique, unpublished); [12]Laxmyr (1984); [13]Juorio and Sloley (in preparation); [14]Juorio (1971); [15]Juorio and Killick (1972); [16]Anton and Sayre (1964); *p-OA and m-OA determined jointly; n.d., none detected; n.a., estimates not available.

Rat Brain Regional Distribution of the Trace Amines

β-Phenylethylamine. The highest concentrations of PE
were observed in some parts of the mesolimbic system such
as the olfactory tubercles and the nucleus accumbens followed by the globus pallidus (Table 2) while concentrations
in the hypothalamus, cerebellum, brain stem and striatum,
ranged between 2.1 and 1.6 ng g^{-1} (Table 2).

Tyramines. The striatum (caudate putamen and globus
pallidus) and some parts of the mesolimbic system (olfactory tubercles and nucleus accumbens) contain the highest
concentration of p- and m-TA, where p-TA in the predominant
isomer (Table 2). Of the other mesolimbic areas investigated, only the septal nuclei contain some detectable p-TA
while in the rest, the levels are below the limits of
sensitivity of the methods (Table 2). The levels of m-TA
in the amygdaloid nucleus and the nucleus tractus diagonal-
is are higher than those of p-TA (Table 2).

In hypothalamic areas, brain stem and cerebellum, the
concentration of the p- and m-isomers of TA are quite low
and sometimes below the limits of sensitivity of the method
(Table 2). In the nucleus arcuatus, both p- and m-TA are
present but the concentration of the m-isomer is higher
(Table 2).

Octopamines. The highest concentration of p-OA was
observed in the hypothalamus. The values for Wistar rats
were 3.4 ng g (Table 2) but higher values (4.9-12.1 ng g^{-1})
were observed for the Roman strain (David and Delacour,
1980). The levels of m-OA were below the limits of sensitivity the method for the hypothalamus of Wistar rats
(Table 2) and somewhat higher of the Roman strain (David
and Delacour, 1980).

Tryptamine. The concentrations are highest in the
caudate putamen, olfactory tubercles and nucleus accumbens
and lowest in the globus pallidus, brain stem and cerebellum (Table 2).

Effect of Lesion of Selected Brain Nuclei on
Brain Trace Amine Concentration

β-Phenylethylamine. Since PE-containing terminals may
exist in the caudate nucleus in a manner similar to the

TABLE 2

The rat brain distribution of PE, p-TA, m-TA, T and 5-HT in selected nuclei

Region	PE ng g^{-1}	p-TA ng g^{-1}	m-TA ng g^{-1}	p-OA ng g^{-1}	m-OA ng g^{-1}	T ng g^{-1}
Striatum						
Caudate-putamen	1.3[1]	10.9[1]	3.7[1]	n.d.[5]	n.d.[5]	1.1[1]
Globus pallidus	3.6[1]	3.2[2]	1.2[2]	n.a.	n.a.	0.4[1]
Mesolimbic System						
Olfactory tubercles	5.3[2]	5.1[2]	2.4[2]	n.a.	n.a.	0.9[1]
Nucleus accumbens	4.8[2]	4.9[2]	1.6[2]	n.a.	n.a.	0.7[1]
Septal nuclei	n.a.	1.2[2]	0.9[2]	n.a.	n.a.	n.a.
Amygdaloid nuclei	n.a.	<0.4[2]	1.2[2]	n.a.	n.a.	n.a.
N. tractus diagonalis	n.a.	<0.5[2]	0.7[2]	n.a.	n.a.	n.a.
Hippocampus (dorsalis)	n.a.	<0.4[2]	0.6[2]	n.a.	n.a.	n.a.
Hypothalamic Areas						
Anterior part	2.1[3]*	<0.5[1]	<0.5[1]	3.4[5]*	n.d.[5]*	n.a.
Posterior part	n.a.	0.7[1]	<0.5[1]	n.a.	n.a.	n.a.
Nucleus arcuatus	n.a.	0.9[1]	1.3[1]	n.a.	n.a.	n.a.
Brain Stem						
Whole	1.6[3]	0.4[3]	0.1[3]	1.1[5]	n.d.[5]	0.2[4]
Cerebellum						
Whole	2.1[3]	0.1[3]	0.01[3]	n.a.	n.a.	0.3[4]

Values are in ng g^{-1} of fresh tissue. The superscripts indicate the source of reference: [1]Juorio, unpublished; [2]Sardar, Juorio and Boulton, 1987; [3]Philips (unpublished, cited by Boulton, 1976); [4]Philips, Durden and Boulton, 1974b; [5]Danielson, Boulton and Robertson, 1977; n.d., none detected; n.a., estimates not available; * whole hypothalamus.

presence of dopamine-containing neurons that project from
the substantia nigra to the caudate nucleus (Andén et al.,
1964), the effects of unilateral electrolytic or 6-hydroxy-
dopamine lesions of the left substantia nigra were assessed
on the striatal concentrations of PE. Since the endogenous
levels of PE are normally close to the limits of sensitivity of the mass spectrometric method employed, the metabolism of PE was blocked by the administration of a small dose
of a monoamine oxidase inhibitor (deprenyl). The use of
deprenyl (2 mg kg^{-1}) does not detract from the validity of
the PE depletion, mainly because the dose employed had
little or no effect on the levels of dopamine, 5-hydroxy-
tryptamine and their metabolites (Greenshaw et al., 1986)
or affects the unequal regional distribution of PE in the
striatum or the hypothalamus (Greenshaw et al., 1985). The
chronic (7 days) nigral electrolytic or 6-hydroxydopamine
legions produced a marked depletion (to 55%-63% of the
control side) of the striatal PE concentration while no
changes in striatal PE were observed in the sham operated
rats (Tables 3 and 4); either lesion type produce a substantial dopamine depletion (to about 30% of the control
side) but did not affect 5-hydroxytryptamine (Tables 3 and

TABLE 3

The effect of chronic (7 days) electrolytic lesions
of the substantia nigra on rat striatal PE, DA and 5-HT
in deprenyl-treated (2 mg kg^{-1}, 2 hours) rats.

	PE ng g^{-1}	DA ng g^{-1}	5-HT ng g^{-1}
		Sham Controls	
Right side	12.6	10020	620
Sham side	12.2	9510	620
		Electrolytic Lesions	
Right side	11.6	9330	710
Lesion side	7.3*	2970*	690

Values are in ng g^{-1} of fresh tissue. *Significantly
different from control side. Results are from Greenshaw,
Juorio and Nguyen, 1986.

TABLE 4

The effect of chronic (7 days) 6-hydroxydopamine (8 μg)
lesions of the rat substantia nigra (left side)
on rat striatal PE, DA and 5-HT in
deprenyl-treated (2 mg kg^{-1}, 2 hours) rats.

	PE ng g^{-1}	DA ng g^{-1}	5-HT ng g^{-1}
Sham Controls			
Right side	11.7	10220	630
Sham side	11.4	9260	640
6-Hydroxydopamine Lesions			
Right side	14.9	10360	630
Lesion side	8.2*	3060*	640

Values are in ng g^{-1} of fresh tissue. *Significantly
different from control side. Results are from Greenshaw,
Juorio and Nguyen, 1986.

4). In contrast, chronic (7 days) electrolytic median and
dorsal raphé lesions produced no changes in striatal PE
levels (Table 5).

The observation that the accumulation of PE in respons
to deprenyl decreased following nigral but not raphé
lesions suggests that a significant source of PE is intraneuronal and that PE-containing cell bodies are present in
the substantia nigra. It may be possible that the differential effects of nigral and raphe lesions could relate
directly to the extent of decarboxylase loss in the striatum; lesions of the substantia nigra result in decreased
activity of this enzyme, as indicated by a decrease in the
accumulation of dopamine following administration of L-DOPA
(Schlosberg et al., 1979; Greenshaw et al., 1986). Nevertheless, equivalent lesions of the midbrain raphe nuclei
also result in a decrease of brain decarboxylase activity

TABLE 5

The effect of chronic (7 days) electrolytic lesions
of the dorsal and median raphé nuclei (RN) on the rat
striatal concentration of PE, DA and 5-HT in
deprenyl treated (2 mg kg^{-1}, 2 hours) rats.

	PE ng g^{-1}	DA ng g^{-1}	5-HT ng g^{-1}
Sham lesions	13.0	9960	680
RN lesions	12.7	10530	350*

Values are in ng g^{-1} of fresh tissue. *Significantly
different from controls. Results are from Greenshaw,
Juorio and Nguyen, 1986.

as indicated by a decrease in the accumulation of 5-
hydroxytryptamine after administration of 5-hydroxytrypto-
phan but do not affect PE (Schlosberg et al., 1979;
Greenshaw et al., 1986).

These results together with the recent observation
that PE stimulated the release of dopamine on 5-hydroxy-
tryptamine from the striatum perfused by a push-pull cannula (Philips and Robson, 1983; Bailey et al., 1987) or
inhibited the electrically evoked release of labelled acetylcholine from brain slices (Baud et al., 1985) provides
considerable support for a possible functional role of PE
in the regulation of neuronal activity.

p- and m-TA. The p- and m-isomers of TA are normal
constituents of the brain and the concentrations of both
amines are reduced by reserpine or intraventricularly
administered 6-hydroxydopamine suggesting an intraneuronal
location in storage granules (Boulton et al., 1977). In an
attempt to elucidate the location of the cell bodies that
give use to terminals containing either p- or m-TA unilateral electrolytic lesions of the rat substantia nigra were
carried out. At survival times of 2, 11 and 25 days the
concentrations of p-TA, m-TA and DA were significantly

alterations in amine levels are most likely a consequence of the degeneration of the nigro-striatal axonal projections. These findings are supported by the fact that electrical stimulation of the substantia nigra for 1 hour induced a substantial increase in dopamine turnover in the rat caudate nucleus; concurrently, there was an increase in striatal m-TA and a reduction in p-TA (Jones et al., 1983) and agree with earlier experiments that have shown that the increases in dopamine turnover produced by neuroleptics is accompanied by a reduction in p-TA and an increase in m-TA (see Juorio, 1982 for a review).

More recently it has been shown that electrolytic lesions of the dorsal and median raphe nuclei produced no significant change in the striatal concentrations p-TA, m-TA or dopamine (Table 7), in contrast, the expected reduction in 5-hydroxytryptamine was observed suggesting that striatal TA-containing terminals do not originate or pass through the dorsal or median raphé nuclei.

TABLE 6

The effect of intraventricular cerebral administration of 6-hydroxydopamine (6-OHDA) (250 µg mL^{-1}) or electrolytic lesions of the substantia nigra (left side) on rat striatal p-TA, m-TA, DA and 5-HT

Treatment	Time days	p-TA ng g^{-1}	m-TA ng g^{-1}	DA ng g^{-1}	5-HT ng g^{-1}
Vehicle	10	10.6[1]	2.1[1]	8440[1]	450[1]
6-OHDA	10	3.6[1]*	0.9[1]*	3150[1]*	410[1]
Right side	11	14.8[2]	3.5[2]	8890[2]	-
Lesion side	11	4.1[2]*	2.2[2]*	3130[2]*	-

The asterisks indicate that the results are statistically significantly different from the corresponding controls and the superscripts indicate the source of reference.
[1]Boulton, Juorio, Philips and Wu, 1977; [2]Juorio and Jones, 1981.

TABLE 7

The effect of chronic (7 days) dorsal and median raphe (RN) lesions on rat striatal p-TA, m-TA, DA and 5-HT

	p-TA ng g^{-1}	m-TA ng g^{-1}	DA ng g^{-1}	5-HT ng g^{-1}
Sham lesions	9.90	4.0	9730	720
RN lesions	11.4	3.9	9980	350*

Values are in ng g^{-1} of fresh tissue. *Significantly different from controls. Results are from Juorio and Greenshaw, 1985.

Octopamine. It has been known for some time that the administration of divided doses of 6-hydroxydopamine (200 mg kg^{-1}, 48 hours) produced marked reductions in heart spleen and salivary gland DA concentrations (Molinoff and Axelrod, 1969). In these experiments OA was determined by a radioenzymatic method that does not resolve between the p- and m-isomers. Nevertheless, it supported that, at least in some peripheral organs, p-and/or m-OA are associated with nerve terminals. The hypothalamus contains the highest concentration of p-OA (Danielson et al., 1977) but the location of the corresponding OA-containing cell bodies has as yet to be determined. More recent experiments have shown that the subcutaneous administration of the noradrenaline neurotoxin DSP-4 (50 mg kg^{-1}, 7 days) produced a significant reduction in mouse hypothalamic p-OA concentrations [controls 9.6 ng g^{-1} ± 0.6 (3), treated 5.3 ng g^{-1} ± 0.7 (4); P<0.01] suggesting that p-OA is co-localized in noradrenergic terminals or that if p-OA neurons exist in mouse hypothalamus, they are sensitive to DSP-4 (A.V. Juorio, in preparation).

Tryptamine. The possible existence of T-containing neurons originating in the midbrain raphe is suggested by several reports of T-mediated responses to electrical stimulation of the raphe nuclei (Jones and Broadbent, 1982; Jones, 1982a,b; Cox et al., 1981). To assess this hypothesis, the effects of electrolytic lesions of the median and dorsal raphe nuclei on striatal, hypothalamic and hippocampal concentrations of T, 5-hydroxytryptamine and dopamine were investigated. No changes in the concentrations of T were observed at 1 week after lesioning the dorsal and median raphe nuclei at which time the 5-hydroxyindoles were markedly reduced (Juorio and Greenshaw, 1985). Furthermore, no reductions were observed in T concentrations in the striatum of rats pre-treated with a monoamine oxidase inhibitor (Table 8). The results indicate that T concentrations are independent of the integrity of 5-hydroxytryptamine-containing neurons of the midbrain raphe nuclei and if they exist their cell bodies are outside these areas.

One obvious possibility for this was the substantia nigra which has been known for some time to contain dopamine cell bodies and whose lesions produced significant reductions in striatal PE, p-TA and TA as well (vide supra). It was found that 7 days following a unilateral electrolytic lesion of the rat substantia nigra there is a significant reduction in the lesion side striatal concentration of T (Table 9); the efficacy of the lesion is

TABLE 8

The effect of chronic (7 days) electrolytic lesions of the dorsal and median raphe nuclei (RN) on the rat striatal concentration of T, DA and 5-HT of pargyline treated rats (200 mg kg^{-1}, 2 h)

Type of Lesion	Time days	T ng g^{-1}	5-HT ng g^{-1}	DA ng g^{-1}
Sham lesion	7	78[1]	600[1]	9730[2]
RN lesion	7	67.5[1]	260[1]*	9980[2]

Values are in ng g^{-1} of fresh tissue. *Significantly different from controls. The superscript indicates the source of reference. [1]Juorio and Greenshaw, 1985; [2]Juorio and Greenshaw, 1986.

TABLE 9

The effects of chronic (7 days) electrolytic lesions of the substantia nigra (left side) on the right and left side striatal concentrations of T, 5-HT and DA in pargyline treated (200 mg kg^{-1}, 2 hours) rats

	T ng g^{-1}	5-HT ng g^{-1}	DA ng g^{-1}
	Sham Controls		
Right side	79.9	1050	13070
Sham side	76.6	1020	12030
	Electrolytic Lesions		
Right side	86.6	1160	14370
Lesion side	48.2*	1200	5520*

Values are in ng g^{-1} of fresh tissue. *Significantly different from control side. Results are from Juorio and Greenshaw, 1986.

confirmed by the concomitant reduction in striatal dopamine and lack of effect on the sham controls (Table 9). Similar reductions in rat ipsilateral striatal T and dopamine were observed after unilateral administration of 6-hydroxydopa-mine into the left substantia nigra (Table 10); in contrast, nigral 5,7-dihydroxytryptamine administration produced no changes in striatal T or dopamine (Table 11).

These findings clearly demonstrate that T levels in the striatum are, under the present conditions, dependent on the integrity of neurons of the nigrostriatal system. In contrast, lesion of the midbrain raphe do not alter the accumulation of T.

Neurophysiological Studies

The iontophoretic studies have given invaluable impulse towards elucidation of the functional role of the

TABLE 10

The effects of chronic (7 days) 6-hydroxydopamine (6-OHDA)
lesions of the substantia nigra (left side) on the right
and left side striatal concentrations of T, 5-HT and DA
in pargyline-treated (200 mg kg^{-1}, 2 hours) rats

	T ng g^{-1}	5-HT ng g^{-1}	DA ng g^{-1}
	Sham Controls		
Right side	80.9	1150	12850
Sham side	77.7	1170	11310
	6-OHDA Lesions		
Right side	86.3	1110	12110
Lesion side	41.8*	1310	3200*

Values are in ng g^{-1} of fresh tissue. *Significantly
different from control side. Results are from Juorio,
Greenshaw and Nguyen, 1987.

trace amines. The iontophoretic administration of PE onto
striatal or cortical neurons of the rat brain reduced the
rate of firing of the neurons, in fashion similar to the
catecholamines (Henwood, Boulton and Phillis, 1979); in
addition, the administration of PE at subthreshold currents
does not produce any effect by itself but markedly potentiates the decrease in the firing rate produced by dopamine
(Table 12). Rat cortical neurons were either excited or
depressed by the iontophoretic application of TA (presumably the p-isomer) (Bevan, Bradshaw, Pun, Slater and Szabadi,
1978) while larger amounts produced mainly a depression
(Henwood, Boulton and Phillis, 1979). Subthreshold application of p-TA produced a marked potentiation of either
dopamine or noradrenaline responses and similar potentiation was observed between m-TA and dopamine (Table 12).

TABLE 11

The effects of chronic (7 days) 5,7-dihydroxytryptamine (5,7-DHT) lesions of the substantia nigra on the right and left side striatal concentrations of T, 5-HT and DA in the striatum of pargyline-treated (200 mg kg^{-1}, 2 hours) rats

	T ng g^{-1}	5-HT ng g^{-1}	DA ng g^{-1}
	Sham Controls		
Right side	79.9	1080	13100
Sham side	68.8	1080	11910
	5,7-DHT Lesions		
Right side	79.5	1100	13320
Lesion side	69.9	580*	10330*

Values are in ng g^{-1} of fresh tissue. *Significantly different from control side. Results are from Juorio, Greenshaw and Nguyen, 1987.

The application of OA (presumably the p-isomer) to rat cortical neurons produced either depression or excitation of the cells (Hicks and McLennan, 1978). The application of p-OA with weak iontophoretic currents markedly enhanced both depressant and excitatory responses to noradrenaline (Table 12). No studies have yet been performed using m-OA. The application of 5-hydroxytryptamine excited or depressed an almost equal number of neurons (Jones and Boulton, 1980b). The depressant effects of 5-hydroxytryptamine were markedly enhanced by subthreshold applications of T while the excitatory responses to 5-hydroxytryptamine were consistently reversed into depressant responses by subthreshold applications of T (Table 12).

TABLE 12

Summary of the main effects following the subthreshold
iontophoretic administration of PE, p-TA, m-TA, p-OA and T
on the neuronal firing rate effects produced by dopamine
(DA), noradrenaline (NA) and 5-hydroxytryptamine (5-HT)

Amine	DA	NA	5-HT
PE	$+I^1$	$+I^4$ $+E^4$	no effect[4]
p-TA	$+I^1$	$+I^1$	n.a.
m-TA	$+I^1$		
p-OA	no effect[2]	$+I^2$ $+E^2$	no effect[2]
T	n.a.	n.a.	$+I^3$ $-E^3$

The superscripts indicate the source of reference: [1]Jones
and Boulton, 1980a; [2]Jones, 1982b; [3]Jones and Boulton,
1980b; [4]Paterson, 1987; +, potentiation; -, reduction; I,
inhibitory response; E, excitatory response; n.a., results
not available.

CONCLUSIONS

1. The trace amines are present in neural tissues of
vertebrates and invertebrates; the concentrations of p-TA,
p-OA and T are substantially higher in some invertebrates
while levels in the vertebrate brain are generally lower.

2. The regional distribution studies indicate that
the highest concentrations of PE are found in areas of the
mesolimbic system while both isomers of TA are highest in
the striatum, p-OA in the hypothalamus and T in the
caudate-putamen and olfactory tubercles.

The results of the lesion studies suggest that PE, p-TA, m-TA or T co-exist with dopamine in the nigral neurons or that PE, p-TA, m-TA or T-containing neurons are or their axons pass through the substantia nigra. So far, either of these three alternative pathways are possible. Hypothalamic p-OA is reduced following systemic administration of a noradrenaline neurotoxin suggesting that p-OA is co-localized with noradrenaline or that p-OA neurons are sensitive to the neurotoxin.

4. The finding that subthreshold doses of trace amines enhance or decreases the responses to dopamine, noradrenaline and/or 5-hydroxytryptamine support that they act as modulators in the central nervous system.

ACKNOWLEDGEMENT

I thank Dr. A.A. Boulton for helpful advice and Saskatchewan Health for financial support.

REFERENCES

Anden N.E., Carlsson A., Dahlstrom A., Fuxe K., Hillarp N.A. and Larsson K. (1964) Demonstration and mapping-out of nigrostriatal dopamine neurones. Life Sci. 3, 523-530.

Anton A.A. and Sayre D.F. (1964) The distribution of dopamine and Dopa in various animals and a method for their determination in diverse biological material. J. Pharm. Exp. Ther. 145, 326-336.

Bailey B.A., Philips S.R. and Boulton A.A. (1987) In vivo release of endogenous dopamine, 5-hydroxytryptamine and some opf their metabolites from rat caudate nucleus by phenylethylamine. Neurochem. Res. 12, 173-178.

Baud P., Arbilla S., Cantrill R.C., Scatton B. and Langer S.Z. (1985) Trace amines inhibit the electrically evoked release of [^{3}H]acetylcholine from slices of rat striatum in the presence of pargyline: Similarities between β-phenylethylamine and amphetamine. J. Pharm. Exp. Ther. 235, 220-229.

Bertler A. and Rosengren E. (1959) Occurrence and distribution of catecholamines in brain. Acta Physiol. Scand. 47, 350-361.

Bevan P., Bradshaw C.M., Pun R.Y.K., Slater N.T. and
Szabadi E. (1978) Comparison of the responses of single
cortical neurons to tyramine and noradrenaline: effects
of desipramine. Br. J. Pharmac. 63, 651-657.
Bogdanski D.F., Pletscher A., Brodie B.B. and Udenfriend S.
(1956) Identification and anay of serotonin in brain.
J. Pharm. Exp. Ther. 117, 82-88.
Boulton A.A. (1976) Identification, distribution, metabol-
ism and function of meta and para-tyramine, phenylethyl-
amine and tryptamine in the brain. Adv. Biochem.
Psychopharmacol. 15, 57-67.
Boulton A.A. and Juorio A.V. (1982) Brain Trace Amines.
In, Handbook of Neurochemistry, 2nd Edition, Volume 1
(A. Lajtha, ed.), Plenum Press, New York, pp. 189-222.
Boulton A.A., Juorio A.V., Philips S.R. and Wu P.H. (1977)
Effects of reserpine and 6-hydroxydopamine and the levels
of some arylalkylamines in rat brain. Br. J. Pharmac.
59, 209-214.
Cox B., Davis A., Juxon V., Lee T.F. and Martin D. (1981)
Core temperature changes following electrical stimulation
of the midbrain raphe and intrahypothalamic tryptamine in
the rat. Br. J. Pharmac. 74, 216.
Danielson T.J., Boulton A.A. and Robertson H.A. (1977)
m-Octopamine, p-octopamine and phenylethanol amine in
mammalian brain: a sensitive, specific anay and effect
of drug. J. Neurochem. 29, 1131-1135.
David J.C. and Delacour J. (1980) Brain contents of
phenylethanolamine, m-octopamine and p-octopamine in the
Roman strain of rats. Brain Res. 195, 231-235.
Durden D.A., Philips S.R. and Boulton A.A. (1973) Identification and distribution of β-phenylethylamine in the
rat. Can. J. Biochem. 51, 995-1002.
Greenshaw A.J., Juorio A.V. and Boulton A.A. (1985)
Behavioural and neurochemical effects of deprenyl and
β-phenylethylamine in Wistar rats. Brain Res. Bull. 15,
183-189.
Greenshaw A.J., Juorio A.V. and Nguyen T.V. (1986) Depletion of striatal β-phenylethylamine following dopamine
but not 5-HT denervation. Brain Res. Bull. 17, 477-484.
Henwood R.W., Boulton A.A. and Phillis J.W. (1979) Iontophoretic studies of some trace amines in the mammalian
CNS. Brain Res. 164, 347-351.
Hicks T.P. and McLennan H. (1978) Comparison of the
actions of octopamine and catecholamines in single
neurones of the rat cerebral cortex. Br. J. Pharmacol.
64, 485-491.

Jones R.S.G. (1982a) Responses of cortical neurones to stimulation of the nucleus raphe medianus: A pharmacological analysis of the rate of indoleamines. Neuropharmacol. 21, 511-520.

Jones R.S.G. (1982b) Noradrenaline-octopamine interactions on cortical neurones in the rat. Eur. J. Pharmacol. 77, 159-162.

Jones R.S.G. and Boulton A.A. (1980a) Interaction between p-tyramine, m-tyramine or β-phenylethylamine and dopamine on single neurons of the cortex and caudate nucleus of the rat. Can. J. Physiol. Pharmacol. 58, 222-227.

Jones R.S.G. and Boulton A.A. (1980b) Tryptamine and 5-hydroxytryptamine actions and interactions on cortical neurons in the rat. Life Sci. 27, 1849-1956.

Jones R.S.G. and Broadbent J. (1982) Further studies on the rates of indoleamines in the responses of cortical neurones to stimulation of nucleus raphe medianus: Effect of indoleamine precursor loading. Neuropharmacol. 21, 1273-1277.

Jones R.S.G., Juorio A.V. and Boulton A.A. (1983) Changes in levels of dopamine and tyramine in the rat caudate nucleus following alterations of impulse flow in the nigrostriatal pathway. J. Neurochem. 40, 396-401.

Juorio A.V. (1971) Catecholamines and 5-hydroxytryptamine in nervous tissues of cephalopods. J. Physiol. 216, 213-226.

Juorio A.V. (1976) Presence and metabolism of β-phenylethylamine, p-tyramine, m-tyramine and tryptamine in the brain of the domestic fowl. Brain Res. 111, 442-445.

Juorio A.V. (1982) A possible role for tyramines in brain function and some mental disorders. Gen. Pharmac. 13, 181-183.

Juorio A.V. and Greenshaw A.J. (1985) Tryptamine concentrations in areas of 5-hydroxytryptamine terminal innervation after electrolytic lesion of midbrain raphe nuclei. J. Neurochem. 45, 422-426.

Juorio A.V. and Greenshaw A.J. (1986) Tryptamine depletion in the rat striatum following electrolytic lesions of the substantia nigra. Brain Res. 371, 385-389.

Juorio A.V., Greenshaw A.J. and Nguyen T.V. (1987) The effect of intranigral administration of 6-hydroxydopamine and 5,7-dihydroxytryptamine on rat brain tryptamine. J. Neurochem. 48, 1346-1350.

Juorio A.V. and Jones R.S.G. (1981) The effect of mesencephalic lesions on tyramine and dopamine in the caudate nucleus of the rat. J. Neurochem. 36, 1898-1903.

Juorio A.V. and Kazakoff C.M. (1984) The presence of β-phenylethylamine, p-tyramine, m-tyramine and tryptamine in ganglia and food muscle of the garden snail (Helix aspersa). Experientia 40, 549-551.

Juorio A.V. and Killick S.W. (1972) Monoamines and their metabolism in some molluscs. Comp. Gen. Pharmacol. 3, 283-295.

Juorio A.V. and Philips S.R. (1976) Arylalkylamines in octopus tissues. Neurochem. Res. 1, 501-509.

Juorio A.V. and Robertson H.A. (1977) Identification and distribution of some monoamines in tissues of the sunflower star, Pycnopodia Helianthoides (Echinodermata). J. Neurochem. 28, 573-579.

Laxmyr L. (1984) Biogenic amines and dopa in the central nervous system of decapod crustaceans. Comp. Biochem. Physiol. 77C, 139-143.

Molinoff P. and Axelrod J. (1969) Octopamine: Normal occurence in sympathetic nerves of rats. Science 164, 428-429.

Paterson I.A. (1987) An interaction between β-phenylethylamine and noradrenaline: an iontophoretic study in the rat cerebral cortex. This Symposium.

Philips S.R., Davis B.A., Durden D.A. and Boulton A.A. (1975) Identification and distribution of m-tyramine in the rat. Can. J. Biochem. 53, 65-69.

Philips S.R., Durden D.A. and Boulton A.A. (1974a) Identification and distribution of p-tyramine in the rat. Can. J. Biochem. 52, 366-373.

Philips S.R., Durden D.A. and Boulton A.A. (1974b) Identification and distribution of tryptamine in the rat. Can. J. Biochem. 52, 447-451.

Philips S.R. and Robson A.M. (1983) In vivo release of endogenous dopamine from rat caudate nucleus by phenylethylamine. Neuropharmacol. 22, 1297-1301.

Reader T.A., Diop L., Gottberg E., Kolta A., Brière R. and Grondin L. (1987) Dopamine in the neocortex: A classical monoamine neurotransmitter as a trace amine. This symposium.

Sardar A., Juorio A.V. and Boulton A.A. (1987) The concentration of p- and m-tyramine in the rat mesolimbic system: its regional distribution and effect of monoamine inhibition. Brain Res. 412, 370-374.

Schlosberg A.V. and Harvey J.A. (1979) Effects of L-DOPA and 5-hydroxytryptophan on locomotor activity of the rat after selective or combined destruction of central catecholamine and serotonin neurons. J. Pharm. Exp. Ther. 211, 296-304.

DOPAMINE IN THE NEOCORTEX: A CLASSICAL MONOAMINE NEUROTRANSMITTER AS A TRACE AMINE

T.A. Reader, E. Gottberg, A. Kolta,
R. Brière, L. Diop and L. Grondin

Centre de recherche en sciences neurologiques, Département de
physiologie, Université de Montréal, Montréal, Qué., H3C 3J7

INTRODUCTION

The mammalian cerebral cortex receives catecholamine-
containing afferents, and at least two such systems have been
very well documented: 1] the noradrenergic (NA) projections
originating from the **Locus coeruleus** or A6 region, and 2] the
cortical dopaminergic (DA) fibers arising from the **Substantia
nigra** and the adjacent ventral tegmental area, or A9 and A10
regions (Dahlström and Fuxe, 1964; Ungerstedt, 1971). The
presence of significant numbers of catecholamine-containing
axonal varicosities, presumed to be the terminals and release
sites, was first demonstrated by fluorescence histochemistry.
Based on these anatomical findings it was soon implied that
they could be acting as conventional neurotransmitters in
cerebral cortex. This was supported by the **in vivo** release
studies of cortical DA and NA (Reader et al., 1976), as well as
by numerous electrophysiological investigations (Bunney and
Aghajanian, 1976; Krnjević and Phillis, 1963; Reader, 1978,
1980; Reader et al., 1979a; Stone, 1973) showing effects of
these CA upon cortical neurons. In the case of NA-containing
terminals and varicosities, although present in all cortical
areas, they were found to be infrequently associated with
differentiated synaptic specializations (Beaudet and Descarries, 1978, 1984). In the case of the cortical DA afferents,
the early histofluorescent studies restricted the terminal
fields of innervation to a few cortical regions, such as the
frontal cortex (anteromedial and suprarhinal systems), the

supragenual and perirhinal systems, as well as the piriform and
entorhinal cortices (Berger et al., 1974; Hökfelt et al., 1974;
Lindvall et al., 1974; Lindvall and Björklund, 1984; Thierry et
al., 1973). Because of the difficulties inherent to the histofluorescence microscopy, and in spite of the fact that several
biochemical studies (Kehr et al., 1976; Palkowitz et al., 1979;
Reader 1981; Reader et al., 1976, 1979b; Versteeg et al., 1976)
had documented more extensive DA projections than those originally described for the rat brain, the existence of DA-containing nerve fibers in the occipital region has often been
questioned. It is only recently that new dopaminergic terminal
fields have been described in the visual cortex of young and
adult rats, using a combined histofluorescence and immunocytochemical approach (Berger et al., 1985).

BIOCHEMICAL SURVEYS OF CORTICAL DOPAMINE DISTRIBUTION

Dissections of the visual cortex.   This study was
performed with adult male Sprague-Dawley rats (275-325 g),
which were decapitated, their brains quickly removed and placed
on ice. The visual cortex was dissected out following stereotaxic coordinates from the occipital region, between IA (interaural) 3.0-1.2 mm, and between V (vertical) 2.5 mm and the
dorsal surface of the hemispheres. Towards the midline it was
delimited at L (lateral) 2.5 mm (Fig. 1).

HPLC assay of the monoamines. The tissue samples were
homogeneized in 450 µl of ice cold monochloroacetic acid
0.1 M. Aliquots of 10 µl of ascorbate oxidase (1 mg/ml in
0.1 M monochloroacetic acid, pH 3.3) were added to the homogenates which were then centrifuged (12,500 rpm for 45 min at
4°C). The pellets were dissolved overnight in 1 ml of 1 N NaOH
to determine protein content (Lowry et al., 1951). The supernatants, filtered (0.22 µm mesh) were injected into an RP18
(5-µm) column. The HPLC conditions were: flow = 0.6-1.2 ml/
min; temperature = 32-38°C; electrochemical detector at
+ 750 mV; gain at 5 nA full scale (f.s.) for cortical samples
or at 50-200 nA f.s. for neostriatum and adrenals. The mobile
phase was 0.1 M monochloroacetic acid pH 3.30-3.35, with
800 mg/L of Na_2-EDTA, 480 mg/L of octyl sodium sulphate, and
8-12% (v/v) of methanol, filtered and degassed prior to use.
The separation DA, 3,4-dihydroxyphenylacetic acid (DOPAC),
homovanillic acid (HVA), NA, 3-hydroxy-4-methoxyphenylglycol
(MHPG), adrenaline (AD), serotonin (5-hydroxytryptamine; 5-HT)
and 5-hydroxyindole-3-acetic acid (5-HIAA) could be obtained
within 40-60 min (Lakhdar-Ghazal et al., 1986).

IA 3.2 mm   IA 1.7 mm

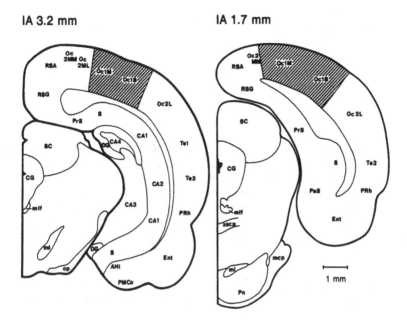

1 mm

Fig. 1. Coronal sections of the rat brain, showing the representative slices that were used to dissect out the occipital cortex, arranged according to their distance in mm from the interaural (IA) plane. The primary visual cortex was composed of both the monocular (Oc1M) and the binocular (Oc1B) parts of visual area 1. The nomenclature and the modified drawings are from the atlas of Zilles (1985). The calibration is 1 mm in both the vertical and the horizontal planes.

Endogenous dopamine contents in rat visual cortex. For the biochemical studies, the primary visual cortex was composed of both the monocular (Oc1M) and the binocular (Oc1B) parts of visual area 1, but did not include the lateral visual area 2 (Oc2L). Towards the midline, since it was delimited at L 2.5 mm (see Fig. 1), it did not include the retrosplenial regions (RSG and RSA). It could however have been contaminated in some cases by the border of the mediomedial (Oc2MM) and mediolateral (Oc2ML) components of visual area 2. For comparison purposes, the endogenous contents in monoamines for the visual cortex as above-defined is given in Table 1, together with values for the hypothalamus, neostriatum and adrenal glands.

TABLE 1. Concentration of endogenous monoamines in the visual cortex, hypothalamus, neostriatum and total adrenal glands of the rat.

Compound n	Visual cortex [16]	Hypothalamus [8]	Neostriatum [9]	Adrenals [9]
DA	0.023±0.003	3.12±0.26	79.33±6.05	28.9± 3.2
DOPAC	0.060±0.011	2.87±0.24	82.35±0.56	14.7± 0.8
HVA	0.023±0.003	0.47±0.13	14.70±0.80	0.3± 0.1
NA	1.489±0.065	15.94±0.88	3.94±0.88	718.7± 57.2
MHPG	0.147±0.024	17.60±2.60	2.58±0.61	42.4± 3.9
AD	0.020±0.003	0.40±0.08	7.52±1.61	3,391.2±231.1
5-HT	1.623±0.106	4.00±0.32	1.63±0.13	6.1± 0.4
5-HIAA	2.507±0.135	12.49±0.62	6.44±0.39	0.2± 0.06

The values represent the means ± SEM, in nanograms per milligram of protein (ng/mg p.). The number of samples assayed [n] is indicated between brackets.

Significance of endogenous dopamine in cortical tissue.
The primary visual cortex dissected in this study did not include the most medial part of the cortex, so that the prolongations of the cingulate cortex towards the occipital lobe were excluded. Therefore, the DA levels were lower than those measured in previous studies for the occipital cortex (Diop et al., 1987; Kehr et al., 1976; Reader 1981) which included part of the cingulate cortex besides the primary visual area. Although endogenous DA has repeatedly been measured in the occipital (visual) cerebral cortex of both the cat (Reader, 1980; Reader and Quesney, 1986; Reader et al., 1976, 1979b; Tork and Turner, 1981) and the rat (Kehr et al., 1976; Palkowitz et al., 1979; Reader 1981; Versteeg et al., 1976), the existence of a true dopaminergic afferent pathway has often been questioned. It is only recently that combined immunocytochemical and fluorescent investigations have shown a DA innervation of the primary visual cortex in the rat (Berger et al., 1985). Furthermore, the presence of DA terminals in this visual area has also been confirmed in radioautographic studies (Descarries et al., 1987), and preliminary quantitative data have estimated this innervation at 6,000 varicosities per cu.mm. of tissue.

ELECTROPHYSIOLOGY OF DOPAMINE IN THE VISUAL CORTEX

Electrophysiological methods. For the microiontophoretic studies, adult male Sprague-Dawley rats (250-300 g) were anesthetized with urethane (1.25-1.5 g/kg, i.p.) and placed in a stereotaxic frame. The bone overlying the occipital cortex was removed, the dura mater retracted and the surface of the cerebral cortex covered with 2% agar in saline. Five- or seven-barrel micropipettes (overall tip diameter 5-8 μm) were filled with the following drugs at pH 4.0: dopamine hydrochloride (DA, 0.5 M); apomorphine hydrochloride (APO, 0.1 M); noradrenaline hydrochloride (NA, 0.5 M) and gamma-aminobutyric acid (GABA, 0.5 M). The compounds DA, NA and APO were dissolved in 0.1% ascorbic acid. The central barrel (2-3 M NaCl; resistance of 2-6 megohms) was used for recording, while one of the side barrels (2 M NaCl) served as balancing channel and for backing currents (-10 nA) when not ejecting drugs. Extracellular unitary activity was amplified, displayed on an oscilloscope and the spikes discriminated from background noise by a spike amplitude discriminator. Only neurons with a relatively stable rate of discharge were studied. On-line analyses of visually-driven cells were performed with a 6502 microprocessor and digitized with an A/D board (Adalab).

Microiontophoretic studies with DA in visual cortex. Two populations of cells were sampled in the visual cortex; i.e.: spontaneously active (SA) units and neurons that could be synaptically activated by using a visual stimulation (VD; visually-driven cells). In order to differentiate the VD from SA neurons we used the peri-stimulus histogram (PSH) generated in response to a photic stimulation. However, it cannot be ruled out that SA units were not visual neurons since the stimulus used may not have been the most adequate to drive them specifically. Non-responsive SA units could also be participating in information processing by modifying the excitability of the synaptically-driven VD neurons. For the majority of cortical neurons (73% of SA and 69% of VD) sampled throughout this study DA exerted a long-lasting inhibitory effect (Fig. 2) and in only a few cases excitations or biphasic responses could be documented (Table 2). The microiontophoretic application of the dopaminergic agonist apomorphine also produced long-lasting inhibitions in more than 60% of the neurons tested.

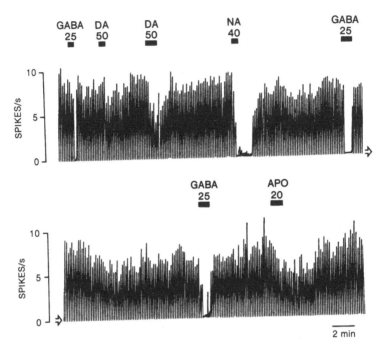

FIG. 2. Continuous ratemeter recordings of the firing rate of a spontaneously active (SA) neuron in the occipital cortex. The duration of drug applications is indicated by the horizontal bars, with the dosage in nA. The frequency was integrated over 10 s intervals.

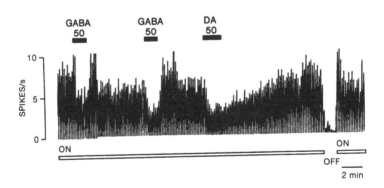

FIG. 3. Continuous ratemeter recordings of the photically-evoked firing rate of a visually driven (VD) neuron.

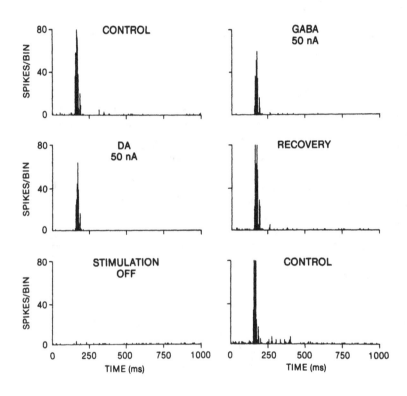

Fig. 4. Peri-stimulus histograms of the cell shown in Fig.3 generated by 60 successive sweeps of a duration of 1,024 ms each. The PSH represent the firing of the neuron for a control period, during the applications of GABA and DA, during a recovery period, while the photostimulator was OFF and for a final control.

Reader et al.

TABLE 2. Effects of dopamine and apomorphine on the firing rate of spontaneously active (SA) and visually-driven (VD) neurons in rat occipital cortex.

DRUG, type of NEURON and response parameters		Number of cells N	Inhibited	Excited	Biphasic	No response
DOPAMINE		50				
SA	n	15	11	2	1	1
Delay	(s)		20 ± 7	n.d.	n.d.	---
T_{max}	(s)		80 ± 23	n.d.	n.d.	---
$_{max}R$	%		71 ± 7	n.d.	n.d.	---
Duration	(s)		206 ± 39	n.d.	n.d.	---
VD	n	35	24	3	0	8
Delay	(s)		29 ± 10	22 ± 4	---	---
T_{max}	(s)		102 ± 22	41 ± 3	---	---
$_{max}R$	%		73 ± 5	45 ± 10	---	---
Duration	(s)		211 ± 33	42 ± 4	---	---
APOMORPHINE		17				
SA	n	5	3	2	0	0
Delay	(s)		14 ± 3	n.d.	---	---
T_{max}	(s)		74 ± 31	n.d.	---	---
$_{max}R$	%		73 ± 13	n.d.	---	---
Duration	(s)		164 ± 60	n.d.	---	---
VD	n	12	8	0	0	4
Delay	(s)		30 ± 8	---	---	---
T_{max}	(s)		63 ± 10	---	---	---
$_{max}R$	%		79 ± 9	---	---	---
Duration	(s)		143 ± 23	---	---	---

The delay was measured from the start of the drug ejection. T_{max} was the time at which the maximum response ($_{max}R$) was recorded, and the duration was the time for the neurons to recover to 90% ± 10% of their initial firing rate. The maximum response ($_{max}R$) is expressed as the percentage of firing rate recorded at T_{max} as compared to the initial firing frequency.

DOPAMINE RECEPTORS IN THE VISUAL CORTEX

The dopamine D_1 receptor. Two distinct dopamine receptors have been described in the CNS: 1] dopamine D_1 receptors, coupled to adenylate cyclase and 2] dopamine D_2 receptors, either negatively coupled or not coupled at all to this 2nd. messenger mechanism (Kebabian and Calne, 1979; Creese et al., 1983). The availability of tritium labelled SCH23390, a very potent and selective dopamine D_1 antagonist (Hyttel 1983; Iorio et al., 1983) has allowed the biochemical and pharmacological characterization of such binding sites in the neostriatum (Billard et al., 1984; Schulz et al., 1985). In addition, the use of [^{3}H]SCH23390 has permitted the radioautographic localization of dopamine D_1 receptors (Dawson et al., 1986) and these anatomical studies have shown a specific although sparse labelling in regions with a faint DA innervation, such as the cerebral cortex.

Dopamine D_1 receptor binding assays. To investigate the possible presence of dopamine D_1 receptors in the cerebral cortex, equilibrium binding determinations were performed with tissue samples from total cortex (saturation curves) and compared to neostriatum. The tissues were homogenized (Polytron, 15 s) in 40-100 volumes (w/v) of cold 50 mM Tris buffer pH 7.4 and centrifuged at 20,000 g, 10 min (4°C). The pellets were washed once and resuspended in the incubation buffer: 50 mM Tris pH 7.4 with 120 mM NaCl, 5 mM KCl, 2 mM $CaCl_2$, and 1 mM $MgCl_2$ (Billard et al., 1984). Saturation curves were performed with 10-12 concentrations of [^{3}H]SCH23390 (New England Nuclear, specific activity 80.4 Ci/mmol). Final assay volumes were of 1,000 µl; i.e.: 100 µl of membrane preparation, 100 µl of radioligand and 800 µl of buffer, with or without 30 µM of (±)SKF38393 (for nonspecific binding). After incubating at 25°C for 45 min, specific binding was measured by rapid filtration under reduced pressure over Whatman GF/C glass fiber filters and two washes (<10s) with 5 ml cold buffer. Protein concentrations were determined (Lowry et al., 1951) and in the final incubations ranged between 0.15-0.20 mg/ml for cerebral cortex, or between 0.08-0.15 mg/ml for neostriatum (Brière et al., 1987). For the neostriatum the B_{max} was of 999 ± 87 fmol/mg p with a K_d of 0.213 ± 0.011 nM (n=6), and for total cerebral cortex 131 ± 11 fmol/mg p with a K_d of 0.439 ± 0.075 nM (n=7). To measure the dopamine D_1 receptor density in the primary visual cortex the specific binding of [^{3}H]SCH23390 was determined and the maximum binding capacity estimated

TABLE 3. Endogenous catecholamine concentrations compared to their receptor densities (B_{max}) and affinities (K_d) in the rat primary visual cortex.

	Concentration [pmol/g]	B_{max} [pmol/g]	K_d [pM]
DOPAMINE	15.0± 1.9		
	90.1± 18.3 **(a)**		
	137.1± 32.6 **(b)**		
D_1 receptors		11.0±1.2	439± 75
	[I-V]	14.7±2.0 **(c)**	
	[VI]	23.7±3.6 **(c)**	
D_2 receptors	[I-V]	10.7±4.6 **(c)**	
	[VI]	3.6±2.0 **(c)**	
NORADRENALINE	880.1± 38.4		
	1,155.0±239.4 **(a)**		
	1,542.7±206.9 **(b)**		
α_1 receptors		15.9±1.4 **(b)**	178± 55
α_2 receptors		15.3±1.7 **(b)**	2,198±433
β receptors		10.3±0.6 **(b)**	1,363±229

For comparative purposes, the original values (means ± S.E.M.) of endogenous noradrenaline and dopamine contents as well as the receptor densities (B_{max}) were converted into picomoles per gram (pmol/g) of tissue wet weight. The picomolar (pM) values for the dissociation constants (K_d) are given when available. The cortical layers are in roman numerals between brackets. The original data used for the calculations was taken from Table 1 or from **(a)** Reader 1981; **(b)** Diop et al., 1987; and **(c)** Boyson et al., 1986.

according to the following formula: B_{max} = Bound x $(K_d + L)/L$, where L is the concentration of radioligand (5 nM). For this region the maximum binding capacity was calculated at 110 ± 12 fmol/mg p (n=6), in agreement with the values (147 ± 20 to 237 ± 36 fmol/mg p) previously estimated from radioautographic studies using the analog [³H]SKF83566 (Boyson et al., 1986).

The dopamine D_2 receptor in visual cortex. In radioligand binding studies with [³H]spiperone (Creese et al., 1983), [³H]spiroperidol (Fields et al., 1977) or [³H]sulpiride (Theodorou et al., 1979) dopamine D_2 receptors have been identified in the CNS. More recently, using [³H]spiroperidol, Boyson et al. (1986) have localized by radioautography such binding sites in cerebral cortex, including the primary visual area or striate cortex (Table 3). The densities of D_2 sites were lower (36 ± 20 to 107 ± 46 fmol/mg p) than those measured in the same study with [³H]SKF83566 (an analog of SCH23390 and marker of D_1 receptors). In addition, this study revealed a differential distribution of D_1 and D_2 receptors: the highest labelling of D_1 sites was found in layer VI (237 ± 36 fmol/ mg p), while the D_2 receptors were more concentrated in layer I (107 ± 46 fmol/mg p).

When the endogenous NA content in primary visual cortex is compared to the number of receptors (Table 3) it can be seen that they exceed by 20- to 30-fold the receptor densities. Since there are few NA-varicosities (5-10%) which actually make true synaptic junctions (Beaudet and Descarries, 1978, 1984), and most NA molecules have to diffuse for relatively long distances, and could be uptaken or oxidized so that only a small percentage reach and activate the specific adrenoceptors, it can be proposed that this excess of endogenous neurotransmitter is required to assure the function of this system in the cerebral cortex. If the same comparison is made for DA, the relationship between endogenous levels and receptor number is much smaller (1 to 3-4 times) so for this system to be functional it would be required that a greater percentage of DA-varicosities establishes true or 'classic' synaptic contacts.

CONCLUDING REMARKS

The present study confirms the existence of a dopaminergic system in the primary visual cortex of the rat. Although the biochemical assays demonstrate low (trace) levels of endogenous DA in this region, the coexistence of the metabolites DOPAC and HVA warrant that DA is not present as a mere precursor of NA, but that it is released from dopaminergic fibers. In addition, the responses to the microiontophoretic application of exogenous DA and of the agonist apomorphine also favour a dopaminergic innervation, which inevitably calls upon neurotransmitter-specific receptors. Although dopamine D_1 receptors in the visual cortex are readily measured and are more abundant than D_2 sites, the functional role of such DA receptor subtypes in the cortical processing of visual information as well as in higher mental functions deserves further elucidation.

Acknowledgements: Supported by the Medical Research Council of Canada (MT-6967) and the University of Montreal. Personal support was from the Fonds de la recherche en santé du Québec to Dr. T.A. Reader (Chercheur-boursier Senior) and Mr. R. Brière (Studentship), from the Centre de recherche en sciences neurologiques (CRSN) to Dr. L. Diop (Herbert H. Jasper Fellow) and from both the CRSN and the Universidad Central de Venezuela to E. Gottberg.

REFERENCES

Beaudet A. and Descarries L. (1978) The monoamine innervation of rat cerebral cortex: synaptic and nonsynaptic axon terminals. Neuroscience **3**, 851-860.

Beaudet A. and Descarries L. (1984) Fine structure of monoamine axon terminals in cerebral cortex. In, MONOAMINE INNERVATION OF CEREBRAL CORTEX (Descarries, L., Reader, T.A., and Jasper, H.H., eds) pp 77-93, Alan R. Liss, New York.

Berger B., Tassin J.P., Blanc G., Moyne M.A. and Thierry A.M. (1974) Histochemical confirmation for dopaminergic innervation of the rat cerebral cortex after destruction of the noradrenergic ascending pathways. Brain Res. **81**, 332-337.

Berger B., Verney C., Alvarez C., Vigny A. and Helle K.B. (1985) New dopaminergic terminal fields in the motor, visual (area 18b) and retrospenial cortex in the young and adult rat. Immunocytochemical and catecholamine histochemical analyses. Neuroscience **15**, 983-998.

Billard W., Ruperto V., Crosby G., Iorio L.C. and Barnett A. (1984) Characterization of the binding of ^{3}H-SCH 23390, a selective D-1 receptor antagonist ligand, in rat striatum. Life Sci. **35**, 1885-1893.

Boyson S.J., McGonigle P. and Molinoff P.B. (1986) Quantitative autoradiographic localization of the D_1 and D_2 subtypes of dopamine receptors in rat brain. J. Neurosci. **6**, 3177-3188.

Brière R., Diop L., Gottberg E., Grondin L. and Reader T.A. (1987) Stereospecific binding of a new benzazepine [^{3}H]SCH23390 in cortex and neostriatum. Can. J. Physiol. Pharmacol. (in press).

Bunney B.S. and Aghajanian G.K. (1976) Dopamine and norepinephrine innervated cells in rat prefrontal cortex. Pharmacological differentiation using microiontophoretic techniques. Life Sci. **19**, 1783-1792.

Creese I., Sibley D.R., Hamblin M.W. and Leff S.E. (1983) The classification of dopamine receptors: Relationship to radioligand binding. Ann. Rev. Neurosci. **6**, 43-71.

Dahlström A. and Fuxe K. (1964) Evidence for the existence of monoamine containing neurons in the central nervous system. I. Demonstration of monoamines in the cell bodies of brain stem neurons. Acta Physiol. Scand. **62** (Suppl. 232), 1-55.

Dawson T.M., Gehlert D.R., McCabe R.T., Barnett A. and Wamsley J.K. (1986) D-1 dopamine receptors in the rat brain: a quantitative autoradiographic analysis. J. Neurosci. **6**, 2352-2365.

Descarries L., Lemay B., Doucet G. and Berger B. (1987) Regional and laminar density of the dopamine innervation in adult rat cerebral cortex. Neuroscience (in press).

Diop L., Brière R., Grondin L. and Reader T.A. (1987) Adrenergic receptor and catecholamine distribution in rat cerebral cortex: binding studies with [^{3}H]prazosin, [^{3}H]idazoxan and [^{3}H]dihydroalprenolol. Brain Res. **402**, 403-408.

Fields J.Z., Reisine T.D. and Yamamura H.I. (1977) Biochemical demonstration of dopaminergic receptors in the rat and human brain using [^{3}H]spiroperidol. Brain Res. **136**, 578-584.

Hökfelt T., Fuxe K., Johansson O. and Ljungdhal A. (1974) Pharmacohistochemical evidence for the existence of dopamine nerve terminals in the limbic cortex. Eur. J. Pharmacol. **25**, 108-112.

Hyttel J., (1983) SCH 23390 - The first selective dopamine D-1 antagonist. Eur. J. Pharmacol. **91**, 153-154.

Iorio L.C., Barnett A., Leitz F.H., Houser V.P. and Korduba C.A. (1983) SCH23390, a potential benzazepine antipsychotic with unique interactions on dopaminergic systems. J. Pharmacol. Exp. Ther. **226**, 462-468.

Kebabian J.W. and Calne D.B. (1979) Multiple receptors for dopamine. Nature (London) **277**, 93-96.

Kehr W., Lindqvist M. and Carlsson A. (1976) Distribution of dopamine in the rat cerebral cortex. J. Neural Transm. **38**, 173-180.

Krnjević K. and Phillis J.W. (1963) Actions of certain amines on cerebral cortical neurons. Br. J. Pharmacol. Chemother. **20**, 471-490.

Lakhdar-Ghazal N., Grondin L., Bengelloun W.A. and Reader T.A. (1986) Alpha-adrenoceptors and monoamine contents in the cerebral cortex of the rodent **Jaculus orientalis**: effects of acute cold exposure. Pharmacol. Biochem. Behav. **25**, 903-911.

Lindvall O., Björklund A., Moore R.Y. and Stenevi U. (1974) Mesencephalic dopamine neurons projecting to neocortex. Brain Res. **81**, 325-331.

Lindvall O. and Björklund A. (1984) General organization of cortical monoamine systems. In, MONOAMINE INNERVATION OF CEREBRAL CORTEX (Descarries, L., Reader, T.A., and Jasper, H.H., eds) pp 9-40, Alan R. Liss, New York.

Lowry O.H., Rosebrough N.J., Farr A.L. and Randall R.J. (1951) Protein measurements with Folin phenol reagent. J. Biol. Chem. **193**, 265-275.

Palkowitz M., Zaborsky L., Brownstein M.J., Fekete M.I.K., Herman J.P. and Kanyicska B. (1979) Distribution of norepinephrine and dopamine in cerebral cortical areas of rat brain. Brain Res. Bull. **4**, 593-601.

Reader T.A. (1978) The effects of dopamine, noradrenaline and serotonin in the visual cortex of the cat. Experientia **34**, 1586-1587.

Reader T.A. (1980) Microiontophoresis of biogenic amines on cortical neurons: amounts of NA, DA and 5-HT ejected, compared with tissue content. Acta physiol. latinoam. **30**, 291-304.

Reader T.A. (1981) Distribution of catecholamine and serotonin in the rat cerebral cortex: absolute levels and relative proportions. J. Neural Transm. **50**, 13-27.

Reader T.A. and Quesney L.F. (1986) Dopamine in the visual cortex of the cat. Experientia **42**, 1242-1244.

Reader T.A., De Champlain J. and Jasper H.H. (1976) Catechol-amines released from cerebral cortex in the cat: decrease during sensory stimulation. Brain Res. **111**, 95-108.

Reader T.A., Ferron A., Descarries L. and Jasper H.H. (1979a) Modulatory role for biogenic amines in the cerebral cortex. Microiontophoretic studies. Brain Res. **160**, 217-229.

Reader T.A., Masse P. and De Champlain J. (1979b) The intra-cortical distribution of endogenous biogenic amines in the cat. Brain Res. **177**, 499-513.

Schulz D.W., Stanford E.J., Wyrick S.W. and Mailman R.B. (1985) Binding of [^{3}H]SCH23390 in rat brain: regional distribution and effects of assay conditions and GTP suggest interactions at a D_1-like dopamine receptor. J. Neuro-chem. **45**, 1601-1611.

Stone T.W. (1973) Pharmacology of pyramidal tract cells in the cerebral cortex: noradrenaline and related substances. Naunyn-Schmiedeberg's Arch. Pharmacol. **278**, 333-346.

Theodorou A., Crockett M., Jenner P. and Marsden C.D. (1979) Specific binding of [^{3}H]sulpiride to rat striatal prepara-tions J. Pharm. Pharmacol. **31**, 424-426.

Thierry A.M., Blanc G., Sobel A., Stinus L. and Glowinski J. (1973) Dopaminergic terminals in the rat cortex. Science **182**, 499-501.

Tork I. and Turner S. (1981) Histochemical evidence for a catecholaminergic (presumably dopaminergic) projection from the ventral mesencephalic tegmentum to visual cortex in the cat. Neurosci. Lett. **24**, 215-219.

Ungerstedt U. (1971) Stereotaxic mapping of the monoamine pathways in the rat brain. Acta Physiol. Scand. Suppl. **367**, 1-48.

Versteeg D.H.G., Van Der Gugten J., De Jong W. and Palkowitz M. (1976) Regional concentrations of noradrenaline and dopamine in rat brain. Brain Res. **113**, 563-574.

Zilles K. (1985) The Cortex of the Rat. A Stereotaxic Atlas, Springer-Verlag, Berlin and Heidelberg.

INVOLVEMENT OF NEWLY SYNTHETIZED CATECHOLAMINES IN THE MODULATION BY TRACE AMINES OF THE RELEASE OF NEUROTRANSMITTERS

Salomon Z. Langer and Sonia Arbilla

Department of Biology
Laboratoires d'Etudes et de Recherches Synthélabo
(L.E.R.S.) 58, rue de la Glacière
F-75013 Paris, France

INTRODUCTION

The receptor mediated modulation of the release of ^{3}H-neurotransmitters from slices of selected brain areas or from peripheral tissues is a well-established model to study interactions between neurotransmitters (Langer, 1981; Langer and Arbilla, 1981).

In non-catecholaminergic neurons, in addition to the existence of presynaptic autoreceptors, which modulate transmitter release, heteroreceptors sensitive to dopamine or noradrenaline have been described. The evoked release of ^{3}H-acetylcholine (ACh) in the corpus striatum is modulated by an inhibitory dopamine receptor of D_2 type which is the target for dopamine receptor agonists (Lehmann and Langer, 1983; Arbilla et al., 1985) exogenous (Arbilla et al., 1985) or endogenous released dopamine (Baud et al., 1985). The evoked release of ^{3}H-serotonin (5HT) in the rat substantia nigra can be inhibited by endogenous dopamine or by activation of dopamine receptors of D_1 type (Benkirane et al., 1987). On the other hand, the release of ^{3}H-5HT in the hippocampus is inhibited by $\alpha_2$adrenoceptor agonists (Benkirane et al., 1986b), while adrenoceptors of the α_1 subtype are involved in the modulation of the stimulated release of ^{3}H-ACh from rat atrial slices (Benkirane et al., 1986a).

Under conditions in which the activity of monoamine oxidase (MAO) is inhibited by pargyline or deprenyl, exposure to ß-phenylethylamine (ß-PEA) facilitates the spontaneous outflow of ^{3}H-dopamine from rat striatal slices (Langer et al., 1985) as well as the release of ^{3}H-noradrenaline from rat hippocampus and rat atrial slices (Benkirane et al., 1986a). Presynaptic release inhibitory dopamine receptors of D_1 or D_2 type as well as adrenoceptors of α_1 or α_2 type can be activated by the released dopamine or noradrenaline, respectively. Therefore, the presynatic modulation of the stimulation evoked release of ^{3}H-neurotransmitters represents a useful model to evaluate changes in catecholaminergic transmission induced by a releasing agent such as ß-PEA.

The present article describes changes in dopaminergic and noradrenergic transmission induced by ß-PEA. These changes are explored as a function of the inhibition of both MAO activity and tyrosine hydroxylase activity.

EFFECTS OF ß-PEA ON THE EVOKED RELEASE OF ^{3}H-TRANSMITTERS : INFLUENCE OF THE INHIBITION OF MONOAMINE OXIDASE ACTIVITY

Under our experimental conditions of electrically-evoked release of ^{3}H-ACh from slices of corpus striatum at 1 Hz, or release of ^{3}H-5HT from the substantia nigra or hippocampus at the frequency of 3 Hz, the concomitant release of endogenous catecholamines does not activate the corresponding release modulatory receptors (Baud et al., 1985; Benkirane et al., 1986b, 1987). Similarly, α_1-adrenoceptors involved in the modulation of the release of ^{3}H-ACh form the rat atria elicited by exposure to 100 mM potassium are not activated by endogenous released noradrenaline (Benkirane et al., 1986a). Therefore, these experimental conditions of stimulation of ^{3}H-transmitter release are suitable to study the D_2 and D_1 receptor mediated effects of dopamine released by ß-PEA, as well as the α_1 and α_2-adrenoceptor mediated effects triggered by the noradrenaline releasing action of ß-PEA.

In the absence of MAO inhibition, ß-PEA does not modify the electrically-evoked release of ^{3}H-ACh from the striatum or ^{3}H-5HT from the hippocampus (Table 1). In contrast, ß-PEA inhibits the electrically-evoked release of ^{3}H-5HT in the substantia nigra through the activation of D_1 receptors as this effect is blocked by 0.1 µM SCH 23390 (data not shown).

When deprenyl 1 µM is included in the Krebs'medium to protect ß-PEA from metabolism by MAO-B (Yang and Neff, 1973), exposure to ß-PEA 3 µM inhibits the evoked release of ^{3}H-ACh from the striatum as well as the release of ^{3}H-5HT from the hippocampus (Table 1). In addition, deprenyl potentiates the concentration-dependent inhibitory action of ß-PEA on the release of ^{3}H-5HT from the substantia nigra (Table 1). The inhibitory effect of ß-PEA on the release of ^{3}H-5HT in the substantia nigra is observed with lower concentrations of ß-PEA than those necessary in the striatum to inhibit ^{3}H-ACh release and or in the hippocampus to inhibit ^{3}H-5HT release. This action of ß-PEA in the substantia nigra is present even in the absence of deprenyl. The low activity of MAO-B in the substantia nigra (Westlund et al., 1985) may explain this regional effect of ß-PEA.

³H-TRANSMITTER Central Structure	EXPERIMENTAL GROUP	S₂/S₁		
		MAO INTACT	DEPRENYL 1 μM	DEPRENYL 1 μM + α-MPT
³H-5HT Substantia nigra	Control	0.97±0.02 (20)	0.96±0.03 (30)	0.85±0.11 (5)
	β-PEA 0.3 μM	1.05±0.10 (4)	0.44±0.03 (16)*	N.T.
	β-PEA 1 μM	0.58±0.02 (12)*	0.34±0.04 (16)*	0.71±0.03 (3)
³H-ACh Corpus Striatum	Control	0.92±0.01 (6)	0.79±0.02 (25)	0.85±0.07 (4)
	β-PEA 3 μM	0.83±0.07 (3)	0.28±0.01 (10)*	0.80±0.03 (5)
³H-5HT Hippocampus	Control	0.99±0.05 (9)	0.93±0.03 (8)	0.91±0.10 (7)
	β-PEA 3 μM	0.85±0.03 (3)	0.53±0.03 (11)*	0.89±0.13 (4)

Table 1 : Influence of monoamine oxidase inhibition on the effects of β-PEA at modulating the electrically-evoked release of ³H-transmitters in different regions of the rat central nervous system. Two periods of electrical stimulation (S_1 and S_2) were applied to elicit the release of ³H-transmitters. S_2/S_1 represents the ratio of transmitter release elicited by the two periods of stimulation under control conditions and after 20 min of exposure, before S_2, to β-PEA. The Krebs medium was either drug free (MAO intact) or contained deprenyl 1 μM. When indicated, α-MPT (300 mg/kg i.p.) was administered 2 hrs before the experiment and was present in the medium at 100 μM. Each value is the mean ± S.E.M. of the number of experiments shown in parentheses. N.T. not tested. * $p < 0.05$ when compared with the corresponding control. For experimental details see Baud et al. (1985) and Benkirane et al. (1986a, 1987).

In contrast to central areas, the release of
^{3}H-ACh evoked by high potassium in the atria is not
affected by exposure to ß-PEA. This lack of effect of
ß-PEA is observed in the presence of deprenyl as well as
in the presence of the non-selective MAO inhibitor
pargyline (Table 2). It should be stressed that under
both conditions, exogenous noradrenaline inhibits
^{3}H-ACh release in the atria (Table 2) and that this
effect is mediated through the activation of
α_1-adrenoceptors (Benkirane et al., 1986a).

EXPERIMENTAL GROUP	S_2/S_1	
	DEPRENYL 1 µM	PARGYLINE 10 µM
Control	0.76+0.02 (8)	0.75+0.07 (9)
Noradrenaline 10 µM	0.45+0.06 (4)*	0.43+0.05 (4)*
ß-PEA 30 µM	0.66+0.08 (10)	0.85+0.09 (3)

Table 2 : Effects of noradrenaline and ß-PEA on the
potassium-evoked release of ^{3}H-ACh from rat atrial
slices under conditions of monoamine oxidase inhibition
by deprenyl or pargyline. Two 2 min periods (S_1 and
S_2) of exposure to potassium 100 mM were applied to
elicit ^{3}H-ACh release. S_2/S_1 represents the ratio
of transmitter release elicited by the two periods of
stimulation under control conditions or after 20 min of
exposure, to noradrenaline 10 µM or ß-PEA 30 µM
before S_2. When indicated, the Krebs'medium contained
either deprenyl 1 µM or pargyline 10 µM. Shown are
mean ± S.E.M. of 4-9 experiments per group. * p <
0.05. For experimental details, see Benkirane et al.
(1986b, 1987).

INFLUENCE OF TYROSINE HYDROXYLASE INHIBITION
ON THE EFFECTS OF ß-PEA ON ^{3}H-TRANSMITTERS RELEASE

Inhibition of tyrosine hydroxylase activity was achieved by pretreatment of the rats with α-methyl-P-tyrosine 300 mg/kg i.p. 2 h before the experiment. The enzyme inhibitor was also present in the Krebs'medium at 100 μM. Under conditions in which MAO-B is inhibited with deprenyl, exposure to ß-PEA inhibits the release of ^{3}H-ACh from the striatum and ^{3}H-5HT from the nigra and hippocampus respectively (Table 1), however these inhibitory effects of ß-PEA are abolished by the inhibition of tyrosine hydroxylase activity. Thus, when ß-PEA is protected by the selective inhibition of MAO-B activity, this trace amine exerts its inhibitory effect on the release of ^{3}H-ACh from the striatum or ^{3}H-5HT from the substantia nigra through the indirect activation of D_2 and D_1 dopamine receptors by releasing preferentially newly synthesized dopamine. Similarly, these results support the view that newly synthesized noradrenaline released by ß-PEA is at the origin of the inhibitory action of this amine on the release of ^{3}H-5HT in the hippocampus.

The lack of effects of ß-PEA at modulating ^{3}H-ACh or ^{3}H-5HT release when the synthesis of catecholamines is inhibited, indicates that ß-PEA is devoid of intrinsic activity at the receptors involved in the modulation of the evoked release of the ^{3}H-transmitters that we examined. Our results also exclude the possibility that ß-PEA recognition sites may have a transmitter release modulatory role. This view was also confirmed after chemical noradrenergic denervation in the hippocampus with DSP4 (Benkirane et al., 1986b) as well as dopaminergic denervation with 6OHDA of the striatum (Baud et al., 1985) or the substantia nigra (Benkirane, Arbilla and Langer, unpublished observations).

CONCLUDING REMARKS

When MAO-B is inhibited by deprenyl, ß-PEA releases dopamine and activates indirectly inhibitory D_2 and D_1 receptors that mediate respectively inhibition of the evoked release of ^{3}H-ACh in the striatum and ^{3}H-5HT in the substantia nigra. In addition, ß-PEA releases noradrenaline that activates indirectly α_2-adrenoceptors which inhibit the release of ^{3}H-5HT in the hippocampus. However, ß-PEA fails to activate α_1-adrenoceptors modulating the release of ^{3}H-ACh in rat atria. Our experimental evidence suggests that the inhibitory effects of ß-PEA on transmitter release occurs in the central nervous system but may not occur in the periphery.

When tyrosine hydroxylase activity is inhibited by α-MpT, exposure to ß-PEA does not modify transmitter release, indicating the involvement of newly synthesized catecholamines in its mechanism of action.

In the absence of inhibition of MAO activity, ß-PEA retains its transmitter release inhibitory activity only in the substantia nigra, indicating that the action of ß-PEA may differ in each brain area depending on the degree of MAO activity.

REFERENCES

Arbilla S., Nowak J.Z. and Langer S.Z. (1985) Rapid desensitization of presynaptic dopamine autoreceptors during exposure to exogenous dopamine. Brain. Res. 337, 11-17.

Baud P., Arbilla S. and Langer S.Z. (1985) Inhibition of the electrically evoked release of ^{3}H-acetylcholine in rat striatal slices: an experimental model for drugs that enhance dopaminergic neurotransmission. J. Neurochem. 44, 331-337.

Benkirane S., Arbilla S. and Langer S.Z. (1986a) Phenyl-ethylamine-induced release of noradrenaline fails to stimulate α_1-adrenoceptors modulating ^{3}H-acetylcholine release in the rat atria, but activates α_2-adrenoceptors modulating ^{3}H-serotonin release in the hippocampus. Naunyn-Schmiedeberg's Arch. Pharmacol. 334, 149-155.

Benkirane S., Arbilla S. and Langer S.Z. (1986b) Newly synthesized noradrenaline mediates the α_2-adreno-ceptor inhibition of ^{3}H-5-hydroxytryptamine release induced by ß-phenylethylamine in rat hippocampal slices. Eur. J. Pharmacol. 131, 189-198.

Benkirane S., Arbilla S. and Langer S.Z. (1987) A functional response to D$_1$ dopamine receptor stimu-lation in the central nervous system: inhibition of the release of ^{3}H-serotonin from the rat substantia nigra. Naunyn-Schmiedeberg's Arch. Pharmacol. 335, 502-507.

Lehmann J. and Langer S.Z. (1983) Dopamine receptors modulating ^{3}H-acetylcholine release in slices of the cat caudate: effects of (-)N-(2-chloroethyl)nora-pomorphine. Eur. J. Pharmacol. 90, 393-400.

Langer S.Z. (1981) Presynaptic regulation of the release of catecholamines. Pharmac. Rev. 32, 337-362.

Langer S.Z. and Arbilla S. (1981) Presynaptic receptors and modulation of the release of noradrenaline, dopa-mine and GABA. Postgraduate Medical Journal 57, 18-29.

Langer S.Z., Arbilla S., Niddam R., Benkirane S. and Baud P. (1985) Pharmacological profile of ß-phenyl-ethylamine on dopaminergic and noradrenergic neuro-transmission in rat cerebral slices: comparison with amphetamine and other trace amines. In, Neuropsycho-pharmacology of the trace amines (Boulton A.A., Maitre L., Bieck P.R. and Reiderer P., ed) pp 27-88, Humana Press, Inc. Clifton N.J.

Westlund K.N., Denney R.M., Kocherspergen L.M., Rose R.M. and Abell C.W. (1985) Distinct monoamine oxidase A and B population in the primate brain. Science 230, 181-183.

Yang H.Y.T. and Neff N.H. (1973) ß-Phenylehtylamine: a specific substrate for type B monoamine oxidase of brain. J. Pharmacol. Exp. Ther. 187, 365-371.

AN INTERACTION BETWEEN β-PHENYLETHYLAMINE AND NORADRENALINE:

AN IONTOPHORETIC STUDY IN THE RAT CEREBRAL CORTEX.

I.A.Paterson,

Psychiatric Research Division,
University of Saskatchewan,
Saskatoon, Saskatchewan,
Canada, S7N 0W0.

The trace amines are a group of endogenous aromatic amines found in low concentrations in the nervous systems of many invertebrate and vertebrate species. They are a diverse group of compounds and it seems likely that they have diverse functions. It has been postulated that trace amines may act as neuromodulators (Boulton, 1976, 1980), false transmitters (Baldessarini, 1974), co-transmitters (Axelrod and Saavedra, 1977), or that they may be neurotransmitters in their own right (Jones, 1982b, Stoof et al, 1976).

β-Phenylethylamine (PE) is a trace amine with a heterogeneous distribution in the human and the rat brain (Durden et al, 1973; Phillips et al, 1978). It is associated with the brain synaptosomal fraction (Boulton and Baker, 1975), and is released from striatal slices by electrical stimulation in a calcium dependent manner (Niddam et al, 1985). Although it is present in small quantities it has a very rapid turnover rate in the rat brain (Durden and Phillips, 1980). All of this suggests that PE plays some role in neurotransmission.

Boulton (1976, 1980) proposed that trace amines may be released from presynaptic sites to alter the sensitivity of the synapse to the synaptic transmitter. There is increasing evidence for such a modulatory role of the trace amines. Tryptamine, in addition to inhibiting cortical neurones, altered their response to iontophoretically applied 5HT (Jones, 1982b). para-Tyramine (pTA), meta-tyramine (mTA) and PE caused an enhancement of responses to iontophoretically applied DA (Jones, 1981; Jones and Boulton, 1980), and pTA and octopamine enhanced responses to iontophoretically applied NA (Jones, 1982a; Jones and Boulton, 1980). These electrophysiological findings show that there are interactions between the trace amines and the classical monoamines, though the mechanisms of these interactions are not clear.

There is evidence that NA may be involved in some of the actions of PE in the central nervous system. The release of 5HT from hippocampal slices is inhibited by PE. This effect of PE was blocked by α-methyl-$\underline{\text{p}}$-tyrosine (αMPT) and yohimbine, indicating that the effect is mediated by newly synthesised NA acting on α2 receptors (Benkirane et al, 1986). Systemically and iontophoretically applied PE inhibited neurones in the rat locus coeruleus (LC), and these responses were inhibited by pretreatment with reserpine and by the α2 antagonist yohimbine (Lundberg et al, 1985). This suggests that PE may have a modulatory effect on noradrenergic transmission within the LC.

The experiments described in this report have tested the hypothesis that PE may modulate noradrenergic transmission by investigating the effects of iontophoretically applied PE on cortical neurones and their responses to NA. Once such an effect was observed, further experiments were carried out to determine the mechanisms of the interaction.

Methods

Male Wistar rats (220 - 300g), anaesthetised with urethane (1.2 g/kg, i.p.), were used in all experiments. A lateral tail vein was cannulated for the intravenous administration of drugs and maintenance doses of urethane (0.1 g/kg). The rat was placed in the stereotaxic frame, and it's body temperature was maintained at $37 \pm 0.5^{\circ}C$ by an electric heating blanket with feedback control.

All recordings were made through six-barrelled glass microelectrodes with a tip diameter of 4-7 μm. The recording barrel was filled with 3M NaCl saturated with Fast Green, giving a resistance of 2-5 MΩ. A second barrel was filled with 3M NaCl and was used to balance the iontophoretic currents. Barrels 3 and 4 always contained ($\pm$)-noradrenaline hydrochloride (0.2M, pH 5.0) and β-phenyl-ethylamine hydrochloride (0.2M, pH 5.0). The remaining barrels each contained one of the following: acetylcholine chloride (ACh, 0.2M, pH 5.0), γ-aminobutyric acid (GABA, 0.2M, pH 3.5), 5-hydroxy-tryptamine bimaleate (0.05M, pH 5.5), monosodium glutamate (0.2M, pH 8.0) or ($\pm$)-noradrenaline hydrochloride (0.2M, pH 5.0). The resistances of the drug barrels were 20-80 MΩ. All drugs were retained in the barrels with currents of 20 nA.

The microelectrodes were introduced into the cerebral cortex through a burr hole in the skull at AP 3.0 mm, LAT ±1.5 mm. The dura was torn and deflected, and the preparation was covered with liquid paraffin to prevent the surface of the cortex from drying. All units were spontaneously active, and were recorded 400-1200 μm below the pial surface. Action potentials were amplified and passed through a variable voltage window discriminator which was set to detect action potentials of a single amplitude. The signal from the discriminator was passed to a rate meter, the output of which was passed to a chart recorder to provide a continuous record of the cell firing rate.

When a neurone was located, control responses to iontophoretically applied NA (and sometimes one or more of the other drugs, except PE) were established. Once stable responses were obtained a weak continuous application of PE was initiated without interrupting the regular applications of the agonists. When a change in the response to NA was seen the PE application was terminated and the NA applications were continued until recovery was seen. If there was no change in the NA response the PE ejection current was turned up until either a change in the response was seen or the baseline firing rate changed. Finally, the response to larger applications of PE was established. The responses to iontophoretic applications were quantified by measuring the area of the response and converting to total spike number (Bradshaw et al, 1974).

A number of rats were subjected to one of the following pretreatments: α-methyl-p-tyrosine (250 mg/kg, i.p., 30 minutes prior to anaesthesia), reserpine (10 mg/kg, i.p., 24 hours) or vehicle (20% ascorbic acid, 2 ml/kg, i.p., 24 hours) or unilateral electrolytic lesions of the Locus Coeruleus 14-21 days before recording. Lesions were made by passing 1 mA, 10s through a monopolar electrode at AP -9.3, LAT +1.1, VERT -7.1 (w.r.t. bregma, Paxinos and Watson, 1982). The position and the extent of the lesions were verified histologically.

Results

Responses to PE

Cortical neurones responded to iontophoretic applications of both PE (30-100 nA, 20-60 s) and NA (30-66 nA, 20-60 s) though fewer cells responded to PE (65%) than to NA (87%). Typical responses to PE and NA are shown in Figure 1. The responses to PE and NA of 46 neurones are shown in Table 1. When NA was excitatory 92% (11/12) of the neurones were excited by PE, and when NA was inhibitory 60% (15/25) of the neurones were inhibited by PE. Three cells gave a biphasic response to NA and all 3 were excited by PE. Only one cell gave a response to PE but not to NA. In general, the responses to PE were weaker than the responses to NA.

Table 1   The responses of 46 neurones to iontophoretic applications of PE and NA. Cells could respond with an excitation (+), a biphasic response (+/-), an inhibition (-), or fail to respond (0).

PE response

NA response	+	+/-	-	0
+	11			1
+/-	3			
-			15	10
0	1			5

The effects of LC lesions and pretreatment with reserpine on responses to PE are shown in Table 2. Reserpine abolished responses to PE although both excitatory and inhibitory responses to NA were seen. The vehicle (20% ascorbic acid) had no effect. When recordings were made in the cerebral cortex ipsilateral to LC lesions only 1 cell (out of 23) responded to PE, while in the contralateral cortex normal responses to PE and NA were seen.

Effects of PE on responses to NA

The effects of PE on responses to NA were tested by applying PE with currents of 0-12 nA (median 3 nA) for 3-20 minutes (median 8 minutes) during regular applications of NA. This was done on 24 cells in control animals (see Table 3). In 3 of the studies PE had no effect, but in 21 studies PE caused an increase in the response to NA to between 130% and 565% of control values. Both excitatory responses (n=3, Figure 2) and inhibitory responses(n=16, Figure 3) were enhanced by

<u>Table 2</u>. The effects of reserpine and LC lesions on the responses of cortical neurones to iontophoretically applied PE and NA. *P<0.001 in comparison to control, Chi-squared test.

Pretreatment	Number of cells observed	% of cells responding to	
		PE	NA
None (control)	46	66	87
Reserpine	18	0 *	78
Vehicle	17	58	82
LC lesion (ipsilateral)	23	4 *	78
LC lesion (contralateral)	22	73	95

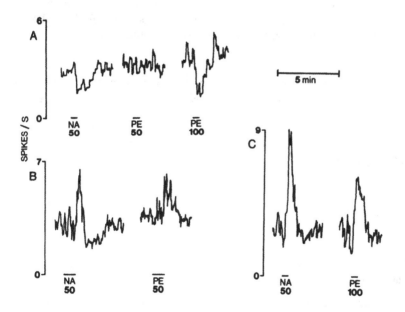

<u>Figure 1</u>. Excerpts from ratemeter records of three cortical neurones showing their responses to NA and PE. Bars indicate the iontophoretic application and the numbers show the current in nA. A] Neurone inhibited by both NA and PE. B] Neurone giving a biphasic response to NA, and excited by PE. C] Neurone excited by both NA and PE.

PE. In two studies PE enhanced both the excitatory and inhibitory components of biphasic responses to NA. The latency of this effect was 1-8 minutes (median 2 minutes), though in all but two cases the first NA response was larger than control, so the latency is largely a function of the time between the start of the PE application and the next NA application. The enhancement of the NA response invariably continued after the PE application was terminated (3-39 minutes, median 12 minutes). No effect of PE was seen on responses to ACh , GABA, or glutamate (Table 3). In one study PE caused a small inhibition in the response to 5HT but in five others it had no effect on 5HT responses.

The enhancement of NA responses by PE could be repeated after the NA responses returned to control values (n=2). On four occasions, after an interaction between PE and NA had been seen, the animal was given an injection of desimipramine (5 or 10 mg/kg, i.v.). The response of the cell to NA was followed until it stabilised, then PE was applied a second time. On all four occasions PE enhanced the NA response, though the magnitude and the duration of the effect were slightly different. In a number of electrodes two barrels were filled with NA to test the effect of a small application of NA (0-3 nA, 4-7 minutes) on NA responses. The small application of NA did not alter responses to NA without changing the baseline firing rate.

Table 3. The effects of small applications of PE on cortical neurone responses to iontophoretically applied NA, ACh, GABA, Glutamate and 5HT.

	Increase	decrease	no effect
NA excitations	3	0	1
NA inhibitions	16	0	2
NA biphasic	2	0	0
ACh	0	0	3
GABA	0	0	6
Glutamate	0	0	4
5HT (inhibitions)	0	1	5

Pretreatment with αMPT, ascorbic acid and electrolytic lesions of the LC did not alter the ability of PE to enhance cortical neurone responses to NA, either in the magnitude or the duration of the effect.

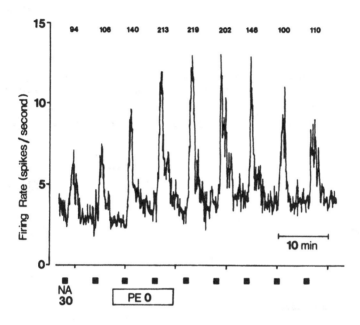

Figure 2. A continuous ratemeter record of a cortical neurone giving excitatory responses to NA. The bars under the trace indicate iontophoretic applications and the numbers show the ejection current in nA. The numbers above the trace show the size of the response as a percentage of the control responses.

Thus with αMPT (n=5) and ascorbic acid (n=4) pretreatments, and with cortical neurones ipsilateral (n=6) and contralateral to the LC lesion (n=4), the magnitude and duration of the enhancement of NA by PE were within the range of values from the non-pretreated animals. Studies in animals pretreated with reserpine also showed an enhancement of NA responses by PE, but there were differences from the control animals. The cortical neurones in the reserpine treated rats showed a slight supersensitivity to NA and with repeated applications of NA there was a marked desensitisation of the response over the first 5 or 6 applications. Once stable responses were obtained, PE was applied in the normal manner. The response to NA was increased to between 170% and 345% of control values but there was a long latency (9-13 minutes). Four studies were carried out but recovery of the NA response to control levels was seen in only one study, after 60 minutes.

In the remaining 3 studies the response of the neurone to NA was followed for 60-90 minutes (when the cell was lost) without seeing recovery of the NA response.

Table 4. The effect of various pretreatments on the action of PE on responses to NA. None of the pretreatments altered the either the magnitude or the duration of the effect of PE, with the exception of reserpine which increased the duration of the effect of PE (see text).

	Effect of PE		
Pretreatment	Increase	Decrease	No Effect
None	21	0	3
αMPT	5	0	0
Reserpine	4	0	0
Vehicle	4	0	0
Ipsilateral LC lesion	6	0	0
Contralateral LC lesion	4	0	0

Discussion and conclusions

Fewer cells responded to PE than to NA, and the responses to PE were weaker than the responses to NA. When responses were obtained to both compounds the responses were qualitatively similar, with the exception of cells which gave biphasic responses to NA and excitatory responses to PE. Pretreatment with reserpine abolished cortical neurone responses to PE, as did ipsilateral LC lesions. In the cortex contralateral to the LC lesions the PE responses were not affected, presumably because the noradrenergic innervation to the cortex arises almost exclusively from the ipsilateral LC (Ungerstedt, 1971). These results suggest that cortical neurone responses to PE are indirect, relying on intact noradrenergic presynaptic terminals and granular stores of NA. This is similar to the effects of PE in the LC (Lundberg et al, 1985). It is unlikely that PE is acting directly on postsynaptic noradrenergic receptors to exert its effect as none of the pretreatments altered responses to NA, and in the sole report on PE binding sites it was found that PE binding was not displaced by α- or β-adrenergic ligands (Hauger et al, 1982). These results do not indicate what the

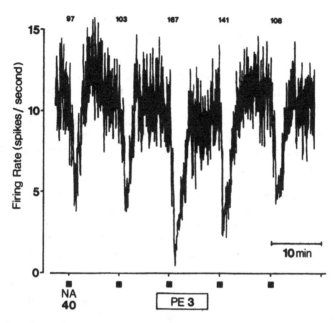

Figure 3. A continuous ratemeter record of a cortical neurone which is inhibited by NA. Details as in figure 2.

mechanisms of the PE responses are, only that they depend on functioning noradrenergic transmission. It could be that PE inhibits the reuptake of NA <u>in vivo</u> (Horn and Snyder, 1973; Raiteri <u>et al</u>, 1978) or it could be releasing NA from the presynaptic terminals (Fuxe <u>et al</u>, 1967; Raiteri <u>et al</u>, 1978). Finally, in view of the possible postsynaptic action of PE (see below), it could be that the responses to PE arise from the postsynaptic enhancement of the efficacy of endogenous NA.

Applications of PE with small iontophoretic currents, which had no effect on the baseline firing rate, caused an increase in the size of responses to NA. This effect was seen with inhibitions, excitations and biphasic responses, and was specific to NA in that responses to GABA, glutamate, ACh and 5HT (5/6) were not affected. On one occasion PE caused a small inhibition of the 5HT response, but it is unlikely that this is a significant effect. The responses to NA were increased to between 130% and 565% of control responses. The effect invariably lasted beyond the termination of the PE application, up to 39 minutes. The effects are similar in magnitude to the reported effects of p-TA, m-TA and PE on responses to DA, and of p-TA and OA on responses to NA (Jones, 1981; Jones, 1982a; Jones and Boulton, 1980). However, the duration of the effect was much longer than any previously reported interaction between trace amines and catecholamines, where the effects of the trace amines rarely outlasted the iontophoretic application. The long duration of the effect is somewhat surprising since PE is rapidly oxidised by MAO-B and exogenous PE has an estimated half-life of between one and five minutes, depending on the brain region (Wu and Boulton, 1975). Since no MAO inhibitors were used in these experiments, it would seem that either PE initiates a mechanism which continues after the PE has disappeared, or that PE is getting into some cellular or extracellular compartment where it is protected from MAO-B to exert it's effect.

The mechanism of the interaction between iontophoretically applied PE and NA is not likely to involve endogenous NA or noradrenergic presynaptic terminals. The finding that reserpine, αMPT and lesions of the LC failed to block the interaction appears to rule out an involvement of endogenous NA. The inability of Desimipramine and lesions of the LC to block the interaction makes it unlikely that the interaction is due to an inhibition of NA reuptake systems, though it can not be ruled out that inhibition of reuptake may contribute to the effect. While other possibilities remain, the most likely explanation of the effect is that PE is acting postsynaptically to enhance the efficacy of NA.

These findings provide the first evidence that PE may have postsynaptic as well as presynaptic effects within the CNS, and provide further evidence to support the hypothesis that trace amines may modulate catecholamine transmission (Boulton, 1976, 1980).

Acknowledgement

I wish to thank Dr. A.A.Boulton for his support and advice.

References

Axelrod, J. and Saavedra, J.M., (1977) Octopamine, Nature, 265, 501-504.

Baldessarini, J., (1974) Release of catecholamines. In Handbook of Psychopharmacology, Vol. 3 (I.I.Iverson, S.D.Iverson and S.H.Snyder, Eds), p37-114, Plenum, New York and London.

Benkirane, S., Arbilla, S. and Langer, S.Z., (1986) Newly synthesised noradrenaline mediates the α2-adrenoreceptor inhibition of [3H]-5-hydroxytryptamine release induced by β-phenylethyl -amine inrat hippocampal slices, Eur. J. Pharmacol., 131, 189-198.

Boulton, A.A., (1976) Cerebral aryl alkyl aminergic mechanisms. In Trace Amines and the Brain (E.Usdin and M.Sandler, Eds), p22-39, Marcel Dekker, New York.

Boulton, A.A., (1980) The trace amines: Neurohumours, Behav. Brain Sci., 7, 418.

Boulton, A.A. and Baker, G.B., (1975) The subcellular distribution of β-phenylethylamine, p-tyramine and tryptamine in the rat brain, J. Neurochem., 25, 477-481.

Bradshaw, C.M., Roberts, M.H.T. and Szabedi, E., (1974) Effects of imipramine and desimipramine on responses of single cortical neurones to noradrenaline and 5-hydroxytryptamine, Br. J. Pharmacol., 52, 349-358.

Durden, D.A. and Phillips, S.R., (1980) Kinetic measurements of the turnover rates of phenylethylamine and tryptamine in vivo in the rat brain, J. Neurochem., 34, 1725-1732.

Durden, D.A., Phillips, S.R. and Boulton, A.A., (1973) Identification and distribution of β-phenylethylamine in the rat, Can. J. Biochem., 51, 995-1002.

Fuxe, K., Grobecker, H. and Jonsson, J., (1967) The effects of β-phenylethylamine on central and peripheral monoamine containing neurones, Eur. J. Pharmacol., 2, 202-207.

Hauger, R.L., Skolnick, P. and Paul, S.M., (1982) Specific [3H]-β-phenylethylamine binding sites in the rat brain, Eur. J. Pharmacol., 83, 147-148.

Horn, A.D. and Snyder, S.H., (1973) Steric requirements for catecholamine uptake by rat brain synaptosomes: studies with rigid analogues of amphetamine, J. Pharmacol. Exp. Ther., 180, 523-530.

Jones, R.S.G., (1981) Specific enhancement of neuronal responses to catecholamine by p-tyramine, J. Neurosci. Res., 6, 49-61.

Jones, R.S.G., (1982a) Noradrenaline-octopamine interactions on cortical neurones in the rat, Eur. J. Pharmacol., 77, 159-162.

Jones, R.S.G., (1982b) Tryptamine: a neuromodulator or neurotransmitter in mammalian brain?, Prog. Neurobiol., 19, 117-139.

Jones, R.S.G. and Boulton, A.A., (1980) Interactions between p-tyramine, m-tyramine or β-phenylethylamine and dopamine on single neurones in the cortex and caudate nucleus of the rat, Can. J. Physiol. Pharmacol., 58, 222-227.

Lundberg, P-.A., Oreland, L. and Engberg, G., (1985) Inhibition of Locus Coeruleus neuronal activity by β-pheylethylamine, Life Sci. , 36, 1889-1896.

Niddam, R., Arbilla, S., Baud, P. and Langer, S.Z., (1985) [3H]-β-phenylethylamine but not [3H]-(±)-amphetamine is released by electrical stimulation from perfused rat striatal slices, Eur. J. Pharmacol., 110, 121-124.

Phillips, S.R., Rozdilsky, B. and Boulton, A.A., (1978) Evidence for the presence of m-tyramine, p-tyramine, tryptamine and phenyethylamine in the rat brain and several areas of the human brain, Biol. Psychiat., 13, 51-57.

Raiteri, M., Del Carmine, R., Bertollini, A. and Levi, G., (1977) Effect of sympathomimetic amines on the synaptosomal transport of noradrenaline, dopamine and 5-hydroxytryptamine, Eur. J. Pharmacol., 41, 133-143.

Stoof, J.C., Liem, A.L. and Mulder, A.H., (1976) Release and receptor stimulating properties of p-tyramine in rat brain, Arch. Int. Pharmacodynam., 220, 62-71.

Ungerstedt, U., (1971) Stereotaxic mapping of the monoamine pathways in the rat brain, Acta Physiol. Scand. Suppl., 367, 1-48.

Wu, P.H. and Boulton, A.A., (1975) Metabolism, distribution and disappearance of injected β-phenylethylamine in the rat, Can. J. Biochem., 53, 42-50.

A CONTRIBUTION TO THE NEUROMODULATORY/NEUROTRANSMITTER ROLE OF TRACE AMINES

J. Harris, S. Trivedi and B. L. Ramakrishna
Arizona State University, Department of
Chemistry, Tempe, Arizona 85287, USA

INTRODUCTION

A role of the trace amines as neuromodulators or even neurotransmitters has been suggested (Boulton 1976, Sabelli et al. 1978) by several investigators but remains an open question.

The view is supported in part from evidence obtained by iontophoretic administration of small amounts of the trace amines which produce no change in neuronal firing rate, but, potentiate changes in cell firing rate induced by iontophoresing a classical transmitters such as DA, NA, or 5-HT (Jones and Boulton 1980, Jones 1984). Such results is in keeping with the concept of a neuromodulator, namely as the ability of a substance to either amplify or attenuate information by a neurotransmitter at synaptic junctions. Using the electrophysiological observations as a point of departure we undertook preliminary experiments to determine whether we could obtain physical biochemical evidence of a modulatory role for trace amines. For the purpose we proposed to use isolated synaptosomes as the model for the study.

A wide array of changes occur when the binding of a substance, among which are neurotransmitters, etc., takes

213

place at its receptor on the cell surface of its target tissue, such as a neuron. Rapid changes take place at the plasma membrane surface, within the membrane and, intracellularly. The events that initiate signal transmission by the receptor across the plasma membrane through the inner membrane into the cytoplasm involve second messengers. Beta adrenergic substances activate within seconds after binding at its specific receptor c-AMP. There are events that appear to be mediated by the existence of other second messengers, e.g., a Ca^{++}, phospholipid and diacylglycerol-activated protein kinase (C-kinase) that regulates phosphatidyl inositol turnover. PE and T appear to operate in the latter manner rather than with c-AMP.

It is becoming increasingly apparent that use of the paramagnetic property via EPR spectroscopy is a powerful means for probing the structure and the dynamics of biological membranes and their phospholipid and protein components. Stable nitroxide spin labels incorporated into biological membranes have afforded a very sensitive probe of two-dimensional lateral and transverse and rotational mobility of the phospholipids and the effect of endogenous or exogenous perturbing substances. Spin labels reveal the state of organization of the phospholipid bilayer of the biological membrane in which the polar head groups are arranged at the intracellular and extracellular surfaces and the fatty-acid acyl chain are stacked in a parallel fashion perpendicular to the plane of the membrane.

The lipid environment is generally described in terms of its fluidity which reflects the degree of order in the lipid acyl chains. A fluidity gradient exists with increased mobility from the outer polar head groups toward the inner terminal methyl groups. This can be determined by using stearic acid labels that have the nitroxide moiety at varying distances along the acyl chain.

Phospholipids in membranes, as well as in liquid dispersions, undergo a thermotropic transition from a gel state to a liquid-crystalline state at a characteristic temperature. The phase behavior is important for its effect on the membrane proteins structure and function. The gel or crystalline state is characterized by a high

degree of order; there is close molecular packing, slow
intra- and intermolecular thermal motion and a low
capacity to dissolve exogenous amphipathes in the
membrane bilayer. The liquid and crystalline state
reflects a disorderly packing with increased molecular
motion including rotations about carbon-carbon bonds,
wagging motion of segments of the molecules, rotation of
phospholipids about their long axis and lateral diffusion
of molecules in the plane of the membrane.

The preceding discussion was intended to provide a
framework for this preliminary study. the
electrophysiological data of the Boulton group in which
the addition of subthreshold amounts of "trace" amine,
e.g., PE, did not produce any effect but markedly
potentiated the effect produced by a neurotransmitter,
e.g., DA decrease in the rate of electrophysiology
response, should have its counterpart in the changes in
the organizational state of synaptosomal membrane
phospholipids. We would anticipate that low
concentrations of PE would not itself alter the
organizational state of the phospholipids but in the
presence of DA should effect an increase in the state
disorder of the synaptosomal membrane reflecting the
message of the receptor to the interior of the membrane
and into the cell.

MATERIALS AND METHODS

Synaptosomal membranes were isolated from adult male
Sprague-Dawley rats by the method of Jones and Matus
(1974).

The doxyl stearic acid spin-label isomers with the
nitroxide group situated at C-4, C-12, or C-16 position
of the hydrocarbon chain were purchased from Molecular
Probes Inc. (Eugene, OR, USA). The spin-labels were used
from stock solutions of 1 mg/ml in $CHCl_3$ stored at -20°C.

Spin-labelling of the synaptosomal membrane was
obtained in the following manner. A film of the spin-
label was produced by placing an aliquot of stock
solution in a small glass tube, the organic solvent
removed, first in a stream of O_2-free N_2, then by vacuum
dessication for at least 3 hr. The isolated synaptosomal

pellet in Krebs-Ringer buffer, pH 7.4, was added to dry
film of spin-label, then shaken and incubated at 4°C for
10 min. Spin-label of 1 or 2 mol% to membrane
phospholipid was used. The spin-labelled synaptosomes
then were incubated at 4°C with the respective biogenic
amine accordingly: PE or DA for 3 min, and PE for 3 min,
then DA for 2 min. After incubation, the spin-labelled
synaptosome with biogenic amine(s) were transferred to a
100-200 µl glass capillary tube.

ESR spectra were obtained using a Brüker ER20000,
9.3 GHz spectrometer equipped with a variable temperature
control unit. The sample capillaries were inserted into
a standard 4 mm quartz ESR tube containing light silicon
oil for thermal stability. Sample temperature was
measured with a thermo-couple placed in the silicon oil
just above the ESR cavity. ESR spectra were obtained in
the absorbance mode for conventional ESR measurements at
a modulation amplitude of 0.125 mT and microwave power of
30 mW. The ESR spectrometer was connected to a computer
which allowed accumulation and integration of the
spectra.

The rate of spin-label rotational motion of lipids
in biological membranes was determined from spectra with
spin-label androstanol and measuring the relative line-
height and line-widths according to the method of
Polnaszek et al. (1981) as applied to synaptosomes
(Harris et al. 1983).

Hyperfine splittings, maximum A_{11} and minimum $A^1{\perp}$
were measured to calculate order parameters and isotropic
splitting factors. Relevant equations are detailed in
Marsh and Watts (1981).

RESULTS

Typical ESR spectra were obtained with the
positional isomers of the nitroxide spin-labelled stearic
acids. The spectra are characteristic for such spin-
labels undergoing fast anisotropic motion in lipid
bilayers, a hyperfine splitting of the absorption
spectrum with three lines at increasing magnetic field.
The shape of the spectrum reflects the orientation and
motion of the spin label and the polarity of its

environment.

Correlation time, which gives a measure of the rate
of rotational motion of the long axis of the lipids in
biological membranes was rapid, of the order of 1 nsec,
over the temperature range 37° to -10°C, for PE, DA and
PE & DA, respectively. At temperatures below -10°C,
correlation time was longer, increasing to 5 to 6 nsec at
-28°C. Table 1 correlation time data show that spin-
labelled synaptosomes remain unchanged by the addition of
PE (0.5 μM). The addition of DA (0.5 μM) slightly
lowered the correlation time while the presence of PE
(0.5 μM) and DA (0.5 μM) reduced the correlation time
considerably. Thus, the acyl chain motion about the long
molecular axis, i.e., wobbling, was unchanged by PE, but
it markedly potentiated the DA induced motion.

The outer hyperfine splitting, A_{max}, which gives a
relative measure of the acyl chain segmental motion,
decreases as the temperature increases reflecting the
increased mobility at higher temperatures. For spin-
labels positioned deeper into the membrane, increased
segmental motion occurs than in regions closer to surface
of the plasma membrane. The temperature dependence of
the outer hyperfine splitting, A_{max}, of the three stearic
acid spin-labels at C-4, C-12 and C-16 effected these
results are summarized in Table 2. The C-4 and C-12
stearic acid spin labels mobilities in the synaptosome
was unchanged by PE, DA, and PE & DA, respectively, at
the range of temperatures from -28° to 37°C. The C-16
stearic acid spin label mobility in the membrane also was
unaffected by PE & DA, but significantly decreased A_{max}
(increased mobility) over the control.

This temperature dependence experiment with the
spin-label stearic acids indicate that the potentiating
effect of PE on DA occurred at a depth of penetration at
the C-16 methylene segment of stearic acid, an
approximate distance of about 20 Å in the synaptosomal
membrane.

The order parameter, S, reflecting the average
amplitude of the membrane lipid acyl chain motion, is
given in Fig. 1. The results of the order parameter
parallel those obtained with measurement of the outer
hyperfine splitting, A_{max}. PE has no effect over the

TABLE 1

Correlation Time (nsec)

16 SASL	Temperature			
	-23°C	-18°C	-12°C	+27°C
Synaptosomes	5.0	2.78	1.56	0.757
Synaptosomes + PE (.5 μM)	5.0	3.1	1.49	0.757
Synaptosomes + DA (.5 μM)	4.4	2.74	1.55	0.757
Synaptosomes + PE (.5 μM) + DA (.5 μM)	2.36	1.23	1.02	0.683

Spin-labelled synaptosomes were incubated with PE or DA for 3 min, then DA for 2 min.

TABLE 2

Outer Hyperfine Splitting A_{max} (mT)

SASL Temp	C-4		C-12		C-16	
	-12°C	27°C	-12°C	27°C	-12°C	27°C
Synaptosomes	3.37	2.82	3.1	2.32	2.44	1.8
Synaptosomes + PE (.5 μM)	3.3	2.8	3.1	2.3	2.4	1.8
Synaptosomes + DA (.5 μM)	3.4	2.84	3.1	2.3	2.4	1.8
Synaptosomes + PE (.5 μM) + DA (.5 μM)	3.4	2.83	3.1	2.33	2.1	1.68

SASL = stearic acid spin label. Spectra were recorded after incubation of spin labelled synaptosomes (in Kreb's ringer buffer, pH 7.4) with PE or DA for 3 min and PE for 3 min, then DA for 2 min.

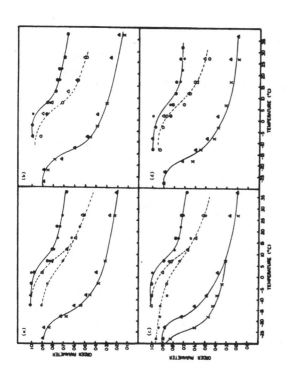

Fig.1.Effect of phenylethyl amine,dopamine and phenylethyl amine + dopamine
on membrane fluidity, order parameter, S,of isolated rat brain synaptosomes over
the temperature range of -28 c to +37 c.The order parameter were calculated for
synaptosomes labeled with stearic acid spin labels having deoxyl group positioned
at carbon 4 (●——●), 12 (▲--▲) and 16 (▲——▲).Effect of (a) phenylethyl amine
.5uM (b) dopamine .5uM (c) phenylethyl amine .5uM + dopamine .5uM (d) phenylethyl
amine .05uM + dopamine .5uM on order parameter of synaptosomes labeled with st
stearic acid having deoxyl group at 4 (●——●), 12 (0--0) and 16 (x——x)

temperature range studied with the spin-labelled stearic acid isomers, C-4, C-12 and C-16. PE potentiated the decreased S induced by DA with C-16 SASL from 0° to -25°C, a reflection of a higher state of disorganization (greater fluidity).

The versatility of lipid spin-label chemistry has been useful in obtaining independent data supporting the electrophysiological results. Correlation times, outer hyperfine splitting and order parameter uniformly parallel the electrophysiological data in PE alone having no effect but in the presence of DA, greatly enhances the DA effect. The orientational information and the rates of molecular motion becomes evident when the rapid motion at higher temperatures are decreased by the decrease in temperature.

This preliminary study supports the suggestion that trace amines as PE can play a modulatory role with classical neurotransmitters such as DA. Work is in progress to clarify the mebrane dynamics concerning the interaction of trace amines and neurotransmitter amines in synaptosomes.

ACKNOWLEDGEMENT

The studies reported here were funded in part by the Chemistry Department, Arizona State University and by Dr. A. L. Bieber, support we gratefully acknowledge. We thank Dr. A. A. Boulton for stimulating discussions.

REFERENCES

Boulton, A. A. (1977) Cerebral aryl alkyl aminergic mechanisms, in Trace Amines in theBrain (Usdin, E. and Sandler, M., eds.), Marcel Dekker, New York.
Boulton, A. A. (1979) Trace amines in the central nervous system. Int. Rev. Biochem., 26, 179-206.
Harris, J., Power, T. J., Biebr, A. L. and Watts, A. (1983) An ESR spin-label study of purified Mojave toxin with rat brain synaptosomal membranes. Eur. J. Biochem., 131, 559-565.
Jones, D. H. and Matus, A. J. (1974) Isolation ofsynaptic plasma membrane from brain by combined flotation-

sedimentation density gradient centrifugation. Biochim. Biophys. Acta, 356, 176-287.

Jones, R. S. G. and Boulton, A. A. (1980) Interactions between p-tyramine, m-tyramine or β-phenylethylene and dopamine on single neurons in the cortex and candate nucleus of the cat. Can. J. Physiol. Pharmac., 58, 222-227.

Jones, R. S. G. (1984) Electrophysiological studies of the possible role of trace amines in synaptic function, in Neurobiology of the Trace Amines (Boulton, A. A., Baker, G. B., Deuhurst, W. G. and Sandler, M., eds.), pp. 205-223, Humana Press, N.J., USA.

Marsh, D. and Watts, A. (1981) in Liposomes from Physical Structure to Thereapeutic Application (Knight, G. G., ed.), pp. 139-188, Elsevier, Amsterdam.

Polnaszek, C. F., Marsh, D. and Smith, I. C. P. (1981) Simulation of the EPR spectra of the cholestane spin probe under conditions of slow axial rotation: Applications to gel phase dipalmitayl phosphatidyl. J. Magnetic Res., 43, 54-64.

Sabelli, H. C., Borison, R. L., Diamond, B. I., Havdala, H. S. and Narasimhachari, N. (1978) Pheneythylamine and brain function. Biochem. Pharmacol., 27, 1707-1711.

RELEASE OF SOME ENDOGENOUS TRACE AMINES FROM

RAT STRIATAL SLICES IN THE PRESENCE AND ABSENCE

OF A MONOAMINE OXIDASE INHIBITOR

Lillian E. Dyck

Neuropsychiatric Research Unit
CMR Building
University of Saskatchewan
Saskatoon, Saskatchewan, Canada S7N 0W0

INTRODUCTION

One of the ways of determining whether a particular
compound has a role in synaptic transmission is to study
the way in which isolated nerve endings accumulate its
radiolabelled congener and subsequently release it under
resting or stimulated conditions. In the past, this strategy has been used to examine the uptake and release of
radiolabelled m-tyramine (mTA), p-tyramine (pTA), and
tryptamine (T) (Dyck, 1984; Baker and Dyck, 1986). Most
previous studies of the uptake and release of labelled PE
and T, in general, have found that these two amines are
poorly taken up, poorly retained and that they are released
mainly by diffusion rather than by a depolarizing stimulus. In contrast, the more polar amines, mTA and pTA, are
actively accumulated, stored and released by depolarizing
stimuli in a calcium-dependent manner; i.e., labelled mTA
and pTA behave in a manner consistent with the appellation,
neurotransmitter. Because radiolabelled congeners do not
necessarily give an accurate reflection of the release of
endogenous neurochemicals, recent studies attempt to measure basal or stimulated release of endogenous compounds.
For example, we have measured the release of endogenous
mTA, pTA and dopamine (DA) from   striatal slices
obtained from rats pretreated with an MAOI which increased
the levels of mTA and pTA to levels which were readily

223

measurable (Dyck et al., 1982). The release of mTA and pTA
was found to be similar to that of DA; i.e., their release
was neurotransmitter-like. In the following study, we have
extended our release studies to include the release of
endogenous PE and T from striatal slices obtained from rats
pretreated with an MAOI to increase the levels of these
amines to more easily measurable amounts. Furthermore, we
have refined the release measurements somewhat by examining
the release of endogenous pTA from striatal slices obtained
from control rats.

MATERIALS AND METHODS

Animals

Male Wistar rats (200-250 g, Charles River, Canada)
were used. They were housed in hanging wire cages with ad
libitum access to food and water.

Chemicals

Dansyl chloride, veratridine, PE, T and tranylcypromine (TCP) were purchased from Sigma Chemical Co. (St.
Louis, MO), pTA from Aldrich (Milwaukee, WI) and mTA from
Vega Biochemicals (Tucson, AZ). The internal standards for
mass spectrometry ([$^2H_4$]mTA, [$^2H_4$]pTA, ^{2}[H_4]PE and [$^2H_4$]T
were synthesized by Dr. B.A. Davis as described previously
(Davis and Boulton, 1980). All solvents utilized were of
HPLC grade and glassware was cleaned in chromic acid to
ensure low blank values in the mass spectrometric analyses. Precoated thin-layer chromatography (TLC) plates
(silica gel G60, E. Merck) were purchased from Brinkmann
Instruments (Ont.).

Measurement of mTA, pTA, T and PE Release

Male Wistar rats were injected intraperitoneally with
10 mg/kg TCP two hours prior to sacrifice. The rats were
killed by cervical dislocation, the brain removed rapidly
and the corpus striata rapidly dissected out on ice.
Striata from two rats were pooled, weighed (about 120 mg)
and chopped into 0.5 mm thick slices using a McIlwain
tissue chopper. The slices were placed in a transfer

apparatus containing 4.0 ml Krebs buffer (see Dyck, 1978 for composition) pH 7.4, mixed gently to separate the slices and incubated at 37°C with continuous gassing with O_2/CO_2 (95%/5%). The slices were transferred at 2 min intervals through five tubes. To measure basal release, Krebs buffer was present in all five tubes. To measure stimulated release, a 50 mM K^+ Krebs buffer dissolved in Krebs buffer was placed in tube 3.

The amounts of mTA, pTA, T and PE present in the five release fractions were quantified simultaneously using well established mass spectrometric techniques (Durden et al., 1973; Philips et al., 1974a; 1974b; Durden, 1985). The amounts of these amines left in the tissue slices at the end of the experiment were also quantified by this technique.

In brief, known amounts of the tetradeutero internal standards ($[^2H_4]$mTA, $[^2H_4]$pTA, $[^2H_4]$PE and $[^2H_4]$T) were added to the release fractions and to homogenates (0.1 N perchlorate) of the tissue. The amines therein were then converted to their dansyl derivatives with dansyl chloride and separated unidimensionally on two or three TLC plates. mTA, pTA, PE and T were quantified by comparing the relative signals due to the molecular ions of the dansyl derivatives of each amine and its deuterated isotopomer.

Measurement of pTA Release

The release of endogenous pTA from corpus striata pooled from two control rats was also measured. The striatal slices were prepared and incubated as described above with the following modification. The slices were transferred from tube 1 to tube 2 after a 2 min interval, but thereafter they were transferred after 3 min intervals. Furthermore, a total of 6 release fractions (rather than 5) were collected. Basal and stimulated release were measured as described above, but in this case tube 4 contained the releasing stimulus (50 mM K^+ Krebs buffer or 10 μM veratridine dissolved in Krebs buffer). pTA released into the various tubes and left in the slices was quantified as described above.

Calculation of Release

The amount of amine released into each tube or fraction was expressed either as ng of amine or as a percentage of the total amount available (i.e., the sum of the amounts present in all of the fractions plus that left in the slices). This was done so a comparison of the release of the four amines (which were present in different amounts) could be made.

To determine whether 50 mM K^+ or veratridine stimulated amine release, the amine release in the stimulated fraction (tube 3 or 4) was compared to that in the basal conditions (normal Krebs buffer in tube 3 and 4) using Student's t-test.

RESULTS

Effect of TCP on Striatal Amine Levels

The amounts of mTA, pTA, T and PE present in the striata pf rats injected with 10 mg/kg TCP for 2 h is shown in Table 1. The amounts are shown as ng present in 4 striata and ng/g wet tissue wt. It can be seen that about 8 ng of pTA, T and PE were present in the 4 striata, while about 2 ng of mTA was present.

As a check on the reliability of the quantification of the amounts of amines released, the amounts of each amine present in the tissue just before the commencement of the release procedure was compared to the total of the amounts of each amine present in the release fractions and left in the striatal slices at the end of the release procedure. As can be seen in Table 1, there were no differences in the total amounts of each amine measured in tissue that had been subjected to a release procedure compared to the amounts in tissue that had not been subjected to a release procedure.

PE Release

The basal (Con) and 50 mM K^+-stimulated releases of PE from rat striatal slices are shown in Figure 1. It can be seen that about 3 ng of the total amount of PE (about 9 ng)

TABLE 1

Concentration of mTA, pTA, T and PE in Rat Striatum

Amine	In Tissue[1]		In Release Fractions and Tissue[2]	
	ng	ng/g	ng	ng/g
mTA	1.68±.21	12.0±1.2	1.58±0.13	13.2± 0.8
pTA	7.63±.44	55.3±2.8	7.20±0.33	61.4± 5.1
T	9.42±.86	64.7±5.0	8.26±1.05	70.6± 7.8
PE	8.26±.76	59.3±3.6	10.40±1.76	88.5±14.3

Rats were pretreated with TCP (10 mg/kg, 2 h). Values are mean ± S.E.M., n=5. [1]These values are the amounts of amines present in tissue that has been homogenized rather than run through a release procedure. [2]These values were calculated by adding up the amounts of amines present in the release fractions and left in striatal tissue at the end of the release procedure.

washed out of the tissue in the first tube or fraction, and that after fraction 2 the basal release of PE decreased linearly with time. The presence of a depolarizing concentration of K^+ in tube 3 did not significantly increase the release of PE into fraction 3 compared to the control.

T Release

The basal (con) and K^+-stimulated releases of T from striatal slices obtained from TCP-pretreated rats are shown in Figure 2. About 2 ng of the total amount of T (about 9 ng) was released spontaneously in the first fraction. The basal release of T was linear after fraction 2 and the presence of 50 mM K^+ in tube 3 did not stimulate a release of T above basal values.

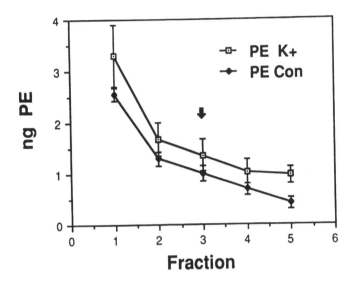

Figure 1. Release of PE from striatal slices obtained from
TCP-treated rats. Values are mean ± SEM (n=5).

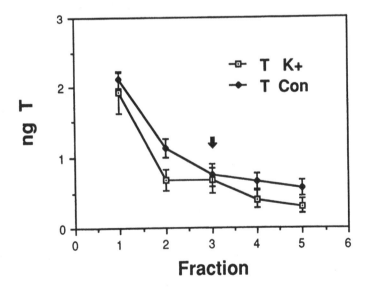

Figure 2. Release of T from striatal slices obtained from
TCP-treated rats. Values are mean ± SEM (n=5).

pTA Release

Figure 3 illustrates the basal (con) and K⁺-stimulated release of pTA from striatal slices obtained from TCP-pre-treated rats. About 1.5 ng of the total amount of pTA (about 7 ng) was washed out in the first fraction, and like T and PE, the basal release of pTA was linear after tube 2; however, unlike PE and T, the presence of 50 mM K⁺ in tube 3 did significantly increase the amount of pTA released compared to the basal condition. The K⁺-stimulated release of pTA was 0.59±11 ng, while the basal release of pTA was 0.17±.03 ng.

mTA Release

The basal (con) and K⁺-stimulated releases of mTA are shown in Figure 4. A total of about 2 ng of mTA was available for release. About 0.3-.4 ng mTA was released in the first fraction, and a linear release of about .02 ng mTA occurred from fraction 2 to fraction 5 in the basal condition. When 50 mM K⁺ was present in tube 3, it stimulated a

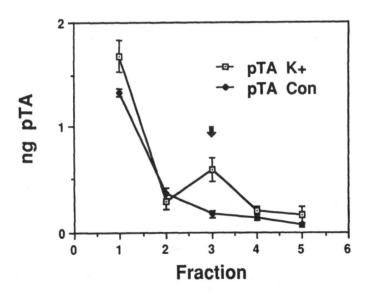

Figure 3. Release of pTA from striatal slices obtained from TCP-treated rats. Values are mean ± SEM (n=5).

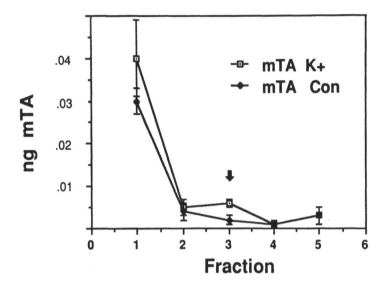

Figure 4. Release of mTA from striatal slices obtained
from TCP-treated rats. Values are mean ± SEM (n=5).

release of mTA (.06 ± .01 ng) compared to the control
experiment (.02 ± .01 ng).

A Comparison of the Basal Release

Since different amounts of PE, T, mTA and pTA were
present in the striatum, in order to compare their basal
releases it was necessary to express the amount of each
amine released into the various fractions as a percentage
of the total amount of each amine available for release.
It can be seen in Figure 5 that basal release of PE was the
most rapid. T also washed out of the striatal tissue
fairly quickly, but the rates of wash out of mTA and pTA
were the slowest of the four amines. These different rates
of wash-out were reflected in the amounts of amine left in
the tissue at the end of the release experiment. Clearly,
those amines which washed out quickly (PE and T) were
present in smaller amounts in the tissue than those amines
that washed out more slowly (mTA and pTA).

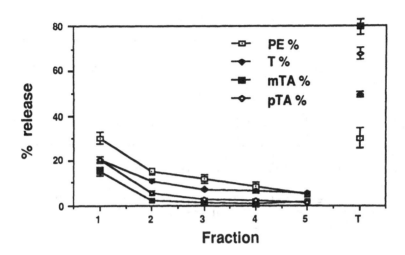

Figure 5. Basal release of mTA, pTA, PE and T from
striatal slices obtained from TCP-treated rats. Values are
mean ± SEM (n=5).

Release of pTA from Control Rats

The amount of pTA available for release from 4 striata
obtained from control rats is only about 2 ng in total (see
Table 2). Of this, about 0.25 ng was released spontaneously into fraction 1 and a linear wash-out of pTA occurred
from fraction 2 through to 6 (see Fig. 6). 50 mM K⁺ did
not stimulate a significant pTA release compared to control; however, 10 μM veratridine released 0.47 ± .04 ng pTA
which was an amount significantly greater than the basal
release (0.10 ± 0.02 ng).

TABLE 2

Concentration of pTA in Rat Striatum

ng	In Tissue[1] ng/g	In Fractions + Tissue[2] ng	ng/g
1.72 ± 0.08	11.92 ± 0.60	1.99 ± 0.27	15.18 ± 1.62

Values are mean ± S.E.M.
[1] This value represents the amount of pTA in 4 control striata (n = 6).
[2] This value represents the amount of pTA released into the various fractions and left in the tissue at the end of the release experiment (n = 13).

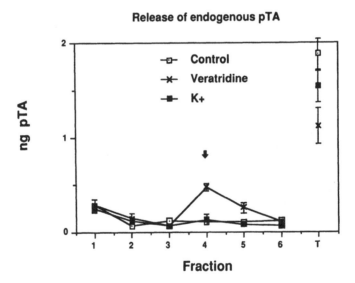

Figure 6. Release of pTA from striatal slices obtained from control rats. Values are mean ± SEM (n=7 for control, n=5 for veratridine and n=10 for 50 mM K+).

DISCUSSION

It is well known that amines such as mTA, pTA, T and PE can mediate both direct and indirect post-synaptic effects (Trendelenburg, 1972). In fact, in order to show direct effects, it may be necessary to preclude the possibility of indirect effects by selecting a mode of administration that favours direct effects or by preventing indirect effects with uptake blockers. Direct actions of mTA (Anden et al., 1970; Ungerstedt et al., 1973), pTA (De La Lande and Waterson, 1968); Trendelenburg, 1972; Miyahara and Suzuki, 1985), PE (Hansen et al., 1980; Antelman et al., 1977) and T (Stollak and Furchgott, 1983) on post-synaptic responses have been reported; thus, one can envisage that these amines have roles as neurotransmitters. The recent reports of high affinity binding sites for T and PE (Kellar and Cascio, 1982; Cascio and Kellar, 1983; Bruning and Rommelspacher, 1984; Wood et al., 1984; Altar et al., 1986; Perry, 1986; Keinzl et al., 1986; Hauger et al., 1982) lend support to this idea, but it is by no means clear yet whether these binding sites represent receptors for chemical messengers. Certainly most previous studies of the uptake and release of the trace amines have concluded that a classical neurotransmitter role is unlikely for T and PE, but that mTA and pTA do exhibit transmitter-like transport (Dyck, 1984; Baker and Dyck, 1986). The findings presented here show that a stimulus (50 mM K^+) which mimics physiologically-induced transmitter release did stimulate mTA and pTA release, but did not stimulate release of PE and T. Thus, these findings agree with previous reports. Similarly, the high rates of wash-out of PE and T, compared to mTA and pTA, agree with previous findings and permit one to conclude that striatal nerve endings do not have specialized systems to store and retain PE and T. This lack of storage is incompatable with conventional concepts of neurotransmission.

The pattern of release of pTA from striatal slices obtained from control rats was not completely identical to the pattern observed with TCP-pretreated rats. In both cases, pTA did not readily wash out of the tissue, as indicated by the large percentage of amine left in tissue slices at the end of the experiment. The spontaneous wash-out of pTA into the fraction 1 was 20.8 ± 1.31% of the total amount in the TCP experiments. In the control situation, the spontaneous wash-out of pTA into fraction 1 was somewhat smaller (10.0 ± 2.3%), thus striatal pTA from control rats appeared to be more tightly retained than striatal pTA from TCP-pretreated rats. While 50 mM K^+ stimulated pTA release from striatal slices obtained from TCP-pretreated rats, it did not stimulate pTA release from striatal slices obtained from control rats. While one might conclude from this that pTA was not released in a transmitter-like fashion in control rats, such a conclusion is inconsistent with many previous studies. As an additional test of this possibility, however, the ability of another depolarizing stimulus (veratridine) to stimulate release of striatal pTA from control rats was investigated. Unlike 50 mM K^+, veratridine (10 μM) did stimulate a significant release of pTA compared to the basal conditions; thus, it can be concluded that striatal pTA can be released in a transmitter-like fashion in control rats. It is not clear why 50 mM K^+ did not stimulate a significant release of pTA. In the TCP-pretreated rats, the K^+-stimulated release of PTA was 8.0 ± 1.2% of the total amount of pTA. If 8% of the total amount of pTA (2 ng) in the control situation were released, then 0.16 ng pTA would have been found in fraction 4. In fact, 0.13 ± 0.05 ng pTA was found in fraction 4, but as can be seen from the S.E.M., a rather large variance occurred. In some samples, pTA appeared to be released while in other samples, it did not. Veratridine, however, apparently had a greater and more consistent releasing effect than 50 mM K^+, and caused a release of 19.9 ± 2.2% of the total amount of striatal pTA from control rats.

The release experiments presented here are an improvement over previous studies in that the release of endogenous amines was measured. A "truly" endogenous pTA release was measured in those experiments using control rats, and an "augmented" endogenous trace amine release was measured in those experiments using TCP-pretreated rats. Furthermore, in the latter experiments, the amounts of mTA, pTA, PE and T in each sample were determined simultaneously. A comparison of the release of each amine was thus facilitated. Philips and Boulton (1979) demonstrated that administration of TCP to rats caused fairly large increases in the concentrations of mTA, pTA, PE and T in rat striatum. Of the many MAOI's they investigated, only TCP increased the levels of all four of these amines, thus TCP was chosen as the most suitable MAOI for a release experiment designed to measure simultaneous release of these four amines.

SUMMARY AND CONCLUSIONS

In the present study, the simultaneous release of mTA, pTA, PE and T from striatal slices obtained from TCP-pretreated rats was investigated. In addition, the release of pTA from control rats was studied. The results are consistent with the view that mTA and pTA can function as neurotransmitters in the conventional sense (are stored and released by a depolarizing stimulus), but that the more lipophilic trace amines, PE and T, do not function as neurotransmitters.

ACKNOWLEDGEMENTS

The author thanks Dr. A.A. Boulton for advice and encouragement, Dr. D.A. Durden for supervising the mass spectrometric analyses, Mr. R.C. Mag-atas and Mr. M. Mizuno for technical assistance, and Saskatchewan Health for financial assistance.

REFERENCES

Altar C.A., Wasley A.M. and Martin L.L. (1986) Autoradiographic localization and pharmacology of unique [^{3}H]tryptamine binding sites in rat brain. Neuroscience 17, 263-273.

Anden N.E., Butcher S.G. and Engel J. (1970) Central dopamine and noradrenaline receptor activity of the amines formed from m-tyrosine, α-methyl-m-tyrosine and α-methyl dopa. J. Pharm. Pharmac. 22, 548-550.

Antelman S.M., Edwards D.J. and Lin M. (1977) Phenylethylamine: evidence for a direct, post-synaptic dopamine-receptor stimulating action. Brain Research 127, 317-322.

Baker G.B. and Dyck L.E. (1986) Neuronal transport of amines in vitro, in Neuromethods Vol. 2 Amines and their metabolites (Boulton A.A., Baker G.B. and Baker J.M., ed.) pp. 457-534. Humana Press, Clifton, New Jersey.

Bruning G. and Rommelspacher H. (1984) High affinity [^{3}H]tryptamine binding sites in various organs of the rat. Life Sci. 34, 1441-1446.

Cascio C.S. and Kellar K.J. (1983) Characterization of [^{3}H]tryptamine binding sites in brain. Eur. J. Pharmacol. 95, 31-39.

Davis B.A. and Boulton A.A. (1980) The metabolism of
    ingested deuterated β-phenylethylamine in a human male.
    Eur. J. Mass Spectrom. Med. Environ. Res. 1, 149-153.
de la Lande I.S. and Waterson J.G. (1968) The action of
    tyramine on the rabbit ear artery. Br. J. Pharmacol.
    34, 8-18.
Dyck L.E. (1978) Uptake and release of meta-tyramine and
    para-tyramine and dopamine in rat striatal slices.
    Neurochem. Res. 3, 775-791.
Dyck L.E. (1984) Neuronal transport of trace amines: an
    overview, in Neurobiology of the Trace Amines: Analyt-
    ical, Physiological, Pharmacological, Behavioral and
    Clinical Aspects (Boulton A.A., Baker G.B., Dewhurst
    W.G. and Sandler M., ed.), pp. 185-204 Humana Press,
    Clifton, New Jersey.
Dyck L.E., Juorio A.V. and Boulton A.A. (1982) The in
    vitro release of endogenous m-tyramine, p-tyramine and
    dopamine from rat striatum. Neurochem. Res. 7, 705-716.
Hansen T.R., Greenberg J. and Mosnaim A.D. (1980) Direct
    effect of phenylethylamine upon isolated rat aortic
    strip. Eur. J. Pharmac. 63, 95-101.
Hauger R.L., Skolnick P.. and Paul S.M. (1982) Specific
    [^{3}H]B-phenylethylamine binding sites in rat brain. Eur.
    J. Pharmac. 83, 147-148.
Kellar K.J. and Cascio C.S. (1982) [^{3}H]-Tryptamine: high
    affinity binding sites in rat brain. Eur. J. Pharmac.
    78, 475-478.
Kienzl E., Riederer P., Jellinger K. and Noller H. (1985)
    A physiochemical approach to characterize [^{3}H] trypt-
    amine binding sites in human brain, in Neuropsychopharm-
    acology of the Trace Amines: Experimental and Clinical
    Aspects) (Boulton A.A., Bieck P., Maitre L. and Riederer
    P., ed.), pp. 469-486 Humana Press, Clifton, New Jersey.
Miyahara H. and Suzuki H. (1985) Effects of tyramine on
    noradrenaline outflow and electrical responses induced
    by field stimulation in the perfused rabbit ear artery.
    Br. J. Pharmac. 86, 405-416.
Perry D.C. (1986) [^{3}H]Tryptamine autoradiography in rat
    brain and choroid plexus reveals two distinct sites.
    J. Pharmacol. Exp. Ther. 236, 548-559.
Philips S.R. and Boulton A.A. (1979) The effect of mono-
    amine oxidase inhibitors on some arylalklamines in the
    rat. Can. J. Biochem. 52, 366-373.

Stollak J.S. and Furchgott R.F. (1983) Use of selective antagonists for determining the types of receptors mediating the actions of 5-hydroxytryptamine and tryptamine in the isolated rabbit aorta. J. Pharmacol. Exp. Ther. 224, 215-221.

Trendelenburg U. (1972) Classification of sympathomimetic amines, in Catecholamines (Blascho H. and Muscholl E., ed.), pp. 336-362, Springer-Verlag.

Ungerstedt U., Fuxe U., Goldstein M., Battista A., OgawaM. and Anagnoste B. (1973) Action of m-tyrosine in experimental models: evidence for possible antiparkinsonian activity. Eur. J. Pharmacol. 21, 230-237.

Wood P.L., Pilapil C., LaFaille F., Nair N.P.V. and Glennon R.A. (1984) Unique [3H]tryptamine binding sites in rat brain: distribution and pharmacology. Arch. Int. Pharmacodyn. 268, 194-201.

PHENYLETHLAMINE AND P-TYRAMINE IN THE EXTRACELLULAR SPACE OF THE RAT BRAIN: QUANTIFICATION USING A NEW RADIOENZYMATIC ASSAY AND IN SITU MICRODIALYSIS

D. P. Henry, W. L. Russell, J. A. Clemens
and L. A. Plebus
Lilly Laboratory for Clinical Research and
Indiana University School of Medicine
Indianapolis, Indiana 46202

ABSTRACT

Sufficiently sensitive and specific methods for the quantitation of p-tyramine (p-tym) and phenylethylamine (PEA) are not generally available since the levels of these substances found in the central nervous sytem tissue are very low. Analytic methodology is stressed even further when attempts are made to quantify these amines in the small samples obtainable from the extracellular space of the brain by in situ microdialysis. Radioenzymatic assays are sensitive analytic procedures that function with small volume samples and therefore match well with the in situ dialysis technique. A brief review of radio-enzymatic assays for trace amines and two new assays are discussed. The first assay based on purified phenylethanolamine N-methyltransferase (PNMT) can be used for trace amines chemically related to phenylethanolamine. The second procedure utilizes tyramine N-methyltrans-ferase (TNMT) which is isolated and purified from germinating barley. The procedure is specific for p-tyramine and phenylethylamine. The TNMT based assay, in conjunction with in situ microdialysis, was used to examine the content of p-tym and PEA in the extracellular space of the rat hypothalamus and striatum. Both amines were detected at levels similar to those observed for more accepted neurotransmitters such as the catechol-amines and histamine. The levels of p-tym and PEA varied rapidly and appeared to also exhibit a diurnal cycle.

239

PEA and p-tym were not released by potassium chloride-induced depolarization. These studies further support a role for p-tym and PEA in the brain.

INTRODUCTION

The availability of analytic techniques for the quantification of some of the trace amines has limited the study of the physiologic function of those compounds in the central nervous system. A new technique, in situ microdialysis, appears to permit direct access to the extracellular space of the brain in conscious ambulatory experimental animals (Ungerstedt, 1984). Samples obtained by this procedure, due both to the low concentration of neuro-transmitters and small sample volume, further extend trace amine analytic methodology. As an extension of our previous studies of radioenzymatic assay (REA) procedures we have developed new techniques with sensitivities that are capable of detecting physiologically relevant concentrations of PEA, p-tym, and octopamine in samples obtained by CNS microdialysis. Both PEA and p-tym were detected in samples obtained from dialysates of the corpus striatum and posterior hypothalamus of the rat. Short term and diurnal variability in the concentration of these amines was observed.

REA'S FOR TRACE AMINE

The REA technique was introduced into the trace amine field (Molinoff, 1969) soon after the procedures were first developed (Snyder, 1966). A number of trace amines have been identified or quantified using the REA including p-tyramine, m-tyramine, p-octopamine (nor-phenylephrine), synephrine, tryptamine and salsolinol.

The characteristics of the first generation assay have been reviewed (Saavedra, 1984). Since many of these techniques used non-specific procedures, such as solvent extraction, to isolate the radiolabeled products formed during the assays, specificity can be questioned. Furthermore, we have found that the purification and optimization of the methyltransferase reactions remarkably increased the sensitivity of the REA for histamine

(Verburg et al., 1983) and the PNMT based assay for
norepinephrine (Henry and Bowsher, 1986). The use of
purified enzymes has, to our knowledge, not previously
been applied to the analysis of trace amines.

REA can be classified based on the enzyme used in
the procedure (Table 1). The most widely used type of REA
has been based on PNMT reaction. This enzyme is present
in high concentrations in the adrenal gland of large
domesticated animals and is thusly relatively available.
The enzyme is highly specific for compounds with an
ethanolamine side chain attached to an aromatic or
hydrophobic substituent. As noted in Table 1, analytes
such as p-octopamine are excellent substrates; but
ethylamine side chain containing compounds such as p-tym
are not substrates for the enzyme. The REA for octopamine
has been modified in order that phenylthanolamine and
tyramine could be quantified (Saavedra, 1974; Tallman et
al., 1976). This was accomplished by first incubating the
sample of interest with dopamine beta-hydroxylase (DBH)
which converts the ethylamines to ethanolamines which are
then substrates for PNMT.

We have modified the REA for p-tyramine and
p-octopamine (Van Huysse, 1987) in the following manner:
(1) The assay utilizes PNMT specifically purified for use
in REA's (Henry and Bowsher, 1986). This enzyme results
in maximal generation of product but minimizes the assay
"blank". (2) The highest specific activity ^{3}[H]-S-
adenosyl-methinone (50-85 Ci/mM) is used. (3) Partially
purified Dopamine-β-hydroxylase obtained from a commercial
source is used. (4) Specificity is increased by the use
of a new TLC system for the separation of the radiolabeled
trace amine products. The sensitivity of the assay for
1.5 and 2.0 pg per assay tube for p-oct and p-tym
respectively which were obtained with a C.V. of 4.8 ± 0.4%
and 8.4 ± 1.2% (VanHuysse, 1987). Despite these
modifications, the assay is complex and labor intensive
and the absolute blank for the p-tym assay is difficult to
define in tissues which contain both p-tym and p-oct. The
revised method, however, is highly effective for the
quantification of the class of trace amines that are
chemically related to phenylethanolamine.

Indoleamine-N-methyltransferase, an enzyme that is only found in high concentrations in rabbit lung has been used in an assay for tryptamine (Saavedra and Axelrod, 1972). This enzyme(s) has little substrate specificity and poor enzyme kinetics. The full scope of methyltransferase activity has not been defined. For example, we recently observed that histamine was a substrate for this enzyme (Herman et al., 1985). Any REA based on this enzyme reaction must be examined closely to determine specificity.

The most recent enzyme to be used in an REA for a group of trace amines has been tyramine-N-methyltransferase (TNMT, EC 2.1.1.27), an enzyme found in the root portion of sprouting barley seeds (Henry and Van Huysse, 1986). It functions in barley to synthesize the allelochemicals, N-methyl-tyramine and N-N-dimethyl-tyramine. The methyl donor is S-adenosyl methinone. The enzyme was first identified in 1963 but has not been studied extensively since then (Mann and Mudd, 1963). In addition to tyramine, the other only aromatic amines that were substrates were phenylethylamine, amphetamine and m-tyramine. The following substances were examined and found not to be substrates: o-tyramine, 4-methoxytyramine, 3-methoxytyramine, tyrosine, phenylalanine, 5-hydroxy-tryptamine, norepinephrine, epinephrine, dopamine, metanephrine, normetanephrine, n-methylphenylethylamine, β-methylphenylethylamine, phenylethanolamine, n-methyl-phenylethanolamine, 4-methoxyphenylethanolamine, histamine, synephrine, benzylamine and methamphetamine. The substrate specificity of this enzyme is then targeted towards a chemically related subset of the trace amines which may be physiologically related since these compounds are synthesized by the same enzyme, aromatic aminoacid decarboxylase (Bowsher and Henry, 1983). The enzyme is extracted from the roots of five day sprouted barley seed and purified by ultracentrifugation, ammonium sulfate fractionation, dialysis, ion exchange chromatography and molecular exclusion chromatography. The assay itself is modeled after standard TLC based assays for catecholamines (Boren et al., 1980). The methylated radiolabeled amine products are isolated by base specific solvent extraction, concentrated in a vacuum centrifuge, and separated by TLC on pre-absorbent, channeled silica gel plates. Finally, the radioactivity is quantified by scintillation counting using an ion pair containing counting solution. The sensitivity of the procedure is 0.8 and 2.8 pg/tube for p-tym and PEA respectively and the C.V. is <5%. The assay can be modified to detect m-tym also.

The major classes of REA's for trace amines are summarized in Table 1. In general, the techniques compare favorably with other analytic methods. The sensitivity of specificity, and simplicity of the TNMT based procedure for the p-tym and PEA make this a potentially useful procedure.

IN SITU BRAIN DIALYSIS

Delgado, Ungerstedt, Pycock and others in the early 1970's attempted to develop, as stated by Ungerstedt in a recent review, "a universally applicable technique to sample the content of the extracellular fluid ...", of the brain (Ungerstedt, 1984). This procedure built on experience gained with other similar techniques such as the push-pull canula and in situ superfusion techniques. The procedure is elegant in its simplicity. Ultrafine (250 micron diameter) cellulose dialysis tubes from commercial sources are inserted into small cannulae which are placed stereotaxically into descrete brain regions. The probes can be permanently attached and are tolerated well by the animals. Dialysis media is then slowly pumped through the tubing at rates of 1 to 10 μl/min. Methodologic aspects of the procedure have been reviewed (Ungerstedt, 1984). Some of the advantages of this system are: (1) the small size of the probes, (2) the lack of exposure of the tissue directly to perfusion media, thereby eliminating mechanical stress and the risk of infection, (3) simplicity of the perfusion apparatus, (4) the bidirectional nature of the process - you can add and sample substances at the same time, (5) the sample obtained is clean and devoid of large molecules. A major shortcoming, however, is that the amount of substance obtainable is very small. The concentration of neurotransmitters in the extracellular space is low and total perfusate volume is in the microliter range. Since most REAs utilize unprocessed samples and function optimally with samples in the 5 to 25 μl range, REA's match well with in situ dialysis.

RESULTS

We have used the TNMT based REA and the microdialysis techniques to evaluate the content of p-tym and PEA in the extracellular space of the rat brain. Probes were placed in either the posterior hypothalamus or corpus striatum based on stereotaxic coordinates from the Atlas of Pellegrino (hypothalamus: 1 mm posterior to bregma, 1 mm lateral to the midsagital sinus, 10 mm ventral to the dura; striatum: 1.5 mm anterior, 3 mm lateral, and 5.8 mm ventral). Three days after surgery, rats were placed in containers which permitted free mobility but maintained alignment of input and output tubing. The striatum was dialyzed with normal saline at a rate of 1 µl/min and the samples were collected and analyzed for PEA and p-tym content. The results of these determinations for a representative animal is shown in Figure 1. The conclusions that we have drawn from this and a number of other similar experiments are as follows: (1) PEA and p-tym are detectable in the extracellular space of the rat brain in both the striatum and hypothalamus. (2) The levels vary markedly between collection periods and between animals. (3) The levels of PEA and p-tym vary independently. This sort of variability has not been seen with the catecholamines or histamine in the same preparation (personal observations).

Next we examined the effect of potassium-induced depolarization by dialysing with 156 mM KCl. In a striated dialysis preparation after four 30 minute collection periods, the dialysis was changed to KCl. Neither PEA nor p-tym levels increased significantly. This procedure has been reported to induce significant release of GABA and DA from rat striatum (Ungerstedt, 1984). The results were unexpected and could be interpreted that p-tym and PEA are not stored and released by depolari-zation in the striatum, that the release of other neurotransmitters by KCl modified the response of PEA and p-tym or that experimental parameters were not optimized.

In another set of experiments using conscious free-moving animals dialyzed continuously for a 24 hour period with samples collected at three hour intervals, we have preliminarily concluded that p-tym and PEA exhibit diurnal cycles that are distinct from simultaneously determined cycles for norepinephrine, epinephrine, dopamine and histamine. The animals were maintained on a 4 AM to 7 PM lights on schedule. The most constant aspect of the cycle was that PEA in the extracellular space of the striatum and hypothalamus and p-tym in the striatum increased during the day, the period of inactivity for the rat, and peaked in the period before the lights were extinguished. The p-tym cycle in the hypothalamus was distinctly different; the lowest levels of p-tym were found just before lights out.

SUMMARY AND CONCLUSIONS

REA's for trace amines were reviewed from the point of view of the enzymes used in the procedures. Two newer assays were introduced - one in which the PNMT used in the procedure was specifically purified for use in REA's and a second in which the enzyme used, TNMT, was isolated from germinating barley seeds. Using the TNMT based assay and in situ microdialysis, p-tym and PEA were detected in the extracellular space of the rat brain at levels which approached those seen for more accepted neurotransmitters such as NE, DA, and histamine. The level of PEA and p-tym varied rapidly over relatively short collection periods (30 minutes). Diurnal cycles were seen in both the hypothalamus and striatum which were both amine and site dependent. The functional correlates of the short and longer term variability are not known at present. Finally, in the striatum neither p-tym nor PEA responded in an expected fashion to KCl-induced depolarization. We conclude that the new REA's for PEA and p-tym and the microdialysis techniques should allow us to approach a number of exciting questions concerning the physiologic role of PEA and p-tym. The cyclic rhythms seen for PEA and p-tym in the extracellular space, though distinct from those seen from other monoamines, support a physiologic role for the PEA and p-tym. The extreme variability between samples and the lack of depolarization induced release certainly are, however, atypical for accepted amine neurotransmitter substances. If we take the current observation together with the documented lower tissue levels and rapid turnover of these substances, we conclude that even though PEA and p-tym are chemically related to the other aromatic aminoacid derived biogenic amines, they are physiologically a unique class of pharmacologically active substances indigenous to the central nervous system.

ACKNOWLEDGMENTS

The authors wish to thank Ms. Dorothy Kahrs for the expert assistance in preparing the manuscript and Dr. R. R. Bowsher for review of its contents.

REFERENCES

Boren K.R., Henry D.P., Selkert E.E. and Weinberger M.H. (1980) Renal modulation of catecholamines and metabolites during volume expansion in the dog. Hypertension 4, 383-389.

Bowsher R.R. and Henry D.P. (1983) Decarboxylation of p-tyrosine: A source of p-tyramine in mammalian tissues. J. Neurochem. 40, 992-1002.

Henry D.P. and Bowsher R.R. (1986) An improved radio-enzymatic assay for plasma norepinephrine using purified phenylethanolamine n-methyltransferase. Life Sci. 38, 1473-1483.

Henry D.P., Van Huysse J.W. and Bowsher R.R. (1986) A new class of radioenzymatic assay for the quantification of p-tyramine and phenylethylamine. Fed. Proc. 45, 158.

Herman K.S., Bowsher R.R. and Henry D.P. (1985) Synthesis of N-π-methylhistamine and N-α-methylhistamine by purified rabbit lung indolethylamine N-methyl-transferase. J. Biol. Chem. 260, 12336-12340.

Mann J.D. and Mudd S.H. (1963) Alkaloids and plant metabolism. IV. The tyramine methylpherase of barley roots. J. Biol. Chem. 238, 381-385.

Molinoff P.B., Landsberg L. and Axelrod J. (1969) An
    enzymatic assay for octopamine and other
    β-hydroxylated phenylethylamines.
    Journ. Pharm. Expt. Therap. 170, 253-261.

Saavedra J.M. and Axelrod J. (1972) A specific and
    sensitive enzymatic assay for tryptamine in tissues.
    Journ. Pharm. Expt. Therap. 182, 363-369.

Saavedra J.M. (1974) Enzymatic isotopic assay for and
    presence of β-phenylethylamine in brain.
    J. Neurochem. 22, 211-216.

Saavedra J.M. (1984) The use of enzymatic radioisotopic
    microassays for the quantification of
    β-phenylethylamine, phenyelethanolamine, tyramine
    and octopamine. In, Neurobiology of the Trace
    Amines (Boulton, A.A., Baker, G.B., Dewhust, W.G.
    and Sandler, M., eds) pp 41-55, Humana Press,
    Clifton, New Jersey.

Snyder SH, Baldessarini RJ, Axelrod J. (1966) A sensitive
    and specific enzymatic isotopic assay for tissue
    histamine. Journ. Pharm. and Expt. Therap. 153,
    544-549.

Tallman J.F., Saavedra J.M. and Axelrod J. (1976) A
    sensitive enzymatic-isotopic method for the analysis
    of tyramine in brain and other tissues.
    J. Neurochem. 27, 465-469.

Ungerstedt U. (1984) Measurement of neurotransmitter
    release by intracranial dialysis. In, Measurement
    of Neurotransmitter Release in vivo (Marsden, C.A.,
    ed) pp 81-105, John Wiley and Sons, New York.

Van Huysse J.W., Bowsher R.R., Willis L.R. and
    Henry D.P. (1987) A radioenzymatic assay for para-
    tyramine and para-octopamine: application to
    mammalian tissues. Anal. Biochem. submitted.

TABLE 1

CLASSIFICATION OF RADIOENZYMATIC ASSAYS FOR TRACE AMINES

	1° Enzyme	2° Enzyme	Procedure	Analytes
I.	Phenylethanolamine N-methyltransferase	None	TLC	p-octopamine m-octopamine phenylethanolamine normetanephrine (synephrine)
	COMMENTS: (1) Limitations due to impure PNMT (2) Inadequate product isolation techniques			
II.	Phenylethanolamine N-methyltransferase	DBH	TLC	phenylethylamine p-Tyramine m-Tyramine
	COMMENTS: (1) Limitations of PNMT assays (2) Undefined blank/manipulative complexity			
III.	Indoleamine N-methyltransferase	None	Solvent Extraction	Tryptamine amphetamine
	COMMENTS: (1) Nonspecific enzyme (2) Poor enzyme kinetics			
IV.	Tyramine N-methyltransferase	None	TLC	p-Tyramine phenylethylamine (m-Tyramine)
	COMMENTS: (1) The most recent assay; unverified			

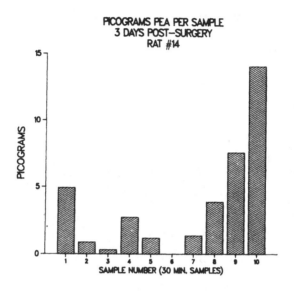

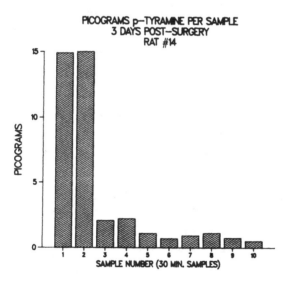

Figure 1. Phenylethylamine and p-tyramine in a rat. Dialysis session was three days after cannula implantation in the corpus striatum. Amine levels are expressed per sample.

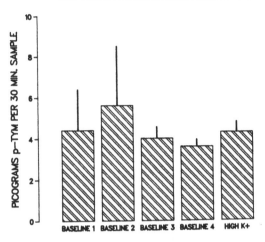

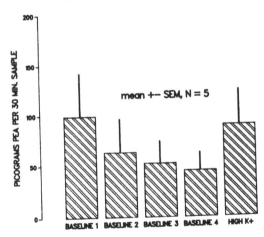

Figure 2. Potassium chloride-induced release of p-tym and PEA in the anesthetized rat. Collection periods were for 30 minutes. KCl was used at 156 mM.

THE EFFECTS OF VARIOUS INDOLES ON THE STIMULATION OF

INOSITOL PHOSPHATE FORMATION IN RAT CORTICAL SLICES

N.N. Osborne and Michelle Hogben

Nuffield Laboratory of Ophthalmology,

University of Oxford,
OXFORD. OX2 6AW. U.K.

Serotonin is a well established neurotransmitter in the CNS and has been implicated in a variety of functions (Osborne 1982, Osborne & Hamon 1988). However, two other indoles, tryptamine and 5-methoxytryptamine, also exist in the brain (Juorio & Greenshaw 1985, Bosin et al. 1979), but in trace amounts. Their roles have still to be elucidated. While binding studies (Wood et al. 1985) provide some evidence for the existence of tryptamine receptors in the brain, it is still not established whether there exist separate functional receptors for serotonin, tryptamine and 5-methoxytryptamine.

Within the past few years a whole host of neurotransmitters (acetylcholine, noradrenaline, serotonin, histamine) have been shown to elicit effects through the inositol lipid receptor mechanism (Berridge et al. 1982, Berridge & Irvine, 1984, Downes 1986). In the case of serotonin it is the 5-HT receptors in brain and retinal tissue which probably utilise inositol triphosphate as a second messenger (Conn & Sanders-Bush 1985, Cutcliffe & Osborne 1987). The aim of our study was therefore to see whether the trace indoles, tryptamine and 5-methoxytryptamine, also stimulate inositol phospholipid turnover and whether such processes were mediated via receptors distinct from 5-HT sites.

EXPERIMENTAL PROCEDURE

Cross-chopped slices (350 x 350 μm) were prepared from

251

whole rat cerebral cortex using a McIlwain tissue chopper, and were incubated in oxygenated physiological buffer pH 7.4 for 30 min. at 37°C in a shaking water bath. This was followed by a 45 min. incubation in fresh oxygenated buffer containing approximately 0.3 µM ^{3}H-inositol (specific activity 16.3 Ci/mmol. from Amersham International). The tissues were then rinsed three times in oxygenated buffer containing 5mM lithium chloride and allowed to settle. Aliquots of 50 µl of packed slices were then placed in 250 µl of oxygenated buffer containing lithium chloride and incubated in a shaking water bath for a further 30 minutes after which substances (10 µl) were added, usually for a further 45 minutes. Incubations were terminated by the addition of a 940 µl chloroform-methanol (1:2 by vol.). Tubes were agitated gently and 310 µl of water and 310 ul chloroform were added. The tubes were left overnight in a refrigerator before extracting ^{3}H-labelled inositol phosphates from the aqueous phase by ion exchange chromatography using Dowex resin in the formate form as described elsewhere (Berridge et al. 1982, Brown et al. 1984). The physiological buffer consisted of: 20 mM HEPES, 1.2 mM Na_2HPO4 5 mM KCl, 1.2 mM $MgSO_4$, 1.2 mM $CaCl_2$, 150 mM NaCl and 11.1 mM glucose. When 5 mM LiCl was added only 145 mM NaCl was present.

RESULTS AND DISCUSSSION

We initially examined the effect of 1 mM each of serotonin, 5-methoxytryptamine, carbachol, noradrenaline, octopamine and tryptamine for their ability to stimulate ^{3}H-inositol phosphate(s) accumulation in rat cerebral cortex slices as well as slices of the cerebellum and hippocampus (Fig.1). The results revealed a number of interesting observations. First, the effects of serotonin and 5-methoxytryptamine in all three areas were very similar. This was in contrast with the effect of tryptamine, which was similar to that of 5-methoxytryptamine in the cerebral cortex and cerebellum, but slightly more intense in the hippocampus. These results for the indole derivatives suggest that tryptamine, at least in the hippocampus, may not elicit its effect through the same sites as serotonin. However, in the case of 5-methoxytryptamine the results can be interpreted as suggesting that its action is the same as serotonin's.

Interestingly, carbachol was barely effective in stimulating ^{3}H-inositol phosphate(s) accumulation in the cerebellum

3

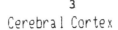

Cerebral Cortex

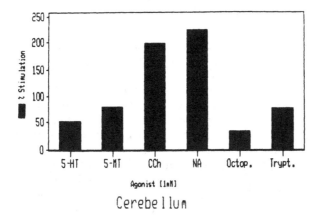

Cerebellum

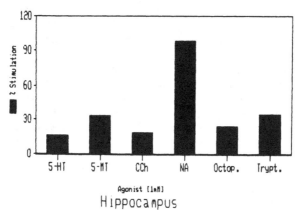

Hippocampus

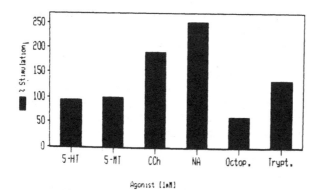

Fig.1.   Stimulation of inositol phosphate(s) by 100μm of
different agonist in three different tissues.   Results are
mean of four different experiments carried out in triplicate.

but was potent to a similar extent in the cerebral cortex
and hippocampus (Fig.1). Octopamine had more or less the
same potency in all three tissues analysed. Noreadrenaline
was the most effective of all the substances tested in stimulating ^{3}H-inositol phosphate accumulation in all three
tissues (Fig.1).

Concentration-effect curves for carbachol, octopamine
and tryptamine in rat cortex slices have previously been reported from our laboratory (Osborne et al.1986). An EC_{50}
value for tryptamine of approximately 650nM was reported.
The shapes of the concentration-effect curves for serotonin
and 5-methoxytryptamine were found to be very similar to
tryptamine, with EC_{50} values between 600 and 700nM (results
not shown). This similarity in the shape (and maximal
effect) in the concentration-effect curves for serotonin,5-
methoxytryptamine and tryptamine, as well as approximately
equal EC_{50} values, provide evidence that the indoles may be
eliciting their effects through similar or identical sites.

While serotonin, 5-methoxytryptamine, tryptamine, carbachol, noradrenaline and octopamine all stimulate ^{3}H-inos-
itol phosphate(s) accumulation, they do not elicit their
influence through the same sites. As shown in Table 1,
atropine and prazosin only antagonised the carbachol and noradrenaline effects respectively. This result is consistent
with other studies which have provided conclusive data to
show that carbachol and noradrenaline act on muscarinic and
α_1-adrenergic receptors respectively to stimulate inositol
phosphate(s) as second messengers (see Downes 1986, Fisher &
Agranoff 1987). Ketanserin, a specific 5-HT_2 receptor antagonist (Conn & Sanders-Bush 1985), in contrast, has equipotent effects on the serotonin, tryptamine and 5-methoxy-
tryptamine-induced stimulation of inositol phosphate(s)
production (Table 1). These results suggest that all three
indoles may influence inositol phosphate production through
the same receptors. This is supported by the fact that
prazosin, atropine, propranolol, yohimbine and mepyramine
do not antagonise the simulatory effects of the three indoles. None of the antagonists used in this study influenced the octopamine-induced stimulation of inositol
phosphate(s) production (see Table 1).

In order to compare further the effect produced by tryptamine, 5-methoxytryptamine and serotonin, three antagonists,
methysergide, ketanserin and MDL 72222 were tested at 5 µM

Table 1

Effect of various substances (1 μM)on the stimulation ^{3}H-inositol phosphate(s) accumulation caused by 100uM agonist

	Octop-amine	Carb-achol	Noradren-aline	5-HT	Trypt-amine	5-MT
Atropine	N.E.	87%*	N.E.	N.E.	N.E.	N.E.
Prazosin	N.E.	N.E.	78%*	N.E.	N.E.	N.E.
Ketanserin	N.E.	N.E.	N.E.	58%*	58%*	55%*
Propranolol	N.E.	N.E.	N.E.	N.E.	N.E.	N.E.
Yohimbine	N.E.	N.E.	N.E.	N.E.	N.E.	N.E.
Mepyramine	N.E.	N.E.	N.E.	N.E.	N.E.	N.E.

Results expressed as mean of three experiments carried out in triplicate. N.E = no effect; * = % reduction.

(Figs.2-4). MDL 72222 is a known 5-HT$_3$ antagonist in peripheral nervous systems (Fozard 1984). It can be seen that at an agonist concentration of 0.2mM, the antagonists methysergide and ketanserin drastically inhibited the inositol phosphate stimulation of all three indoles. In contrast, the 5-HT$_3$ antagonist had a maximal 45±8% inhibition of the tryptamine and serotonin effect, and even less of an effect on the stimulation caused by 5-methoxytryptamine. When the agonist concentrations were 0.5mM, 5μM of methysergide or ketanserin had less of an inhibitory effect, averaging around 50% inhibition for all three indoles. At an even higher concentration of agonist (1mM), ketanserin and methysergide still had an inhibitory effect, but this was less than in the instances where the agonist was only 0.5mM. The 5-HT$_3$ antagonist had only a slight effect when the agonist concentration was 0.5 mM, while at an agonist concentration of 1 mM there appeared to be no inhibition caused by the 5-HT$_3$ antagonists. The results shown in Figs.2-4 suggest that the effects of ketanserin, methysergide and the 5-HT$_3$ antagonist are similar for the stimulation of inositol phosphate(s) for all three indoles. Furthermore, in view of ketanserin and methysergide being 5-HT$_2$ antagonists (though methysergide is not as specific as ketanserin), one may conclude that both tryptamine and 5-methoxytryptamine elicit their effects through the same serotonergic receptors (5-HT$_2$).

Further support for the idea that the three indoles elicit their effects through the same receptors comes from studies on rat cortex tissues from animals of different ages (Table 2). It can be seen that the stimulation of inositol

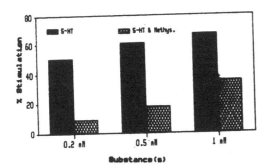

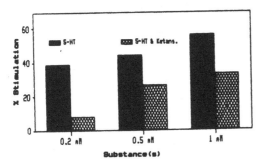

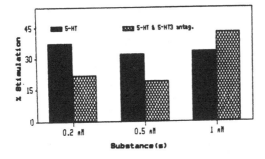

Fig.2.

Effect of ketanserin, methysergide and MDL 72222 at 5μM on the stimulation of inositol phosphate(s) caused by 0.2 mM, 0.5mM and 1mM of serotonin. Results are mean of three separate experiments each carried out in triplicate.

phosphate(s) by the three indoles in cortex tissues of 8 day, 25 day and adult animals is similar in that it remains more or less constant through development.   This also occurs for octopamine and noreadrenaline but not for carbachol.   In the case of carbachol, the stimulation of inositol phosphate(s) at 8 days is at least 4-fold than in the adult (Table 2).

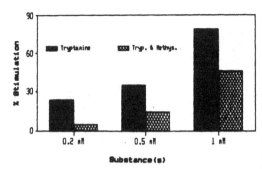

Fig.3.

Effect of ketanserin, methysergide and MDL 72222 at 5µM on the stimulation of inositol phosphate(s) caused by 0.2 mM, 0.5 mM an 1 mM of tryptamine. Results are mean of three separate experiments each carried out in triplicate.

CONCLUSIONS

The present study shows that the trace amines tryptamine, 5-methoxytryptamine and octopamine stimulate inositol phosphate(s) accumulation in rat brain slices. All the evidence points to the mediation of the effects produced by 5-methoxytryptamine and tryptamine are through the $5-HT_2$ serotonergic receptor sub-type. The octopamine effect is not mediated via any of the known receptors which utilise inositol phosphate(s) as a second messenger (viz. muscarinic, α_1-adrenergic or $5-HT_2$). It is clearly necessary to characterise the octo-

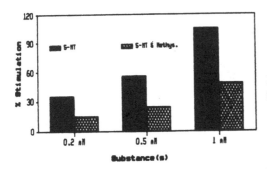

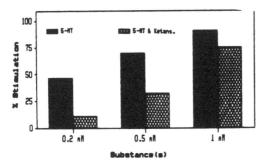

Fig.4.

Effect of ketanserin,
methysergide and MDL
72222 at 5 μM on the
stimulation of inos-
itol phosphate(s)
caused by 0.2 mM,
0.5 mM and 1 mM of
5-methoxytryptamine.
Results are mean of
three separate
experiments each
carried out in tri-
plicate.

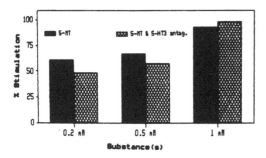

pamine response and demonstrate that it is receptor-mediated
in order to support a "transmitter" role for the amine.

The indication that 5-methoxytryptamine, tryptamine and
serotonin stimulate inositol phosphate(s)production via the
same receptors (i.e. 5-HT$_2$) would at first appear to argue
against the idea that tryptamine and 5-methoxytryptamine are

Table 2

% Increase in Inositol phosphates

Developmental aspects (Frontal cortex or rat)

	8 days[*]	25 days[*]	Adult[@]
5-MT (100 μM)	87	47	62 ± 8
5-MT (1 mM)	159	97	78 ± 5
Tryptamine (100 μM)	55	34	59 ± 4
Tryptamine (1 mM)	168	110	75 ± 9
5-HT (100 μM)	99	81	48 ± 7
5-HT (1 mM)	146	89	52 ± 5
Octopamine (100 μM)	21	25	28 ± 7
Octopamine (1 mM)	34	43	33 ± 5
Carbachol (100 μM)	868	448	176 ± 14
Carbachol (1 mM)	1040	591	199 ± 12
Noradrenaline (100 μM)	216	208	194 ± 15
Noradrenaline (1 mM)	238	210	225 ± 26

[*] = Average of two experiments
[@] = Mean ± S.D. where n = 3)

mediators in the brain.   It is,of course,possible that the "true" receptors for 5-methoxytryptamine and tryptamine in the physiological state do not use inositol phosphate(s) as a second messenger. These two indoles may only stimulate inositol phosphate(s) in the in vitro state because of their similarity in structure to serotonin.   It is also possible that the "true" agonist in the physiological state for the 5-HT$_2$receptors is not serotonin,but one or both of the other two indoles?

Acknowledgements: We are grateful to the Wellcome Trust for financial support. We also thank Dr. Fozard for supplying us with MDL 72222.

REFERENCES

Berridge, M.J. and Irvine R.F.   (1984)   Inositol triphosphate, a novel second messenger in cellular signal transduction.   Nature 312, 316-321.

Berridge, M.J. Downes, C.P. and Hanley M.R.   (1982) Lithium amplifies agonist dependent phosphatidylinositol responses in brain and salivary glands. Biochem.J. 206, 587-595.

Bosin, T.R., Jonsson, G. and Beck, O.   (1979)   On the occurrence of 5-methoxytryptamine in brain.   Brain Res., 173, 79-88.

Brown, E., Kendall, D.A. and Nahorski, S.R.   (1984)   Inositol phospholipid hydrolysis in rat cerebral cortex slices I. Receptor characterisation. J.Neurochem. 42, 1379-1387.

Conn, P.J. and Sanders-Bush, E.   (1985)   Serotonin-stimulated phosphoinositide turnover mediation by S_2 binding site in rat cerebral cortex but not in subcortical regions. J.Pharmacol.Exp.Therap., 234, 195-203.

Cutcliffe, N. and Osborne,N.N.   (1987)   Serotonergic and cholinergic stimulation of inositol phosphate formation in the rabbit retina.   Evidence for the presence of 5-HT_2 and muscarinic receptors. Brain Res. (in press)

Downes, C.P.   (1986)   Agonist-stimulated phosphatidylinositol 4,5-bisphosphate metabolism in the nervous system. Neurochem.Int., 9, 211-230.

Fisher, S.K. and Agranoff, B.W.   (1987)   Receptor activation of inositol lipid hydrolysis in neural tissues. J. Neurochem., 48, 999-1017.

Fozard, J.R.   (1984)   MDL 72222: a potent and highly selective antagonist at neuronal 5-hydroxytryptamine receptors. Naunyn-Schmiedeberg's Arch. Pharmacol., 326 36-44.

Juorio, A.V. and Greenshaw, A.J.   (1985)   Tryptamine concentrations in areas of 5-hydroxytryptamine terminal innervation after electrolytic lesions of midbrain raphe nuclei. J. Neurochem., 45, 422-426.

Osborne, N.N.   (1982)   Biology of serotonergic transmission. John Wiley & Sons, Chichester.

Osborne, N.N. and Hamon, M.   (1988)   Neuronal Serotonin. John Wiley & Sons, Chichester.

Osborne, N.N., Cutcliffe, N. and Peard, A.   (1986)   Trace amines (ethylamine, octopamine and tryptamine) stimulate inositol phospholipid hydrolysis in rat cerebral cortex slices. Neurochem. Res., 11, 1525-1531.

Wood, P.L., Martin, L.L. and Altar, C.A.   (1985)   ^{3}H-Tryptamine receptors in rat brain. In: Neuropsychopharmacology of the trace amines (Boulton, A.A., Bieck, P.R., Maitre, L. and Riederer, P., edits.,), Hamana Press, New Jersey, pp. 101-114.

THE ROLE OF FORM A MONOAMINE OXIDASE IN HEAD-TWITCH

RESPONSE INDUCED BY PARA-HYDROXYAMPHETAMINE

N. Satoh, S. Satoh, Y. Takahashi, T. Oikawa,
A. Yonezawa, M. Shioya, T. Tadano, K. Kisara,
K.Tanno, Y. Arai[*] and H. Kinemuchi[*].
Department of Pharmacology, Tohoku College of Pharmacy,
Sendai 983 and [*]Department of Pharmacology, School of
Medicine, Showa University, Tokyo 142, Japan

An animal test for hallucinogenic activity in man could be a valuable aid in the choice or rejection of psychoactive compounds for clinical use. Since it is impossible to determine whether an animal is experiencing hallucinations or not, an attempt to predict whether a drug, which has not been tested in humans, is likely to be hallucinogenic must be based on correlation between the hallucinogenic activity of known drugs and a measurable specific effect in animals (Corne and Pickering, 1967). Such correlations have been studied for a variety of hallucinogens in many animal species. Previous attempts to study correlation between animal tests and hallucinogenic activity have either required more than one test to encompass the different chemical structures or have been restricted to correlations within a specific chemical series.

HEAD-TWITCH RESPONSE (HTR) AND HALLUCINOGENIC ACTIVITY

Head-twitch, rapid discrete shaking of the head, seems to be in the normal mammalian behavior repertoire. Increase in the frequency of this HTR regarded to be a simple, reliable experimental model of hallucination, since many hallucinogens such as LSD, mescaline, phencyclidine and psilocybin increase this response in mice (Corne and Pickering, 1967; Vetulani et al., 1980). This response has also been elicited by many other compounds such as 5-HT (Nakamura and Fukushima, 1978; Handley and Brown, 1982), 5-HTP (Corne et al.,1963), p-tyramine (Orikasa et al., 1980),

quipazine (Malick et al., 1977; Vetulani et al., 1980), 5-
methoxytryptamine (Nakamura and Fukushima, 1978), and ergo-
metrine (Corne and Pickering, 1967). Several reports suggest that the increased HTR frequency elicited by many
inducers is mainly due to increased activity of the central
5-HT system (Corne et al., 1963; Clineschmidt and Lotti,
1974; Orikasa et al., 1980; Satoh et al., 1985). In addition, HTR induced by e.g. 5-HT (Handley and Brown, 1982)
was greatly changed by central action of selective adrenergic blockers or stimulants. This type of modulation was
also found for HTR elicited by another inducer, para-
hydroxyamphetamine (Tadano et al., in press)(see below).
These studies thus suggest that, in addition to the 5-HT
system, the central NE system or some adrenoceptors present
in presynaptic 5-HT neurons may also be involved in this
response. In addition, some non-selective MAOIs, greatly
enhanced HTR induced by cerebral 5-HTP administration
(Corne et al., 1963). This potentiation of HTR was also
found after tyramine application in combination with a
non-selective MAOI (Sakurada, 1975).

PARA-HYDROXYAMPHETAMINE (p-OHA) AND MOUSE HTR

In rodents, amphetamine (AMPH) is metabolized to p-OHA
and p-hydroxynorephedrine. Both metabolites are found accumulated in the brain after systemic single AMPH administration, but the latter metabolite is less concentrated in
striatal neurons. Since p-OHA cause a depletion of brain CA
stores, it has been implicated in some pharmacological effects of AMPH (Clay et al., 1971; Taylor and Sulser, 1973).
p-OHA, when applied directly to the brain, markedly
induces HTR in mice (Sakurada, 1975; Satoh et al., 1985;
Tadano et al., 1986; 1987). In contrast, another AMPH metabolite, p-hydroxynorephedrine caused no such response
(Sakurada, 1975). It is thus assumed that this p-OHA-induced HTR may be a symptom of hallucination. This p-OHA-
induced HTR was greatly affected by many drugs that act on
NE or 5-HT neurons (Satoh et al.,1985; Tadano et al.,
1987). In addition, when p-OHA was given in combination
with an icv injection of a 5-HT dose that would not cause
HTR, the HTR frequency was markedly increased (Tadano et
al., 1986). Taken together with findings of HTR inhibition
by anti-5-HT drugs (Tadano et al., 1986) and CA depletors
(Satoh et al., 1985), the mechanism by which HTR is induced
by p-OHA appears to be similar to that of some other HTR
inducers.

TWO FORMS OF MAO AND HTR INDUCED BY p-OHA AND 5-HTP

The enzyme, monoamine oxidase (MAO) is divided into two forms, MAO-A and MAO-B, based mainly on difference in inhibitor sensitivities (Kinemuchi et al., 1984). Clorgyline is a selective MAO-A inhibitor (MAO-AI) and l-deprenyl is a selective MAO-BI. Pargyline is usually classified as a selective MAO-BI, but it is by no means selective for MAO-B alone in terms of enzyme inhibition in vivo. In rodent brains, 5-HT is predominantly oxidized by MAO-A, beta-phenylethylamine by MAO-B, and tyramine by both (Kinemuchi et al., 1984). Subcellular studies indicated that MAO is localized mainly, but not exclusively, intraneuronally, whereas MAO-B is more extraneuronal (Oreland et al., 1983). Taken together with findings of the potentiation effect by a non-selective MAOI on HTR frequency (Corne et al., 1963; Sakurada, 1975), it is necessary to determine which form of MAO will preferentially influence HTR induced by p-OHA.

Administration of various icv doses of p-OHA alone markedly increased HTR frequency and this effect continued for 60-80 min. Total head-twitches by p-OHA in 90 min was dose-dependent. The duration of HTR after icv p-OHA agrees with the half-life of this compound after its icv administration (Taylor and Sulser, 1973). Pretreatment with clorgyline greatly increased the total number of head-twitches induced by p-OHA. Similarly, pretreatment with lower doses of clorgyline also markedly increased the total number of HTR. In contrast, l-deprenyl did not change the total number of head-twitches compared to the controls. Pretreatment with pargyline, however, increased the total number of head-twitches. This potentiation by combined treatment with p-OHA and pargyline, however, was much less than that observed by the combination of p-OHA and clorgyline (Tab. 1).

Table 1. Effects of MAOIs on the p-OHA-induced HTR rate

	p-OHA 20 μg	p-OHA 40 μg	p-OHA 80 μg
p-OHA control	3.0±1.1	6.5±1.4	13.2±2.1
p-OHA+Clorgyline (1 mg/kg)	8.3±1.3[*]	22.0±4.1[*]	33.6±6.4[*]
p-OHA+Pargyline (5 mg/kg)	7.3±0.8[*]	15.8±1.5[*]	24.3±2.4[*]
p-OHA+l-Deprenyl (2 mg/kg)	6.5±0.3[*]	9.9±1.3	13.1±1.9

Either MAOI was ip injected 1 hr before p-OHA. HTR, counted for 90 min. Values, expressed as mean±SEM. *, $p < 0.01$

These results indicate that selective inhibition of brain MAO-A may potentiate the frequency of head-twitching

induced by p-OHA. Pargyline, a less selective MAO-BI than
l-deprenyl, also potentiates the response, probably by
inhibiting MAO-A less compared to MAO-B inhibition. To
determine whether this potentiation by MAOI was the same as
that by 5-HTP, we performed similar experiments with the
same MAOIs, and 5-HTP doses (50-100 mg/kg, i.p.) that could
not markedly increase the HTR frequency. As a result,
potentiation of 5-HTP was observed only after pretreatment
with clorgyline, but not after pargyline and l-deprenyl.

The above results led us to conclude that brain MAO-A
may participate in the increase in HTR frequency induced by
p-OHA and 5-HTP (via formation of 5-HT) in mice. Acute
administration of MAOIs elevates levels of NE, DA and 5-HT
and these elevated amines are then available for release
into synapses after neuronal activity. The predominant
presence of MAO-A within the neurons would implicated this
enzyme in regulating the release of these transmitters.
However, when L-tryptophan is injected into rats after
clorgyline, brain 5-HT concentrations increase, but there
is no change in overt behaviour. This might be due to
deamination of elevated 5-HT by intraneuronal MAO-B after
selective inhibition of intraneuronal MAO-A. This would
prevent the behavioural expression of 5-HT. In contrast,
when animals receive non-selective MAOIs, this amine precursor elicits a characteristic behavioural syndrome of
hyperactivity and lateral head-weaving. When both selective MAOIs were given in combination in doses at which each
selectively inhibited MAO-A or MAO-B, the characteristic
behaviours that follow injection of the amine precursors
was observed. If all of this taken together is true for
HTR, the HTR increased after MAO-AI and p-OHA might be
selective inhibition of MAO-B by p-OHA itself. This mechanism, however, is not the case for 5-HTP-induced HTR, since
the 5-HT thus formed does not inhibit MAO-B.

REFERENCES

Clay, G.A. Cho, A.K. and Roberfroid, M. (1971). Effect of
    SKF-525A on the norepinephrine-depleting actions of d-
    amphetamine. Biochem. Pharmacol. 20, 1821-1831.
Clineschmidt, B.V. and Lotti, V.J. (1974). Indole antago-
    nists: Relative potencies as inhibitors of tryptamine and
    5-HTP-induced responses. Br. J. Pharmacol. 50, 311-313.
Corne, S.J. Pickering, R.W. and Warner, B.T.(1963). A met-
    hod for assessing the effects of drugs on the central
    actions of 5-HT. Br. J. Pharmacol. 20, 106-120.

Corne, S.J. and Pickering, R.W. (1967). A possible corre-lation between drug-induced hallucinations in man and a behavioural response in mice. Psychopharmacol. 11, 65-78.

Handley, S.L. and Brown, J. (1982). Effects of the 5-hydroxytryptamine-induced head-twitch of drugs with sele-ctive actions on alpha$_1$- and alpha$_2$-adrenoceptors. Neuro-pharmacol. 21, 507-510.

Kinemuchi, H. Fowler, C.J. and Tipton, K.F. (1984). Sub-strate specificities of the two forms of monoamine oxi-dase. In Monoamine oxidase and disease (Tipton, K.F. Dostert, P. and Strolin Benedetti, M., eds), pp. 53-62. Academic Press, New York.

Malick, J.B. Doren, E. and Barnett, A. (1977). Quipazine-induced head-twitch in mice. Pharmacol. Biochem. Behav. 6, 325-329.

Nakamura, M. and Fukushima, H. (1978). Effects of reser-pine, para-chlorophenylalanine, 5,6-dihydroxytryptamine and fludiazepam on the head-twitches induced by 5-hydro-xytryptamine or 5-methoxytryptamine in mice. J. Pharm. Pharmacol. 30, 254-256.

Oreland, L. Arai, Y. and Stenstrom, A. (1983). The effect of deprenyl (selegyline) on intra- and extraneuronal dopamine oxidation. Acta Neurol. Scand. suppl 95, 81-85.

Orikasa, S. Sakurada, S. and Kisara, K. (1980). Head-twitch response induced by tyramine. Psychopharmacol. 67, 53-59.

Sakurada, S. (1975). Central action of beta-phenylethyla-mine derivatives (6). Head-twitches induced by intracere-braly administered tyramine in isocarboxazid pretreated mice. Folia pharmacol. japon. 71, 779-787.

Satoh, S. Satoh, N. Tadano, T. and Kisara, K. (1985). Enhancement of para-hydroxyamphetamine induced head-twitch response by catecholamine depletion. Res. Commun. Sub. Abuse 6, 213-219.

Tadano, T. Satoh, S. and Kisara, K. (1986). Head-twitches induced by p-hydroxyamphetamine in mice. Japan. J. Phar-macol. 41, 519-523.

Tadano, T, Satoh, S. Satoh, N. Kisara, K. Arai, Y. and Kinemuchi, H. (1987). Involvement of alpha-adrenoceptors in para-hydroxyamphetamine-induced head-twitch response. Psychopharmacol. in press.

Taylor, W.A. and Sulser, F. (1973). Effects of amphetamine and its hydroxylated metabolites on central noradrenergic mechanisms. J. Pharm. Exp. Ther. 185, 620-632.

Vetulani, J. Bednarczyk, B. Reichenburg, K. and Rokosz, A. (1980). Head-twitches induced by LSD and quipazine. Neuropharmacol. 19, 155-158.

EFFECTS OF TRYPTAMINE ON THE HEAD-TWITCH RESPONSE

INDUCED BY 5-HYDROXYTRYPTAMINE

Shuzo Orikasa

Neuropsychiatric Research Unit
CMR Building, University of Saskatchewan
Saskatoon, Saskatchewan, Canada S7N 0W0

The role of conventional amines such as DA, NA, and 5-HT as neurotransmitters in the CNS is now well accepted. Trace amines on the other hand have received relatively little attention and their physiological role at least in mammalian brain is not understood. One possible reason for this is that trace amines show little or no pharmacological activity at doses comparable to those of the conventional amines (Gaddum, 1953; Woolley and Shaw, 1953; Schain, 1961; Ennis and Cox, 1982). This is because their endogenous levels are substantially lower than those of the conventional amines and they are metabolised exceedingly quickly (Boulton, 1976). It has been suggested that they act as modulators of the more conventional amines (Boulton, 1976).

We have recently shown that T as well as 5-HT elicits a head-twitch response (HTR) in mice when the drugs are delivered directly into the brain and it is well known that the HTR is mediated by central serotonergic neurones (Corne et al., 1963).

In the present experiments the possible modulatory action of T on 5-HT function was investigated by examining the effects of T on the 5-HT-induced HTR.

MATERIALS AND METHODS

Male Swiss strain mice (20-24 g) were used. The technique used for intracerebroventricular (i.c.v.) injection

was that described by Haley and McCormick (1957) and modified according to Brittain and Handley (1967). The injection was accomplished by inserting a 30 gauge 1/2 inch
needle, to which 23 gauge stainless steel tubing was glued
as a stopper (3 mm distance), through the soft bone on the
coronal suture about 1 mm lateral to bregma. The mice were
placed in individual plastic cages (25 x 18 x 13 cm) immediately after the icv injection. The number of HTR's was
counted for 1 min at 5 min intervals between 5 and 46 min
after injection. Serotonin and T were hydrochloride salts
purchased from Sigma Chemical Co., Missouri. Additional
doses of 5-HT or T were dissolved in physiological saline
solution containing 0.1% ascorbic acid. Basic doses of
5-HT were then dissolved in this solution.

RESULTS

As shown in Figure 1, the HTR was elicited by i.c.v.
injection of 5-HT at doses of 100-320 nmol/mouse and by T
at doses of 180 and 320 nmol/mouse. The total number of

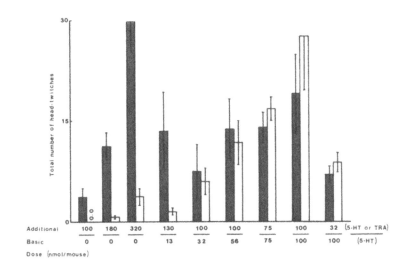

Figure 1. Effects of a combination of various portions of
either additional doses of serotonin (stippled columns) or
tryptamine (open columns) and the basic doses of serotonin
on HTR. Each value is the mean of 5 mice. Vertical bars
show the S.E.M.

head-twitches induced by T was less than that induced by
5-HT and 100 nmol/mouse T did not elicit the HTR.

Figure 1 also illustrates the effects of adding 5-HT
or T to varying basic doses of 5-HT. The total number of
head-twitches induced by adding 130 nmol/mouse T to 13
nmol/mouse 5-HT was less than that induced by adding 130
nmol/mouse 5-HT to 13 nmol/mouse 5-HT. As the ratio of the
additional dose to the basic dose decreased, the potency of
T in potentiating the HTR induced by 5-HT approached that
of 5-HT. On the occasions that the basic dose of 5-HT was
greater than the additional dose, T was equipotent with
5-HT added to the basic doses.

Figure 2 shows the time course of the HTR induced by
combinations of either 5-HT or T and basic doses of 5-HT.
It appears that a combination of additional doses of 5-HT
and basic doses of 5-HT induced a higher peak response than

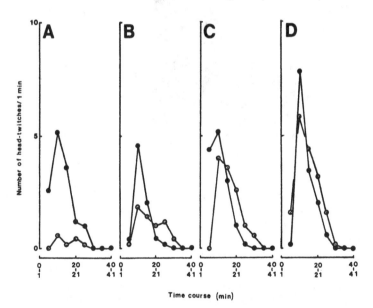

Figure 2. Effects of a combination of various portions of
either additional doses of serotonin (closed circles) or
tryptamine (open circles) and the basic doses of serotonin
on HTR. The combinations of additional doses and the basic
dose are as follows: (A) 130 and 13, (B) 100 and 32,
(C) 100 and 56, and (D) 75 and 75 (doses are given in
nmol/mouse).

a combination of T and basic doses of 5-HT. However, the
duration of effect of the combination of T and basic doses
of 5-HT was longer than that of the combination of
additional doses of 5-HT and basic doses of 5-HT.

As shown in Figure 3, the addition of 5-HT (10-100
nmol/mouse) to a basic dose of 5-HT (180 nmol/mouse)
significantly increased the total number of head-twitches.
A similar potentiating effect was observed when T (10-100
nmol/mouse) was added to the basic dose of 5-HT. Also in
this experiment T and 5-HT added appeared to be equipotent.

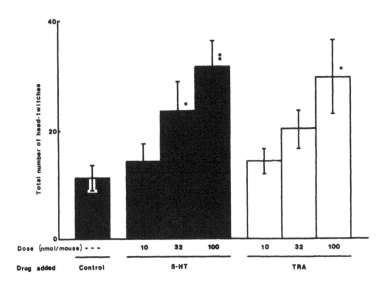

Figure 3. Effects of concurrent injection of additional
doses of serotonin (5-HT, 10-100 nmol/mouse, stippled
columns) and tryptamine (TRA, 10-100 nmol/mouse, open
columns on HTR induced by the basic dose of 5-HT (180
nmol/mouse). The control group (closed column) received
only the basic dose of 5-HT. Each value is the mean of 10
mice. Vertical bars show the S.E.M., *p <0.05, **p <0.01,
when compared with the control group (by Student's t-test).

DISCUSSION

It has been reported that the HTR is mediated by the
central serotonergic system (Corne et al., 1963). In the
present experiment the HTR was induced by i.c.v. injections

of 5-HT. This effect of 5-HT was in agreement with previous reports (Suchowsky et al., 1969; Nakamura and Fukushima, 1978). High, but not low doses of T induced the HTR. Low doses of T, however, significantly potentiated the 5-HT-induced HTR. Therefore, 5-HT initiates the HTR, while low doses of T may act as a modulator of 5-HT in this animal model. The observation that T prolongs the effect of 5-HT would support such speculation.

Tryptamine added to basic doses of 5-HT was equipotent to 5-HT added to basic doses of 5-HT. However, previous reports have shown that T has little or no pharmacological activity in the mammalian brain (Gaddum, 1953; Woolley and Shaw, 1953; Schain, 1961; Ennis and Cox, 1982), and in the present experiments the potency of T administered alone was much less than that of 5-HT. Therefore, it seems unlikely that the potentiation of the 5-HT-induced HTR by T is due to the summation of 5-HT-induced and T-induced responses.

It has been reported that responses of single neurones to iontophoretically applied 5-HT were enhanced by a weak concurrent application of T, which by itself did not alter the baseline cell firing rate (Jones and Boulton, 1980). This is similar to the present experiments where the effects of 5-HT were potentiated by a concurrent injection of small amounts of T, which by themselves do not .elicit the HTR. Those results may suggest that a high concentration of compounds is not always required to regulate the neural function. The rate of production of T is probably quite high since the half-life of T is very short compared to 5-HT (Meek et al., 1970). Therefore, despite the low concentration, T has a rapid turnover and it may be involved in the regulation of serotonergic systems. The results of the present experiments suggest that this animal model may be a useful tool to investigate the regulatory effects of drugs on 5-HT function in the mammalian brain.

ACKNOWLEDGEMENTS

I thank Saskatchewan Health for continuing financial support and the Saskatchewan Health Research Board for the provision of a fellowship.

REFERENCES

Boulton A.A. (1976) Cerebral aryl alkyl aminergic mechanisms. In, Trace Amines and the Brain (Usdin, E. and Sandler M., eds) pp. 21-40, Marcel Dekker, New York.

Brittain R.T. and Handley S.L. (1967) Temperature changes produced by the injection of catecholamines and 5-hydroxytryptamine into the cerebral ventricles of the conscious mouse. J. Physiol. 192, 805-813.

Corne S.J., Pickering R.W. and Warner B.T. (1963) A method for assessing the effects of drugs on the central actions of 5-hydroxytryptamine. Br. J. Pharmacol. 20, 106-120.

Ennis C. and Cox B. (1982) The effect of tryptamine on serotonergic release from hypothalamic slices is medicated by a cholinergic interneurone. Psychopharmacol. 78, 85-88.

Gaddum J.H. (1953) Tryptamine receptors. J. Physiol. 119, 363-368.

Haley T.J. and McCormick W.G. (1957) Pharmacological effects produced by intracerebral injection of drugs in the conscious mouse. Br. J. Pharmacol. 12, 12-15.

Jones R.S.G. and Boulton A.A. (1980) Tryptamine and 5-hydroxytryptamine: actions and interactions on cortical neurones in the rat. Life Sci. 27, 1849-1856.

Meek J.K., Krall A.R. and Lipton M.A. (1970) Psychotropic drugs and the metabolism of intracerebrally injected tryptamine, 5-hydroxytryptamine, and norepinephrine. J. Neurochem. 17, 1627-1635.

Nakamura M. and Fukushima H. (1978) Effect of 5,6-dihydroxytryptamine on the head-twitches induced by 5-HTP, 5-HT, mescaline and fludiazepam in mice. J. Pharm. Pharmacol. 30, 56-58.

Schain R.J. (1961) Some effects of a monoamine oxidase inhibitor upon changes produced by centrally administered amines. Br. J. Pharmacol. 17, 261-266.

Suchowsky G.K., Pegrassi L., Moretti A. and Bonsignori A. (1969) The effect of 4-H-3-methylcarboxamide-1,3-benzoxazine-2-one (FI 6654) on monoaminoxidase and cerebral 5-HT. Arch. Int. Pharmacodyn. 182, 332-340.

Woolley D.W. and Shaw E. (1953) Antimetabolites of serotonin. J. Biol. Chem. 203, 69-70.

EFFECTS OF TRYPTAMINE AND 5-HT MICROINJECTED

INTO THE NUCLEUS ACCUMBENS ON D-AMPHETAMINE

INDUCED LOCOMOTOR ACTIVITY

Paul J. Fletcher

Neuropsychiatric Research Unit
CMR Building
University of Saskatchewan
Saskatoon, Saskatchewan, Canada S7N 0W0

INTRODUCTION

Several reports have demonstrated the presence of specific [^{3}H]-tryptamine (T) binding sites in rat brain (Altar et al., 1987; Cascio and Kellar, 1983; Kellar and Cascio, 1982; Perry, 1986). These [^{3}H]-T binding sites appear to be pharmacologically unique, and distinct from the various 5-HT receptor types. Autoradiographic studies have revealed that [^{3}H]-T binding sites are heterogenously distributed, and that high concentrations occur in areas receiving dense dopaminergic input, such as the nucleus accumbens and caudate-putamen (Altar et al., 1987; McCormack et al., 1986; Perry, 1986). Dopaminergic projections to these areas play a major role in the control of locomotor activity (Pijnenburg and Van Rossum, 1973; Pijnenburg et al., 1976), and it has been suggested that T in the nucleus accumbens may interact with dopamine (DA) to modulate motor activity (Altar et al., 1986). The following experiment was conducted to investigate this possibility.

Amphetamine induced locomotor activity is thought to be mediated primarily by DA release in the nucleus accumbens (Kelly et al., 1975; Kelly and Iversen, 1976). Therefore the effects of T injected into the nucleus accumbens on the locomotor activity induced by the systemic

273

administration of d-amphetamine were examined. The design
of the experiment also allowed an examination of the
effects of T on spontaneous locomotor activity. The
effects of 5-HT were examined also for comparative
purposes.

METHODS

Male Wistar rats (180-200 g) were bilaterally implanted with stainless steel guide cannulae aimed at the anterior nucleus accumbens (AP + 3.4, Lat ± 1.7, V - 5.7 mm,
according to Konig and Klippel, 1963), and then allowed 2
weeks to recover before testing. Testing was conducted in
a plexiglass cage (40x40x23 cm) positioned inside an
infra-red photobeam recording device, consisting of an
array of 12 x 12 photobeams. Horizontal activity was measured by the interruption of any one of these photobeams co-
ordinated ambulatory activity was measured by the interruption of two or more consecutive photobeams. All rats were
habituated to the apparatus for 2h on the 2 days preceding
testing. Twenty-four hours prior to testing all rats were
treated (IP) with 50 mg/kg pargyline HCl. Thirty minutes
before testing rats were injected with either 0.5 mg/kg
d-amphetamine sulphate or 1 ml/kg 0.9% saline. Immediately
before testing rats in the first experiment received bilat-
eral injections of artifical CSF, 50 or 75 µg tryptamine
HCl (1 µl delivered in 3 minutes) into the nucleus accumbens. In the second experiment rats were microinjected with
CSF or 25 µg 5-HT bimaleate. Activity measures were
recorded for 2h at 10 minute intervals. At the end of the
experiment cannulae sites were verified using standard
histological procedures. Only data from those animals with
cannulae placed in the anterior nucleus accumbens were used
for analysis. Behavioural verification was also achieved
in 2 groups of rats by measuring their activity following
microinjection of 10 µg d-amphetamine sulphate, or CSF into
the nucleus accumbens. D-amphetamine significantly
increased motor activity over 1h (mean activity score 3,405
± 402) compared to CSF (843 ± 75).

RESULTS

Figure 1 illustrates the effects of T injected alone,
and in combination with d-amphetamine, on ambulatory

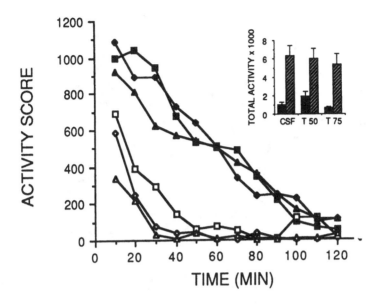

Figure 1. The effects of 50 µg T (squares), 75 µg T (tri-
angles) and CSF (diamonds) injected into the nucleus accumbens on the activity elicited by 0.5 mg/kg d-amphetamine
(solid symbols) or 0.9% saline (open symbols). Inset: The
activity scores summed across the session for each group;
solid bars represent saline, shaded bars represent d-
amphetamine treatment. Each point represents mean activity
scores from 6-8 rats.

activity. Data were analysed by 3-way analysis of variance
(T dose x amphetamine dose x time). d-Amphetamine significantly increased ambulatory activity [F(1,30) = 53.1, p<0.-
001], though this effect declined with time. Neither the
main effect of T, nor the T x amphetamine interaction
reached significance [F(2,30) = 0.5, and 0.8 respectively].
Thus, T at the doses tested failed to alter the ambulatory
response to systemically administered d-amphetamine. T did
not alter spontaneous activity at any of the 10 minute time
bins.

Figure 2 shows the effects of 5-HT injected into the
nucleus accumbens on spontaneous activity, and d-amphetamine induced activity. Again d-amphetamine induced a

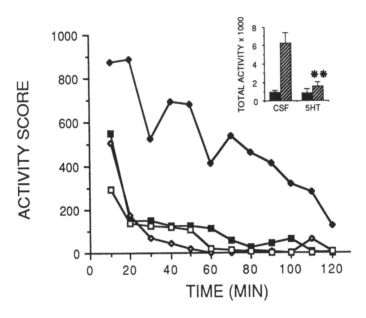

Figure 2. The effects of 25 µg 5-HT (squares) and CSF
(diamonds) injected into the nucleus accumbens on the
activity elicited by 0.5 mg/kg d-amphetamine (solid symbols) or 0.9% saline (open symbols). Inset: activity
scores summed across the session for each group; solid bars
(d-amphetamine), shaded bars (saline). Each point represents mean activity scores from 6-8 rats.

significant increase in activity [F (1,20) = 19.6, p<0.001]
which declined across time. This effect of d-amphetamine
was attenuated by 5-HT as shown by the significant main
effect of 5-HT [F(1,20) = 10.5, p<0.001] and the significant 5-HT x amphetamine interaction [F(1,20) = 7.4,
p<0.012]. This attenuation by 5-HT was of rapid onset and
persisted throughout the test period. The 5-HT alone did
not alter spontaneous motor activity at any point, or as
measured for the whole session.

        In both experiments identical profiles of results were
seen on the measure of horizontal activity (data not
shown).

DISCUSSION

The results show that T injected into the nucleus accumbens does not alter the level of spontaneous locomotor activity. Further, T did not alter the locomotor activity induced by the systemic administration of a low dose of d-amphetamine. Since the effects of amphetamine on motor activity are mediated primarily by DA release in the nucleus accumbens it appears that T does not modulate DA activity in this area. Thus, the functional significance of T binding sites remains unclear.

It is unlikely that the lack of effect of T is the result of an insufficient amount of injected T. The doses employed in this study (50 and 75 µg) are far in excess of the total amount of endogenous T present in rat brain (Philips et al., 1974). Endogenous T is catabolised rapidly by monoamine oxidase (MAO) but this process is slowed markedly by MAO inhibition (Philips and Boulton, 1979). Since the animals used in the present study were pretreated with the MAO inhibitor pargyline, it is unlikely that the failure of T to modify the effects of d-amphetamine is due to the short half-life of this amine.

In contrast to the lack of effect of T, 5-HT injected into the nucleus accumbens abolished the locomotor response induced by d-amphetamine. This finding is consistent with the view that 5-HT exerts an inhibitory influence of 5-HT on dopaminergic activity in the nucleus accumbens (Carter and Pycock, 1978; Costall et al., 1979; Gerson and Baldessarini, 1980; Jones et al., 1981). The results reported here indicate that T and 5-HT have different actions when microinjected into the nucleus accumbens. This is in keeping with the view that these indoleamines may have different pharmacological actions in the CNS (Jones, 1982).

ACKNOWLEDGEMENTS

Supported by Saskatchewan Health and a Fellowship from Saskatchewan Health Research Board.

REFERENCES

Altar C.A., Wasley A.M. and Martin L.L. (1986) Autoradio-
graphic localisation and pharmacology of unique
[^{3}H]tryptamine binding sites in rat brain. Neurosci. 17,
263-273.
Carter C.J. and Pycock C.J. (1978) Differential effects of
central serotonergic manipulation on hyperactive and
stereotyped behaviour. Life Sci. 23: 953-960.
Cascio C.S. and Kellar K.J. (1983) Characterisation of
[^{3}H]tryptamine binding sites in brain. Eur. J.
Pharmacol. 95: 31-39.
Costall B., Hui S.-C. G. and Naylor R. J. (1979) The
importance of serotonergic mechanisms for the induction
of hyperactivity by amphetamine and its antagonism by
intra-accumbens (3,4-dihydroxyphenylamino)-2-imidasoline
(DPI). Neuropharmacol. 18: 605-609.
Gerson S.C. and Baldessarini R.J. (1980) Minireview: Motor
effects of serotonin in the central nervous system. Life
Sci. 27: 1435-1451.
Jones D.L., Mogenson G.J. and Wu M. (1981) Injections of
dopaminergic, cholinergic, serotonergic and GABAergic
drugs into the nucleus accumbens: effects on locomotor
activity of the rat. Neuropharmacol. 20: 29-38.
Jones R.S.G. (1982) Tryptamine: A neuromodulator or neurotransmitter in mammalian brain? Prog. Neurobiol. 19,
117-139.
Kellar K.J. and Cascio C.S. (1982) [^{3}H]Tryptamine: High
affinity binding sites in rat brain. Eur. J. Pharmacol.
78, 475-478.
Kelly P.H., Seviour P.W. and Iversen S.D. (1975) Amphetamine and apomorphine responses in the rat following
6-OHDA lesions of the nucleus accumbens septi and corpus
striatum. Brain Research 94, 507-522.
Kelly P.H. and Iversen S.D. (1976) Selective 6-OHDA-
induced destruction of mesolimbic dopamine neurons:
Abolition of psychostimulant-induced locomotor activity
in rats. Eur. J. Pharmacol. 40, 45-56.
McCormack J.K., Beitz A.J. and Larson A.A. (1986) Autoradiographic localization of tryptamine binding sites in
the rat and dog central nervous system. J. Neurosci. 6,
94-101.
Perry D.C. (1986) [^{3}H]Tryptamine autoradiography in rat
brain and choroid plexus reveals two distinct sites. J.
Pharmacol. Exp. Ther. 236, 548-559.

Philips S.R., Durden D.A. and Boulton A.A. (1974) Identification and distribution of tryptamine in the rat. Can. J. Biochem. 52, 447-451.

Philips S.R. and Boulton A.A. (1979) The effect of monoamine oxidase inhibitors on some arylalkylamines in rat striatum. J. Neurochem. 33, 159-167.

Pijnenburg A.J.J. and Van Rossum J.M. (1973) Stimulation of locomotor activity following injection of dopamine into the nucleus accumbens. J. Pharm. Pharmacol. 25, 1003-1005.

Pijnenburg A.J.J., Honig W.M.M., Van der Heyden J.A.M. and Van Rossum J.M. (1976) Effects of chemical stimulation of the mesolimbic dopamine system upon locomotor activity. Eur. J. Pharmacol. 35, 45-58.

Wu P.H. and Boulton A.A. (1973) Distribution and metabolism of tryptamine in rat brain. Can. J. Biochem. 51, 1104-1112.

EFFECTS OF TRYPTAMINE ON THE ACQUISITION OF

A ONE-WAY ACTIVE AVOIDANCE TASK

Paul J. Fletcher

Neuropsychiatric Research Unit
CMR Building
University of Saskatchewan
Saskatoon, Saskatchewan, Canada S7N OWO

INTRODUCTION

It has been shown previously that tryptamine (T) administered to iproniazid treated rats interferes with avoidance behaviour maintained by electric shock (Vogel and Cohen, 1977). A large body of evidence shows that procedures which enhance central 5-HT activity induce deficits in a variety of avoidance conditioning tasks (Ogren, 1982a). For example, the 5-HT releaser p-chloroamphetamine (PCA) impairs the acquisition and retention of an unsignalled one-way active avoidance task (Ogren, 1982b). These effects appear to be mediated specifically by brain 5-HT mechanisms, and cannot be explained in terms of non-associative effects (Ogren, 1982b; Ogren and Johansson, 1985). Since some of the behavioural effects of T may involve an action on brain 5-HT mechanisms (Jones, 1982) the following experiments were designed to examine the effects of T on the acquisition of an unsignalled one-way active avoidance task, similar to that used by Ogren (1982b).

METHODS

Male Wistar rats (240-270 g) served as subjects. Testing was conducted in a standard shuttle-box divided into 2 identical compartments by a 4 cm high hurdle. The stainless steel bars forming the floor of the box could be electrified. Seven minutes after drug treatment each rat was allowed to explore the box freely for 3 minutes, whereupon it was placed in the "shock" compartment. Five seconds later a scrambled constant current shock (0.5 mA) was delivered to the floor. The shock could be avoided within 5s, or after shock onset, by traversing the hurdle to reach the safe compartment. The rat was then allowed to remain in the safe compartment for 20s, and then transferred to a holding cage for 10s before beginning the next trial. A maximum of 30 trials was conducted; a criterion of 9 out of 10 consecutive avoidance responses was used to define acquisition of the task. Response latencies on each trial were recorded by a hand-held stop watch.

Three experiments were conducted. The first examined the effects of 2.5 and 5 mg/kg tryptamine HCl (IP) on avoidance acquisition. In the second experiment the effects of pretreatment with 1 mg/kg metergoline (1h) and 5 mg/kg methysergide (30 min) on the deficit induced by 5 mg/kg tryptamine were examined. The third experiment examined the effects of pretreatment with PCPA ethyl ester (400 mg/kg, PO, 3 days) and haloperidol (0.1 mg/kg, 1h) on the deficit induced by 5 mg/kg tryptamine. All rats, with the exception of saline controls in experiment 1, were pretreated with 100 mg/kg iproniazid 24h prior to testing.

RESULTS

Iproniazid treatment did not induce any changes in avoidance behaviour compared to saline treatment (Table 1). Therefore any changes observed in animals treated with iproniazid + T must be induced by T alone. At both doses T significantly increased the number of trials required to reach the acquisition criterion, and consequently also increased the number of shocks received. Escape and avoidance latencies were not significantly altered by T. At a dose of 10 mg/kg T induced symptoms resembling the 5-HT syndrome, and those animals tested generally failed to escape from shock (data not shown). Injection of 5 mg/kg 5-HT induced a mild sedation, but did not affect acquisition of the task.

Table 2 illustrates the effects of 1 mg/kg metergoline and 5 mg/kg methysergide on the avoidance deficit induced by 5 mg/kg T. Neither of these 5-HT antagonists altered avoidance behaviour in their own right, but both compounds

TABLE 1

MEDIAN (AND RANGE) RESPONSE PARAMETERS FOR GROUPS IN EXPERIMENT 1

Group	First Avoidance	Trials to Criterion	No. Shocks	Response Failures	Escape[1] Latencies	Avoidance[2] Latencies
Sal + Sal	3 (2-5)	14 (11-21)	3 (2-7)	0 (0)	2.2 (1.1-3.6)	0.9 (0.7-2.1)
Sal + Ipron	4 (2-6)	15 (12-23)	5 (3-8)	0 (0-1)	1.7 (0.7-2)	2.1 (1.4-3.7)
Ipron + 2.5 T	5 (1-8)	20* (13-27)	9** (3-11)	0 (0-6)	1.6 (0.6-2)	1.9 (0.8-2.9)
Ipron + 5 T	4 (3-9)	24** (14->30)	9** (4-18)	0 (0-1)	1.7 (1.3-3.2)	2.1 (0.9-3.4)
Ipron + 5 5-HT	4 (3-8)	16 (12-20)	6 (3-7)	0 (0)	2.5 (1.7-2.6)	1.6 (1.1-2.5)

[1]On trials 1-10.
[2]On all trials; from shock onset.
* p <0.05 compared to Sal + Sal.
** p <0.01 compared to Sal + Sal.
n = 7 per group.

TABLE 2

RESULTS OF EXPERIMENTS 2 AND 3 SHOWING THE MEDIAN NUMBER OF
TRIALS TO CRITERION IN RATS TREATED WITH VEHICLE OR
5 mg/kg T FOLLOWING PRETREATMENT WITH VARIOUS AGENTS

Expt. 2		Veh		Metergoline (1 mg/kg)		Methysergide (5 mg/kg)	
	Veh	T	Veh	T	Veh	T	
	14	25**	13	15	13	15	
Expt. 3		Veh		PCPA (400 mg/kg)		Haloperidol (0.1 mg/kg)	
	Veh	T	Veh	T	Veh	T	
	14	24**	15	21**	16	28**	

** Differs from Veh-Veh condition, p <0.01; n = 5-7 per
group.

clearly blocked the acquisition deficit induced by T. The
effects of pretreatment with PCPA and haloperidol are shown
also in Table 2. Prior depletion of brain 5-HT levels with
PCPA did not affect avoidance behaviour when administered
alone, but failed to reverse the deficit induced by T.
Similarly blockade of dopamine receptors with the neuroleptic haloperidol did not affect the T induced acquisition
deficit.

DISCUSSION

The results described here confirm and extend the previously reported finding that the systemic administration
of T to iproniazid pretreated rats impairs the acquisition
of an active avoidance task. This effect of T may involve
some action at 5-HT receptors since it was reversed by the
5-HT antagonists metergoline and methysergide. However,
treatment with 5-HT, which does not penetrate the blood
brain barrier did not induce an acquisition deficit, and so
the effect of T is probably not the result of an action at
peripheral 5-HT receptors. Pretreatment with PCPA did not
reverse the effect of T. Since PCPA at the dose used here

depletes brain 5-HT to approximately 80-90% of control levels (Fletcher and Sloley, unpublished), the effect of T is unlikely to result from the release of endogenous 5-HT. An action of T on dopaminergic systems can be ruled out also since haloperidol failed to reverse the T-induced avoidance deficit.

Examination of the various parameters of responding (shown in Table 1) revealed that avoidance latencies were not significantly affected by T. Thus, since T treated rats were able to initiate normal avoidance responses a gross motor impairment cannot account for the acquisition deficit. This is supported also by the observation that T treated rats tended to perform chains of 3 or 4 avoidance responses interspersed with 1 or 2 escape responses. Escape latencies were not altered by T, indicating that the acquisition deficit was not caused by an alteration in the animal's sensitivity to the shock stimulus.

Two possible explanations for this behavioural effect of T deserve consideration. Firstly, it is possible that by increasing brain 5-HT neurotransmission T interferes with the memorial storage and/or retrieval processes necessary for the acquisition of the task. This mechanism apparently accounts for the deficit induced by PCA (Ogren, 1982b; Ogren and Johnasson, 1985). Secondly, since correct acquisition of this task involves the formation of an association between the environment in which the shock is delivered and the shock itself, an impairment in attentional or perceptual processes may underlie the T deficit. It is of interest therefore that Vogel and Cohen (1977) noted a similarity in their avoidance task between the effects of T and the effects of LSD and dimethyltryptamine, and that T and LSD-like hallucinogens possess similar pharmacological and physiological action (Martin and Sloan, 1986). Clearly, however, further work is required to examine the behavioural mechanisms by which T impairs the acquisition of avoidance responding.

ACKNOWLEDGEMENTS

Supported by Saskatchewan Health and the Saskatchewan Health Research Board.

REFERENCES

Jones R.S.G. (1982) Tryptamine: A neuromodulator or neurotransmitter in mammalina brain? Progress in Neurobiology 19, 117-139.

Martin W.R. and Sloan J.W. (1986) Relationship of CNS tryptaminergic processes and the action of LSD-like hallucinogens. Pharmacol. Biochem. Behav. 24, 393-399.

Ogren S.-O. (1982a) Central serotonin neurones and learning in the rat. In, Biology of Serotonergic Transmission (Osborne N.N., ed.) John Wiley and Sons Ltd., New York, pp. 317-334.

Ogren S.-O. (1982b) Forebrain serotonin and avoidance learning: Behavioural and biochemical studies on the acute effect of p-chloroamphetamine on one-way active avoidance learning in the male rat. Pharmacol. Biochem. Behav. 16, 881-895.

Ogren S.-O. and Johansson C. (1985) Separation of the associative and non-associative effects of brain serotonin released by p-chloroamphetamine: Dissociable serotoninergic involvement in avoidance learning, pain and motor function. Psychopharmacol. 86, 12-26.

Vogel W.H. and Cohen F. (1977) Effects of tryptamine on the conditioned avoidance response in rats. Brain Res. 122, 162-164.

THE EFFECTS OF p-TA and p-OA ON NORADRENERGIC MECHANISMS

OF THE RAT VAS DEFERENS

S.M. Celuch and A.V. Juorio

Neuropsychiatric Research Unit
Department of Psychiatry
University of Saskatchewan
Saskatoon, Saskatchewan, Canada S7N 0W0

INTRODUCTION

In the prostatic portion of the vas deferens the noradrenaline (NA) which is released by nerve stimulation modulates the twitch responses elicited by the motor neuro-transmitter. These modulatory effects are mediated by activation of inhibitory presynaptic α_2 adrenoceptors, inhibitory postsynaptic β_2 adrenoceptors and excitatory postsynaptic α_1 adrenoceptors (Ambache et al., 1972; von Euler and Hedqvist, 1975; MacDonald and McGrath, 1980). The aim of this study was to investigate the effects of two sympathomimetic amines, p-tyramine (p-TA) and p-octopamine (p-OA) on the twitch responses of the prostatic portion of the rat vas deferens and to compare their effects with those of NA. p-Tyramine is an indirectly acting amine which produces its effects by releasing NA (Burn and Rand, 1958); in addition it was suggested that in some peripheral tissues this amine activates the adrenoceptors directly (Krishnamurty and Grollman, 1972; Morcillo et al., 1984). p-Octopamine is an amine of mixed action with both direct and indirect effects (Trendelenburg et al., 1962; Celuch and Juorio, 1987).

METHODS

The prostatic portions of the vasa deferentia of male Wistar rats (280-320 g body weight) were placed in 5 ml

capacity organ baths containing Krebs solution at 37°C and
bubbled with a mixture of 95% O_2 plus 5% CO_2. Five periods
of electrical stimulation at 0.025 Hz (pulses of 0.5 ms and
supramaximal voltage) were performed in each experiment.
The twitch responses elicited by the electrical pulses were
measured immediately before (pre) and 20 min after (post)
the addition of amines to the organ bath. The ratio between the height of both twitches was defined as ratio
"post/pre". For normal tissues the height of the twitches
was measured 250 ms after the corresponding electrical
pulse. When the animals were pretreated with reserpine the
height of the twitches at 300 ms was considered. A single
concentration of one amine was added during each period of
stimulation in the presence of the MAO inhibitor pargyline
(10 µM). In some experiments the tissues were preincubated
with yohimbine or corynanthine (1.0 µM, 30 min). Statist-
ics were performed by analysis of variance followed by the
Tukey-Kramer test.

RESULTS

Field stimulation of the prostatic portion of the rat
vas deferens at 0.025 Hz induced a discrete contraction
(twitch response) with a maximum at 250 ms after each electrical pulse (Fig. 1). p-Tyramine caused a concentration-
dependent inhibition of the twitch responses and had excitatory effects on the late part of the contraction. The
lowest concentrations of the amine slightly increased or
did not change the height of the twitches (Fig. 1).
p-Octopamine also had excitatory-inhibitory effects. The
effects of NA were mainly excitatory except at the highest
concentration (15.8 µM).

From the corresponding concentration-effect curve
(CEC) the concentrations of amines which caused 30% inhibition of the twitch responses at 250 ms were estimated:
p-TA 1.0 µM (n=8); p-OA 3.4 µM (n=6); NA 15.4 µM (n=5).
Then, the order of potencies for the inhibitory effect was
p-TA>p-OA>NA. p-Tyramine induced more rhythmic activity
than the other two amines. High concentrations of NA
caused sustained contraction of the prostatic portion of
the vas deferens (0.43 ± 0.05 g for NA 15.8 µM; n=11) while
this effect was not observed with p-TA or p-OA.

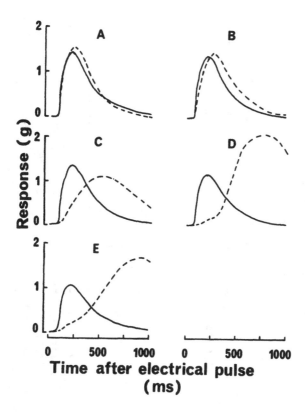

Figure 1. Concentration-effect curve (CEC) for p-TA in the prostatic portion of the rat vas deferens. Record of one experiment. —— twitch response to an electrical pulse previous to the addition of p-TA to the organ bath; ---- after 20 min of stimulation (0.025 H_z) in the presence of the amine. The following concentrations (µM) of p-TA were assayed: A = 0.158; B = 0.527; C = 1.58; D = 5.27; E = 15.8.

In the presence of the α_2-adrenoceptor antagonist yohimbine, the amines increased the height of the twitches along the entire CEC; the maximal excitatory effect of p-TA was greater than that of p-OA or NA (Table 1). The α_1-adrenoceptor antagonist corynanthine potentiated the inhibitory effects of the amines (Table 1). The order of potencies for the three amines in the presence of the antagonists was p-TA>p-OA≈NA.

TABLE 1

Effects of NA, p-TA and p-OA on the twitch responses of the prostatic portion of the rat vas deferens. Effects of specific α-adrenoceptor antagonists.

Experimental Condition	Ratio "post/pre"			
	NA	p-TA	p-OA	Control
No antagonists	0.69±0.10Δ (5)	0.17±0.07ΔΔ (4)	0.31±0.07ΔΔ (6)	0.98±0.02 (6)
Yohimbine 1.0 µM	1.46±0.20** (6)	2.03±0.22** (6)	1.61±0.10** (4)	-
Corynanthine 1.0 µM	0.32±0.06 (6)	0.02±0.01 (4)	0.26±0.04 (3)	-

Shown are means ± S.E.M. of the ratios "post/pre" (ratio between the height of the twitch response in the presence of the exogenous amine and the height of the twitch previous to the addition of the amine; for details see "Methods") for 15.8 µM of the amines. In the control group no amines were added to the organ bath. Number of observations is indicated in parentheses. Δ $p < 0.05$; ΔΔ $p < 0.01$ compared with the control group. ** $p < 0.01$ compared with the corresponding amine in the absence of α-antagonists.

To investigate whether the inhibitory-excitatory effects of p-TA and p-OA depend on direct or indirect actions of the amines, the CEC for both compounds were determined after pretreating the animals with reserpine. The effects of the pretreatment on the concentration of endogenous NA and on the height of the twitch responses are shown in Table 2.

Reserpine pretreatment shifted the CEC for p-TA and p-OA toward higher concentrations of the amines; in tissues with normal contents of NA, a concentration of 1.58 μM p-TA was required to achieve a ratio "post/pre" (for definition see "Methods") of 0.47 ± 0.05 (n=8), but in reserpine pre-treated tissues a 10-fold higher concentration of p-TA was required to produce a similar ratio (0.42 ± 0.12; n=6). Similarly, a higher concentration of p-OA was required in reserpine pretreated tissue (15.8 μM) than in normal tissues (5.27 μM) to produce equivalent ratios. In addition, p-TA and p-OA 1.58 μM increased the height of the twitches after reserpine pretreatment [ratios "post/pre" of 1.41 ± 0.14 (n=6) and 1.34 ± 0.18 (n=5) respectively].

TABLE 2

Twitch responses and concentration of NA in the
prostatic portion of the rat vas deferens
after pretreatment with reserpine

	No Pretreatment	Reserpine (5 mg/kg, i.p., 24h)
Twitch response (g)	1.06±0.04 (59)	0.69±0.08 (21)*
Endogenous NA (μg/g)	13.90±0.90 (5)	0.38±0.09 (6)*

The twitch responses were measured 250 ms and 300 ms after the electrical pulse (0.5 ms, supramaximal voltage) in non-pretreated tissues and in tissues pretreated with reserpine respectively. Noradrenaline was determined by HPLC at the end of the pharmacological experiments. * p <0.05.

DISCUSSION

In the prostatic portion of the rat vas deferens, p-TA and p-OA, as well as NA, have both inhibitory and excitatory effects on the twitch responses elicited by field stimulation. The facilitatory effect of the amines on the twitch responses measured at 250 ms depends on the activation of α_1 adrenoceptors, while the inhibitory action is mediated by α_2 adrenoceptors. The final net effect of each concentration of amine results from the combination of these two actions. It is widely accepted that the α_2 inhibitory adrenoceptors are presynaptic and that the excitatory α_1 adrenoceptors are localized in the smooth muscle (MacDonald and McGrath, 1980; French and Scott, 1983; Illes and Dorge, 1985).

Among the amines assayed the most potent was p-TA followed by p-OA and NA. These results together with the differences among the amines concerning the rhythmic activity and the sustained contraction could indicate differences between populations of receptors activated through exogenous NA and through the direct or indirect actions of p-TA and p-OA.

The effects of p-TA and p-OA after reserpine pretreatment could be due to a direct action of p-TA and p-OA, but an indirect action of the amines mediated by the release of the remaining NA in the tissues cannot be disregarded.

ACKNOWLEDGEMENTS

The authors thank Saskatchewan Health for financial support and the Saskatchewan Health Research Board for a scholarship (to S.M.C.).

REFERENCES

Ambache N., Dunk L.P., Verney J. and Zar M.A. (1972) Inhibition of post-ganglionic motor transmission in vas deferens by indirectly acting sympathomimetic drugs. J. Physiol. 227, 433-456.
Burn J.H. and Rand M.J. (1958) The action of sympathomimetic amines in animals pretreated with reserpine. J. Physiol. 144, 314-336.

Celuch S.M. and Juorio A.V. (1987) Effects of deuterium substitution on the chronotropic responses to some sympathomimetic amines in the isolated rat atria. Naunyn Schmied. Arch. Pharmacol. (in press).

Euler U.S. von and Hedqvist P. (1975) Evidence for an α- and β_2-receptor mediated inhibition of the twitch response in guinea pig vas deferens by noradrenaline. Acta Physiol. Scand. 93, 572-573.

French A.M. and Scott N.C. (1983) The use of neurally released agonist in the measurement of antagonism at α-adrenoceptors. Br. J. Pharmacol. 80, 655-661.

Illes P. and Dorge L. (1985) Mechanism of α_2-adrenergic inhibition of neuroeffector transmission in the mouse vas deferens. Naunyn Schmied. Arch. Pharmacol. 328, 241-247.

Krishnamurty U.S.R. and Grollman A. (1972) Contractile response of rat aorta to norepinephrine and tyramine. J. Pharm. Pharmacol. 182, 264-272.

MacDonald A. and McGrath J.C. (1980) The distribution of adrenoceptors and other drug receptors between the two ends of the rat vas deferens as revealed by selective agonists and antagonists. Br. J. Pharmacol. 71, 445-458.

Morcillo E., Perpina M. and Esplugue S.J. (1984) Responsiveness to tyramine in isolated lung parenchimal strip of guinea-pig and rat. Eur. J. Pharmacol. 97, 13-19.

Trendelenburg U., Muskus A., Fleming W.W. and Gomez Alonso de la Sierra B. (1962) Effect of cocaine, denervation and decentralization on the response of the nictitating membrane to various sympathomimetic amines. J. Pharmacol. exp. Ther. 138, 181-193.

THE USE OF MAO INHIBITORS IN LESION STUDIES:

AN ASSESSMENT OF 'FLOODING' IN TRACE AMINE MEASUREMENTS

T.V. Nguyen, A.V. Juorio and A.J. Greenshaw*

Neuropsychiatric Research Unit
CMR Building, University of Saskatchewan
Saskatoon, Saskatchewan, Canada S7N OWO
and
*Department of Psychiatry, University of Alberta,
Edmonton, Alberta, Canada T6G 2B7

INTRODUCTION

Recent studies have demonstrated that selective lesions of mammalian CNS may result in decreases in trace amine concentrations in selected areas of the brain (see review, Juorio, 1987). In the case of tyramines this has been demonstrated in the absence of MAO inhibitor (Juorio and Jones, 1981). With other trace amines, MAO inhibition has been used to elevate the normally low endogenous levels so that relative decreases in accumulation may be used as an index of lesion effects (PE and T studies: Greenshaw et al., 1986; Juorio et al., 1987). The reason for the latter strategy is to circumvent 'floor effects' that occur when attempting to measure decreases in quantities that are already close to the limits of sensitivity of the available assay systems. Concentrations of trace amines increase greatly after administration of MAO inhibitors (Philips and Boulton, 1979). While the rationale for the use of MAO inhibitors in lesion studies is clear, there is the problem of lipophilicity to be considered with respect to substrate localisation in CNS. The present study was conducted to assess the relative dependence of some reported lesion effects on the type of MAO inhibitor used in this context.

MATERIALS AND METHODS

Male Wistar rats were used. They were housed in group cages under a 12 hr light/dark cycle at 19-21°C. Lesions were made by an injection of 6-OHDA (8 µg in 4 µL) into the left substantia nigra pars compacta. The rats were anaesthetised with sodium pentobarbital (50 mg kg^{-1}) during the lesion procedures. The stereotaxic injection procedures were as described by Juorio et al. (1987).

Six weeks after the 6-OHDA lesion the rats were injected with either (-)deprenyl (2 mg kg^{-1}) or pargyline (200 mg kg^{-1}) and killed 2 hrs later. Neurochemical analyses were conducted using HPLC and mass-spectrometric procedures as described by Greenshaw et al. (1986) and Juorio et al. (1987).

All drug doses are expressed as the salt form, except 6-OHDA which is given as free base. Pargyline hydrochloride was given by intraperitoneal (i.p.) injection and (-)deprenyl hydrochloride by subcutaneous (s.c.) injection in a volume of 1 mL kg^{-1}.

RESULTS

The effects of unilateral 6-OHDA lesions of the substantia nigra on striatal concentrations of amines and metabolites are illustrated by the data in Tables 1 and 2.

In the deprenyl pretreatment groups (Table 1) at six weeks after administration of 6-OHDA, levels of PEA and of p- and m-tyramine were decreased on the lesion side relative to the intact striatum. At this time DA and DOPAC were also significantly decreased on the lesion side. Levels of tryptamine were lower on the lesion side relative to the intact side but this was not statistically significant. Levels of 5-HT and the amino acids p-tyrosine and L-tryptophan were unaffected by the lesions. After pargyline pretreatment (Table 2) PEA levels on the lesion side were not significantly reduced. There were also no changes in relative levels of 5-HT, 5-HIAA, DOPAC and HVA, or of the measured amino acids. Tryptamine, p- and m-tyramine and DA were, however, reduced on the lesion side.

TABLE 1

The Effects of Unilateral 6-OHDA Lesions of the Substantia
Nigra on Striatal Concentrations of Amines and Metabolites
in (-)Deprenyl-treated (2 mg kg^{-1}, 2 hr) Rats (n=7)

	Intact Striatum (ng g^{-1})	Lesioned Striatum (ng g^{-1})	% of Control
β-Phenylethylamine	10.70	5.30*	50
Tryptamine	6.40	2.70	42
p-Tyramine	14.60	2.70*	18
m-Tyramine	6.90	1.70*	25
Dopamine	9090	1500*	17
DOPAC	840	260*	31
HVA	560	<10*	2
5-HT	520	500	96
5-HIAA	630	550	87
p-Tyrosine	11300	14220	126
L-Tryptophan	5100	6060	119

Values are means; statistical comparisons were made by
Student's t-test; * p <0.05.

TABLE 2

The Effects of Unilateral 6-OHDA Lesions of the Substantia
Nigra on Striatal Concentrations of Amines and Metabolites
in Pargyline-treated (200 mg kg^{-1}, 2 hr) Rats (n=7)

	Intact Striatum (ng g^{-1})	Lesioned Striatum (ng g^{-1})	% of Control
β-phenylethylamine	91	84	92
Tryptamine	120	68*	57
p-Tyramine	80	47*	59
m-Tyramine	15.9	7.7*	48
Dopamine	12560	3320*	26
DOPAC	160	190	119
HVA	40	70	175
5-HT	980	870	89
5-HIAA	230	200	87
p-Tyrosine	10430	9850	94
L-Tryptophan	5870	5550	95

Values are means; statistical comparisons were made by
Student's t-test; * p <0.05.

DISCUSSION

We have previously reported decreases in PEA, p- and m-tyramine and in tryptamine levels in the striatum seven days following 6-OHDA lesions of the substantia nigra (Juorio and Jones, 1981; Greenshaw et al., 1986; Juorio et al., 1987). These effects were accompanied by the well-documented decrease in DA at one week post-6-OHDA administration: an effect that was also evident in the present study at six weeks post-lesion (Tables 1 and 2). In the present experiments PEA depletion was only evident after deprenyl pretreatment, indicating probable 'flooding' of this amine after pretreatment with high doses of pargyline. Tryptamine depletion was not significant after deprenyl pretreatment whereas this effect was unmasked by pretreatment with pargyline. This type of differential effect with respect to deprenyl or pargyline pretreatment was not observed with p- and m-tyramine.

The present study demonstrates that, for PEA and tryptamine, lesion effects are observed under selective conditions and it is necessary to choose appropriate doses and type of MAO inhibitors for studies of this kind.

ACKNOWLEDGEMENTS

We thank Professor A.A. Boulton for continuing support and constructive comments. This research is supported by Saskatchewan Health and by the MRC of Canada. T.V. Nguyen is a Medical Research Council Fellow; A.J. Greenshaw is an Alberta Heritage Foundation Medical Scholar.

REFERENCES

Greenshaw A.J., Juorio A.V. and Nguyen T.-V. (1986) Depletion of striatal β-phenylethylamine following dopamine but not 5-HT denervation. Brain Res. Bull. 17, 477-484.

Juorio A.V. (1987) Lesion of selected brain areas as a tool for the demonstration of some trace biogenic amines neural pathways. Gen. Pharmac. 18, 1-5.

Juorio A.V., Greenshaw A.J. and Nguyen T.V. (1987) Effect of intranigral administration of 6-hydroxydopamine and 5,7-dihydroxytryptamine on rat brain tryptamine. J. Neurochem. 48, 1346-1350.

Juorio A.V. and Jones R.S.G. (1981) The effect of mesencephalic lesions on tyramine and dopamine in the caudate
nucleus of the rat. J. Neurochem. 36, 1898-1903.
Philips S.R. and Boulton A.A. (1979) The effect of monoamine oxidase inhibitors on some arylalkylamines in rat
striatum. J. Neurochem. 33, 159-167.

CATECHOLAMINE UPTAKE INTO CULTURED MOUSE ASTROCYTES

John X. Wilson[1] and Wolfgang Walz[2]

[1]Department of Physiology, Health Sciences Centre
The University of Western Ontario
London, Ontario, Canada N6A 5C1
and
[2]Department of Physiology, University of Saskatchewan,
Saskatoon, Saskatchewan, Canada S7N 0W0

Earlier studies have suggested that glial cells may influence neuronal signalling in the brain. In particular, uptake and metabolism by glial cells may terminate the actions of the catecholaminergic transmitters released from neurons (Henn and Hamberger, 1971; Pelton et al., 1981; Kimelberg and Pelton, 1983). Thus, regulation of glial transport and metabolic activities may provide a mechanism for modulating neurotransmission. Ascorbic acid is stored within catecholaminergic neurons and is secreted with neurotransmitters (O'Neill et al., 1984; Kratzing et al., 1985). One function for ascorbic acid is the retardation of oxidative processes that degrade and inactivate catecholamines; in other words, ascorbic acid may serve as a chemical preservative for catecholamines within neurons and in extracellular fluid. However, because the termination of the neurophysiological actions of catecholamines also is effected by cellular uptake, it is of interest to know if ascorbic acid alters the catecholamine uptake process. Therefore the aims of the present study were to characterize the uptake of norepinephrine in mouse cerebral glial cells and to determine if ascorbic acid could affect uptake rates.

Glial cells were cultured from the cerebral hemispheres of neonatal mice, according to the procedure of Hertz et al. (1982). They were grown in Petri dishes containing serum-supplemented modified Eagle's minimum essential medium. These cells stained positively for

glial fibrillary acidic protein by the procedure of Wilson
et al. (1986). They showed the stellation response to
dibutyryl cyclic AMP that is characteristic of astrocytes.
The absence of neurons was confirmed by silver staining
and microscopical examination of representative cultures.

The cells were used for experiemtns after 2 to 3
weeks in culture. The growth medium was discarded and
replaced with serum-free Hepes-buffered incubation medium
(pH 7.4, 37 C). The kinetic characteristics of catecholamine uptake were studied during incubation with levo-
[7-3H(N)]-norepinephrine (NEN Canada; 0.5 μC/ml incubation
medium) in the presence or absence of levo-ascorbic acid
(J.T. Baker Chemical Company and Sigma). At the end of the
incubation period the cultures were rinsed seven times with
ice-cold buffer and the cells were scraped into 1 ml of
water. Cell protein was analyzed by the method of Lowry
et al. (1951). Aliquots of the harvested cells and of the
incubation medium were combined with scintillation cocktail
(Biofluor, NEN Canada) and their tritium contents measured
by liquid scintillation counting. Cellular uptake rates
for tritiated norepinephrine were calculated based on the
specific activity of the amine in the medium and were
expressed as pmol/mg protein/10 min incubation period. A
zero time value, obtained by adding tritiated norepinephrine to the cultures and immediately washing them, was
subtracted from all of the uptake data.

Tritiated norepinephrine was taken up from the
incubation medium and radioactivity accumulated in the cell
cultures (Figure 1). Uptake depended on the extracellular
concentration of total (i.e. tritiated and nonradiolabeled)
norepinephrine. Kinetic analysis of the uptake mechanism
indicated that both saturable and nonsaturable processes
were involved. For the saturable component of uptake the
half-maximal effective concentration of norepinephrine
(Km) was on the order of 0.1 $\underline{\mu}$M and the maximum uptake rate
(Vmax) was approximately 1 pmol/mg protein/10 min
(Figure 1).

Ascorbic acid competitively inhibited the saturable
component of tritiated norepinephrine uptake (Figure 1).
The half-maximal inhibitory concentration (IC50) for
ascorbic acid was approximately 0.2 $\underline{\mu}$M (Figure 2). Maximal
inhibition of saturable norepinephrine uptake was achieved

with about 2 μM ascorbic acid. Ascorbic acid at concentrations of 2 μM or less did not alter the non-saturable component of glial norepinephrine uptake.

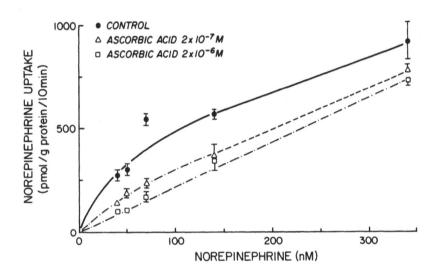

Figure 1. Norepinephrine uptake by mouse cerebral glial cells in the presence and absence of ascorbic acid. Both norepinephrine and ascorbic acid were added to the cells' incubation medium at the beginning of the 10 min incubation period (37 C, pH 7.4). Plotted are the mean ± standard error for triplicate incubations from a representative experiment.

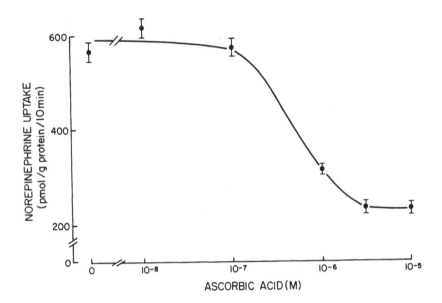

Figure 2.   Effect of ascorbic acid on norepinephrine uptake
by mouse cerebral glial cells.   Both norepinephrine (0.14
uM) and ascorbic acid (0 - 10 μM) were added to the cells'
incubation medium at the beginning of the 10 min incubation
period (37 C, pH 7.4).   Plotted are the mean ± standard
error for triplicate incubations from a representative
experiment.

The results of the present experiments indicate that
mouse cerebral astrocytes can take up norepinephrine from
the extracellular fluid.   The kinetics of saturable uptake
by these murine cells, in the absence of ascorbic acid,
resemble those reported for rat cerebral astrocytes
(Kimelberg and Pelton, 1983).   A novel finding of the

present experiments is that ascorbic acid competitively inhibits the accumulation of tritiated amine in the mouse glial cells. This observation suggests that ascorbic acid may regulate the glial clearance of norepinephrine from the brain's extracellular fluid. Further study is required to learn how this mechanism is involved in regulation of intercellular communication by noradrenergic systems in the central nervous system.

Supported by the Medical Research Council of Canada. The technical assistance of Lea Babcock and Ewa Jaworska is gratefully acknowledged.

Henn F.A. and Hamberger A. (1971) Glial cell function: uptake of transmitter substances. Proc. Natl. Acad. Sci. USA 68, 2686-2690.
Hertz L., Juurlink B.H.J., Fosmark H., and Schousboue A. (1982) Methodological appendix: astrocytes in primary cultures. In, Neuroscience Approached Through Cell Culture, Vol 1 (Pfeiffer S.E., ed), CRC Press, pp. 175-186.
Kimelberg H.K. and Pelton E.W. (1983) High-affinity uptake of (3H)norepinephrine by primary astrocyte cultures and its inhibition by tricyclic antidepressants. J. Neurochem. 40, 1265-1270.
Kratzing C.C., Kelly J.D., and Kratzing J.E. (1985) Ascorbic acid in fetal rat brain. J. Neurochem. 44, 1623-1624
Lowry O.H., Rosebrough N.J., Farr A.L., and Randall R.J. (1951) Protein measurement with the folin phenol reagent. J. Biol. Chem. 193, 265-275.
O'Neill R.D., Fillenz M., Sundstrom L., and Rawlins J.N.P. (1984) Voltametrically monitored brain ascorbate as an index of excitatory amino acid release in the unrestrained rat. Neurosci. Lett. 52, 227-233.
Pelton E.W., Kimelberg H.K., Shiperd S.V., and Bourke R.S. (1981) Dopamine and norepinephrine uptake and metabolism by astroglial cells in culture. Life Sci. 28, 1655-1663.
Wilson G.A.R., Beushausen S., and Dales S. (1986) In vivo and in vitro models of demylinating diseases: XV. Differentiation influences on the regulation of coronavirus infection in primary explants of mouse CNS. Virology 151, 253-264.

Clinical and Metabolic

LOCALIZATION AND INHIBITION OF MONOAMINE OXIDASE ACTIVITIES

K.F. Tipton, J.P. Sullivan, A.-M. O'Carroll, °A. Stenström, °°E. Sundström, °°°L. Oreland, °°°°I. Fagervall, °°°°S.B. Ross and °°°°C.J. Fowler. Department of Biochemistry, Trinity College, Dublin 2, Ireland; °Department of Pharmacology, University of Umeå, S-901 87 Umeå, Sweden; °°Department of Histology, Karolinska Institutet, S-104 01 Stockholm, Sweden; °°°Department of Pharmacology, University of Uppsala, Sweden; and °°°° Research and Development Laboratories, Astra Alab AB, S-151 85 Södertälje, Sweden.

Use of the selective irreversible monoamine oxidase (MAO) inhibitors clorgyline and (-)-deprenyl (selegiline) have allowed the distribution and specificities of the two forms of the enzyme, MAO-A and MAO-B, to be determined in a number of tissues and species (for reviews, see Fowler and Tipton, 1984; Tipton, 1986). The distributions and specificities of MAO-A and -B in a number of human tissue homogenates are summarised in Table 1.

The involvement of MAO-A in the oxidation of 5-hydroxytryptamine (5-HT), adrenaline and noradrenaline (NA) in the brain may provide an explanation for the antidepressant efficacy of MAO-A selective inhibitors such as clorgyline (see e.g. Lipper et al., 1979). In contrast, inhibitors of MAO-B, which would be expected to have little effect on 5-HT oxidation, have been reported to be poor antidepressants, although Pare (1984) has suggested that there may be a sub-class of depressives who do not respond to such inhibitors.

One of the most important problems associated with the use of MAO inhibitors as antidepressants has been the hypertensive crisis that can result after the subsequent ingestion of tyramine-containing foods (the "cheese-reaction", see Blackwell and Marley, 1966). The involvement of both forms of MAO in the oxidation of tyramine led to hopes that selective inhibitors might give rise to a less severe cheese-reaction. However, in both man and rat, the deamination of tyramine by the intestine (the most important tissue with respect to the cheese-reaction) is brought about mainly by MAO-A (Strolin-Benedetti et al., 1983; Hasan et al., 1987). The possible use of reversible MAO-A inhibitors has been considered by a number of workers (for reviews, see Tipton et al., 1984; Fowler and Ross, 1984) on the grounds that high concentrations of tyramine should be able to displace such inhibitors, and thus be metabolised by intestinal MAO, if they were to interact competitively with the enzyme.

Table 1: Contribution of MAO-A to substrate oxidation in human tissue homogenates.

Substrate	Percentage MAO-A activity in:			
	Brain	Liver	Kidney Cortex	Kidney Medulla
(±)-Adrenaline	58	-	-	-
Dopamine	49	-	-	-
5-Hydroxytryptamine	92	97	94	91
(-)-Noradrenaline	55	5	-	-
2-Phenylethylamine	7	5	7	7
Tryptamine	46	76	58	60
Tyramine	55	73	54	55

The percentages were calculated from V_{max} values. (Data of O'Carroll et al. 1983, 1986; Sullivan et al., 1986).

The involvement of both forms of MAO in the oxidation of tryptamine (see Table 1) would suggest that, contrary to earlier suggestions (Bieck et al., 1984), the determination of urinary tryptamine excretion may not provide a simple measure of MAO-A function. However, the possibility that the metabolism of this amine is compartmented in such a way that MAO-A plays a larger role than is suggested by these in vitro studies cannot be excluded.

The compartmentalisation of MAO is of particular relevance with respect to the activities of intra- and extra- neuronal MAO in the brain. In hypotonic brain homogenates, an inaccurate estimate of the contributions of the two enzyme forms may be given if the ratios of MAO-A : -B activities are different inside and outside the nerve terminals. The presence of transport systems for specific amines would be expected to concentrate them in the nerve endings, thus leading to a greater contribution from intra-neuronal MAO than would be expected from the observation that the majority of brain MAO is located extra-symaptosomally (Van der Krogt et al., 1980; Oreland et al., 1983).

Several different methods, such as subcellular fractionation, cell culture, and denervation, have been employed in order to determine the intra - and extraneuronal components of MAO activity (for review, see Fowler et al., 1984). The most versatile method (in that it can be used in post-mortem human brain samples), however, is that first reported by Ross and Renyi (1969) who measured the concentrations of 5-HT and its deaminated metabolite 5-hydroxy-indoleacetic acid (5-HIAA) in brain slices in the absence and presence of the monoamine reuptake inhibitor cocaine. A significant proportion of the deamination was found to be cocaine-sensitive, and this was taken to be the intra-neuronal deamination (Ross and Renyi, 1969). More recent studies have extended this approach to dopaminergic synaptosomes (Demarest et al., 1980; Stenström et al., 1985; Arai et al., 1985), and further refined the method by using selective uptake inhibitors (Ask et al., 1983; Fagervall and Ross, 1986). Validation of this methodology has been provided by lesion studies: thus hemitransection of rats reduces the degree of striatal intrasynaptosomal MAO activity towards dopamine to about 10% of control (and reduces synaptosomal dopamine uptake to the same level), whereas extrasynaptosomal MAO is not reduced (Stenström et al., 1985). Similarly, selective lesions of serotoninergic and noradren= ergic neurones by treatment with the selective neurotoxins PCA (p-chloroamphetamine) and DSP4 (N-(2-chloroethyl)-N-ethyl-2-bromobenzylamine), respectively, reduces the appropriate intrasynaptosomal MAO activity without effect on the extra-synaptosomal activity (Fagervall and Ross, 1986; see Fig. 1).

A further demonstration of the validity of this methodology has been provided by use of the antimitotic agent methylazoxymethanol acetate (MAM), which when administered to pregnant rats on day 15 of gestation at a dose of 25 mg/kg 1.v. produces offspring (here termed 'MAM rats') with forebrain microencephaly accompanied by a relative hyperinnervation of monoamine and acetylcholine terminals (Johnston and Coyle, 1979; Jonsson and Hallman, 1982). The intra- NA and 5-HT- synaptosomal MAO activities were found to be increased in the adult MAM rats to roughly the same extent as the NA and 5-HT concentrations, whereas the extrasynaptosomal activities were not changed (Table 2).

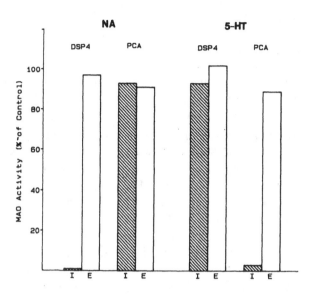

Fig. 1. Deamination of NA and 5-HT within (I) and outside (E) their respective synaptosomes (rat hippocampus) following selective lesion by DSP4 and PCA, respectively. Values are calculated from the data of Fagervall and Ross (1986).

Uptake inhibitors have been used to determine the intra- and extrasynaptosomal components of dopamine deamination in the human post-mortem brain. An initial study, where cocaine was used as uptake inhibitor, demonstrated that whilst MAO-B was responsible for much of the extraneuronal deamination of dopamine, this substrate was deaminated predominantly by MAO-A intrasynaptosomally (Oreland et al., 1983). This finding has subsequently been confirmed in two separate studies, one also using cocaine as uptake inhibitor but using amiflamine instead of clorgyline to define the MAO-A component (Stenström et al., submitted), and one using selective monoamine reuptake inhibitors instead of cocaine (O'Carroll et al., 1987). Since nerve terminal amine uptake proccesses are not completely specific, the intraneuronal metabolism of transmitter amine can occur, after its uptake, both in its own nerve terminals and within those using different transmitters. Sharman (1976) has termed these processes intrahomoneuronal and intraheteroneuronal metabolism, respectively. The sum of these two processes will be obtained if the non-selective uptake inhibitor cocaine is used, whereas only the former is measured when selective uptake inhibitors are used (O'Carroll et al., 1987). The distribution of MAO-A and -B activities found within and outside dopaminergic synaptosomes is rather different from that found in the rat brain (where MAO-A predominates both intra- and extrasynaptosomally) (Oreland et al., 1983) but is rather similar to that found in the pig brain (Stenström et al., submitted) and the guinea-pig brain (Ross, 1987). In the case of noradrenaline, O'Carroll et al. (1987) have shown MAO-B to constitute some 60% of the total activity in noradrenergic terminals from human hypothalamus.

These results are of relevance to the actions of the dopaminergic neurotoxin 1-methyl-4-phenyl-1,2,3,6-tetrahydropyridine (MPTP), which has been shown to be converted to the active neurotoxin, the 1-methyl-4-phenylpyridinium ion (MPP$^+$), by a process involving the action of MAO-B (Heikkila et al., 1984). The absence of significant amounts of that form of the enzyme in dopaminergic nerve terminals would indicate this process to occur extraneuronally, consistent with the observation that MPP$^+$ is actively transported into the dopaminergic nerve terminal (Javitch et al., 1985) and that selective inhibitors of dopamine uptake can protect against the neurotoxic effects of MPTP (Sundström and Jonsson, 1985).

Table 2. Frontal cortical NA, 5-HT and 5-HIAA concentrations and cerebral cortical intra- and extrasynaptosomal MAO activities in adult control and MAM rats.

	Control	MAM
[NA] (ng/g)	547±24	1162±45 (212%)**
MAO vs NA (pmol/g/min):		
Total	225±13	251± 5 (112%)[NS]
Intrasynaptosomal	60± 6	94± 9 (155%)*
Extrasynaptosomal	165±10	158± 9 (96%)[NS]
[5-HT] (ng/g)	586±73	893±52 (153%)*
[5-HIAA] (ng/g)	399±56	541±22 (136%)*
MAO vs 5-HT (pmol/g/min):		
Total	1057± 57	1212±78 (115%)[NS]
Intrasynaptosomal	378±28	612±40 (162%)**
Extrasynaptosomal	679±40	601±46 (89%)[NS]

Data are means ± s.e.m., n=6-8. The NA and 5-HT concentrations for the MAO experiments were 0.25 and 0.1 µM, respectively (methodology similar to that of Fagervall and Ross, 1986, used). The NA, 5-HT and 5-HIAA concentrations in the tissues were determined essentially as described by Hallman et al. (1984). Values in brackets are the mean values for the MAM rats as % of the mean values for the control rats. NS, $p>0.05$;* $p<0.05$** $p<0.01$, two-tailed t-test. (E. Sundström, A. Stetsröm and A.K. Mohammed, unpublished results).

The toxic effects of MPP[+] are believed to arise from the inhibition of mitochondrial oxidative phosphorylation (Nicklas et al., 1985; Ramsay et al., 1986) which is not tissue-specific. The presence of MAO-B in human noradrenergic synaptosomes might thus suggest that MPP[+] could be formed in situ. Similar conclusions might be

drawn for serotoninergic nerve terminals, where
the presence of appreciable quantities of MAO-B
have been demonstrated immunologically (Thorpe
et al., 1987). However, MPTP is a concentration-
dependent inhibitor of both forms of MAO (Tipton
et al., 1986), and studies on the inhibition by
MPTP of the MAO-A activity in human synaptosomes
(which had been treated with (-)-deprenyl to inhibit MAO-B) showed that the degree of inhibition
increased markedly when the synaptosomes were
disrupted (J.P. Sullivan and K.F. Tipton, unpublished results). These results indicate that
MPTP penetrates synaptosomes rather poorly, which
may be a contributory factor to the relative
insensitivity of noradrenergic and serotoninergic
synaptosomes (although in the former case there
may be a species- and strain-dependency) to the
toxic effects of this compound.

REFERENCES.
Arai Y., Stenström, A. and Oreland L, (1985). The
    effects of age on intra- and extraneuronal
    monoamine oxidase -A and -B activities in the
    rat brain. Biogenic Amines 2, 65-71.
Ask A.-L., Fagervall I. and Ross S.B. (1983).
    Selective inhibition of monoamine oxidase in
    monoaminergic neurons in the rat brain.
    Naunyn-Schmiedeberg's Arch. Pharmacol. 324, 79-
    87.
Bieck P.R., Nilsson E., Schick C., Waldmeier P.C.
    and Lauber J. (1984). Urinary excretion of
    tryptamine in comparison to normetanephrine and
    beta-phenylethylamine in human volunteers after
    subchronic treatment with different monoamine
    oxidase inhibitors, in Neurobiology of the
    Trace Amines (Boulton A.A., Baker G.B., Dewhur-
    st W.G. and Sandler M. eds), pp 525-539.
    Humana Press, Clifton.
Blackwell B. and Marley E. (1966). Interactions
    of cheese and its constituents with monoamine
    oxidase inhibitors. Br. J. Pharmacol. Chemo-

ther. 26, 121-141.
Demarest K.T., Smith D.J. and Azzaro A.J. (1980).
The presence of the type A form of monoamine
oxidase within nigrostriatal dopamine-containing neurones. J. Pharmacol. exp. Ther. 215,
461-468.
Fagervall I. and Ross S.B. (1986). A and B forms
of monoamine oxidase within the monoaminergic
neurones of the rat brain. J. Neurochem. 47,
569-576.
Fowler C.J. and Ross S.B. (1984). Selective
inhibitors of monoamine oxidase A and B:
biochemical, pharmacological and clinical
properties. Med. Res. rev. 4, 323-358.
Fowler C.J. and Tipton K.F. (1984). On the
substrate specificities of the two forms of
monoamine oxidase. J. Pharm. Pharmacol. 36,
111-115.
Fowler C.J., Magnusson O. and Ross S.B. (1984).
Intra- and extraneuronal monoamine oxidase.
Blood Vess. 21, 126-131.
Hallman H., Sundström E. and Jonsson G. (1984).
Effects of the noradrenaline neurotoxin DSP4
on monoamine neurons and their transmitter
turnover in rat CNS. J. Neural Transm. 60,
89-102.
Hasan F., McCrodden J.M. and Tipton K.F. (1987).
The involvement of intestinal monoamine oxidase in the transport and metabolism of tyramine
J. Neural Transm., in press.
Heikkila R.E., Manzino L., Cabbat F.S. and Duvoisin R.C. (1984). Protection against the dopaminergic neurotoxicity of 1-methyl-4-phenyl-1,2,
3,6-tetrahydropyridine by monoamine oxidase
inhibitors. Nature (Lond). 311, 467-469.
Javitch J.A., D'Amato R.J., Strittmatter S.M. and
Snyder S.H. (1985). Parkinsonism-inducing
neurotoxin, N-methyl-4-phenyl-1,2,3,6-tetrahy-
dropyridine: uptake of the metabolite N-methyl-
4-phenylpyridine by dopamine neurons explains
selective toxicity. Proc. Natl. Acad. Sci. USA

82 2173-2177.

Johnstone M.V. and Coyle J.T. (1979). Histolog-ical and neurochemical effects of fetal treatment with methyl-azoxymethanol on rat neocortex in adulthood. Brain Res. 170, 135-155.

Jonsson G. and Hallman H. (1982). Effects of prenatal methylazoxymethanol treatment on the development of central monoamine neurons. Dev. Brain Res. 2, 513-520.

Lipper S., Murphy D.L., Slater S. and Buchsbaum M.S. (1979). Comparative behavioural effects on clorgyline and pargyline in man. A preliminary evaluation. Psycho-pharmacology 62, 123-128.

Nicklas W.J., Vyas I. and Heikkila R.E. (1985). Inhibition of NADH-linked oxidation in brain mitochondria by 1-methyl-4-phenyl-1,2,3,6-tetrahydropyridine. Life Sci. 36, 2503-2508.

O'Carroll A.-M., Fowler C.J., Phillips J.P., Tobbia I. and Tipton K.F. (1983). The deamination of dopamine by human brain monoamine oxidase. Specificity for the two enzyme forms in seven brain regions. Naunyn-Schmiedeberg's Arch.Pharmacol. 322, 198-202.

O'Carroll A.-M., Bardsley M.E. and Tipton K.F. (1986). The oxidation of adrenaline and noradrenaline by the two forms of monoamine oxidase from human and rat brain. Neurochem. Int. 8, 493-500.

O'Carroll A.-M., Tipton K.F., Sullivan J.P., Fowler C.J. and Ross S.B. (1987). Intra- and extrasynaptosomal deamination of dopamine and noradrenaline by the two forms of human brain monoamine oxidase. Implications for the neurotoxicity of N-methyl-1,2,3,6-tetrahydropyridine in man. Biogenic Amines 4, 47-60.

Oreland L., Arai Y., Stenström A. and Fowler C.J. (1983). Monoamine oxidase activity and localisation in brain and the activity in relation to psychiatric disorders. Mod. Probl. Pharmacopsychiat. 19, 246-254.

Pare C.M.B. (1984). Clinical studies with monoamine oxidase inhibitors and tricyclic antidepressants in Monoamine Oxidase and Disease
(Tipton K.F., Dostert P. and Strolin Benedetti
M, eds), pp. 469-478. Academic Press, London

Ramsay R.R., Dadgar J., Trevor A.J. and Singer
T.P. (1986). Energy-driven uptake of N-methyl-
4-phenylpyridine in brain mitochondria mediates
the neurotoxicity of MPTP. Life Sci. 39, 581-
588.

Ross S.B. (1987). Distribution of the two forms of
monoamine oxidase within monoaminergic neurons
of guinea pig brain. J. Neurochem. 48, 609-
614.

Ross S.B. and Renyi A.L. (1969). Inhibition of
the uptake of tritiated 5-hydroxytryptamine in
brain tissue. Eur J. Pharmacol. 7, 270-277.

Sharman D.F. (1976). Can the intra- and extra-
homoneuronal metabolism of catecholamines be
distinguished in the mammalian central nervous
system? Ciba Found. Symp. 39, 203-216.

Stenström A., Arai Y. and Oreland L. (1985). Intra
- and extraneuronal monoamine oxidase -A and
-B activities after central axotomy (hemitransection) on rats. J. Neural Transm. 61, 105-
113.

Stenström A., Hardy J. and Oreland L. (1987).
Intra- and extra-dopamine-synaptosomal localization of monoamine oxidase in striatal homogenates from four species. Submitted.

Strolin Benedetti M., Boucher T., Carlsson A. and
Fowler C.J. (1983). Intestinal metabolism of
tyramine by both forms of monoamine oxidase in
the rat. Biochem. Pharmacol. 26, 2337-2342.

Sullivan J.P., McDonnell L., Hardiman O.M., Farr-
ell M.A. and Tipton K.F. (1986). The oxidation of tryptamine by the two forms of monoamine
oxidase in human tissues. Biochem. Pharmacol.
35, 3255-3260.

Sundstrom E. and Jonsson G. (1985). Pharmacolog-

ical interference with the neurotoxic actions
of 1-methyl-4-phenyl-1,2,3,6-tetrahydropyridine
(MPTP) on catecholaminergic neurons in the
mouse. Eur. J. Pharmacol. 110, 293-299.

Thorpe L.W., Westlund K.N., Kochersperger L.M.,
Abell C.W. and Denney R.M. (1987). Immunocytochemical localisation of monoamine oxidase A
and B in human peripheral tissues and brain.
J. Histochem. Cytochem. 35, 23-32.

Tipton K.F. (1986). Enzymology of monoamine oxidase. Cell Biochem. Funct. 4, 79-87.

Tipton K.F., Dostert P. and Strolin Benedetti M.
(eds) (1984). Monoamine Oxidase and Disease.
Academic Press, London.

Tipton K.F., McCrodden J.M. and Youdim M.B.H.
(1986). Oxidation and enzyme-activated irreversible inhibition of rat liver monoamine
oxidase-B by 1-methyl-4-phenyl-1,2,3,6-tetrahy-
dropyridine (MPTP). Biochem. J. 240, 379-383.

Van der Krogt J.A., Koot-Gronsveld E. and Van den
Berg C. (1983). Localisation of rat striatal
monoamine oxidase activities towards dopamine,
serotonin and kynuramine by gradient centrifugation and nigrostriatal lesions. Life Sci. 33,
615-623.

PRODRUGS OF β-PHENYLETHYLAMINE AND TRYPTAMINE: STUDIES IN THE RAT

Glen B. Baker, Ronald T. Coutts, Tadimeti S. Rao,
T. W. Eric Hall and Ronald G. Micetich

PMHAC Research Unit, Department of Psychiatry and Faculty
of Pharmacy and Pharmaceutical Sciences, University of
Alberta, Edmonton, Alberta, Canada   T6G 2B7

INTRODUCTION

A functional deficiency of the trace amines β-phenyl-ethylamine (PE) and tryptamine (T) in depressive disorders has been suggested by a number of researchers (Dewhurst, 1968; Boulton and Milward, 1971; Fischer et al., 1972; Sabelli and Mosnaim, 1974), and one of the most pronounced effects seen in the brains of laboratory animals after administration of most monoamine oxidase (MAO)-inhibiting antidepressants is a marked increase in concentrations of PE and T (Boulton, 1976; Philips and Boulton, 1979; McKim et al., 1980). It has also been demonstrated that administration of either of these amines to rodents produces marked behavioural effects (Dewhurst, 1968; Dourish, 1982, 1984; Greenshaw, 1984).

With the above information about PE and T in mind, it was considered of interest to study the behavioural effects of selective elevation of PE or T. Although administration of MAO inhibitors produces an elevation of levels of these amines, such drugs generally cause an increase in levels of both PE and T as well as of other amines such as noradrenaline (NA), dopamine (DA) and 5-hydroxytryptamine (5-HT). Administration of PE or T themselves, as has been done in many behavioural investigations, suffers from the disadvantage that these two amines are rapidly cleared from the brain after attaining very high initial levels. In order to overcome these problems and to develop models in which the effects of selective elevations of these trace amines could be investigated, we have chosen to use a prodrug approach. A number of N-alkylated analogues of PE and T have been synthesized in the hope that they might be metabolized in the body to give elevated, relatively sustained concentrations of PE or T in brain. The drugs synthesized and tested, in animal models, to date include N-2-cyanoethyl-PE (CEPE), N-3-chloropropylPE (CPPE), N-propargylPE (PGPE), N,N-dipropargylPE (DPGPE) and N-2-cyanoethylT

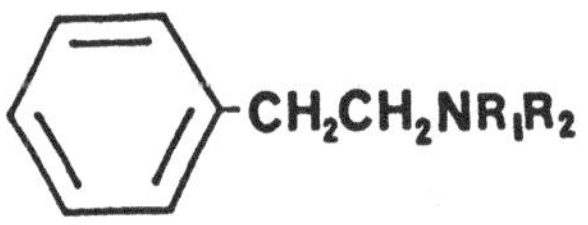

$R_1 = R_2 = H$	PE
$R_1 = H,\ R_2 = CH_2CH_2CN$	CEPE
$R_1 = H,\ R_2 = CH_2CH_2CH_2Cl$	CPPE
$R_1 = H,\ R_2 = CH_2C \equiv CH$	PGPE
$R_1 = R_2 = CH_2C \equiv CH$	DPGPE

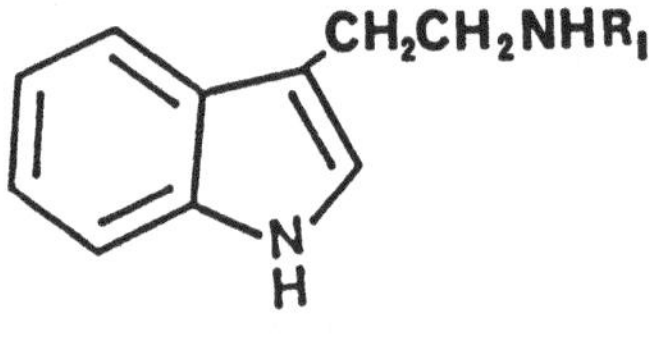

$R = H$	T
$R_1 = CH_2CH_2CN$	CET

Fig. 1. Structures of trace amines and their prodrugs.

(CET) (Coutts et al., 1985; Baker et al., 1987a,b; Rao et al., 1987a,b). The structures are shown in Figure 1.

The N-cyanoethyl and N-3-chloropropyl analogues were chosen because it had previously been demonstrated, using corresponding analogues of amphetamine and tranylcypromine, that these groups could be removed metabolically from phenylalkylamines in humans and/or rats (Beckett et al., 1972; Coutts et al., 1986; Nazarali et al., 1983, 1987a). Similarly, studies with pargyline and deprenyl had indicated that in these drugs the N-propargyl group could be readily removed metabolically (Durden et al., 1975; Pirisino et al., 1978; Reynolds et al., 1978; Coutts et al., 1981; Philips, 1981; Weli and Lindeke, 1985, 1986). The N,N-dipropargyl derivative of PE was formed as a side product in the synthesis of PGPE and was also considered of interest as a potential prodrug.

METHODS

Injection of prodrugs. For studies on the pharmaco-kinetics of the prodrugs and on their effects on MAO activity and concentrations of catecholamines and 5-HT and their metabolites, groups of male Sprague-Dawley rats (175-230 g) were injected intraperitoneally with the appropriate drug (0.1 mmol/kg), and groups were killed at 5, 15, 30, 60, 90, 120, 180 or 240 min after injection. Tissue and blood samples were collected immediately and frozen in isopentane over solid carbon dioxide.

Neurochemical analyses. Tissue levels of the prodrugs and of PE and T were analyzed by the procedure of Baker et al. (1986) or Nazarali et al. (1987b), which utilize extraction and derivatization using pentafluorobenzene-sulfonyl chloride or pentafluorobenzoyl chloride, respectively, followed by separation and quantitation using electron-capture gas chromatography.

Brain concentrations of NA, DA, 5-HT, homovanillic acid (HVA), 3,4-dihydroxyphenylacetic acid (DOPAC) and 5-hydroxyindole-3-acetic acid (5-HIAA) were measured utilizing high-pressure liquid chromatography with electrochemical detection. Samples were homogenized in ice-cold 0.1 M perchloric acid containing 0.05 mmol ascorbic acid and 3,4-dihydroxybenzylamine (50 ng/ml) as internal standard. After centrifuging to remove the protein precipitate, portions of the samples were placed in a Waters Intelligent Sample Processor (WISP) and 20 µl aliquot was used for analysis. The mobile phase was composed of 10 mmol Na_2HPO_4, 0.5 mmol disodium EDTA, 5 mmol SOS, 10% acetonitrile, and 5% methanol; the solution was adjusted to a pH value of 2.5 and was pumped at a flow rate of 1 ml/min through an Econosphere-C_{18} (dimensions 4.6 mm x 250 mm, 5 µm particle size) column.

MAO activity was determined using a modification of the procedure of Wurtman and Axelrod (1963), with ^{14}C-labelled 5-HT and PE being used as substrates for MAO-A and MAO-B, respectively.

In pharmacokinetic studies on brain and liver levels of the drugs and PE and T, semilogarithmic plots of drug or amine concentrations against time were used to determine distribution and elimination half-lives (α and β, respectively) employing the feathering technique (Gibaldi and Perrier, 1982). Concentration-time curves were utilized to determine the area under the curve (AUC) by the trapezoidal rule (Gibaldi and Perrier, 1982). Data were analyzed with analysis of variance followed by independent t-tests in the case of single-pair comparisons or Newman-Keuls tests in the case of multiple pairwise comparisons (α =0.05). All probabilities are two-tailed.

RESULTS AND DISCUSSION

As can be seen from Figure 2, all four N-alkylated analogues of PE produced sustained increases in brain concentrations of PE, indicating that they may be effective prodrugs. The order of effectiveness of the drugs in this regard was PGPE>DPGPE>CEPE>CPPE. The sustained increases are in contrast to the very rapid disappearance of PE from brain following injection of PE itself (Coutts et al., 1985; Baker et al., 1987a).

Figure 3 illustrates the relative availabilities of the various prodrugs of PE in rat brain, blood and liver after administration of an equimolar dose of each. Figure 4 represents a comparison of the availability of PE in brain, blood and liver after administration of equimolar amounts of the drugs.

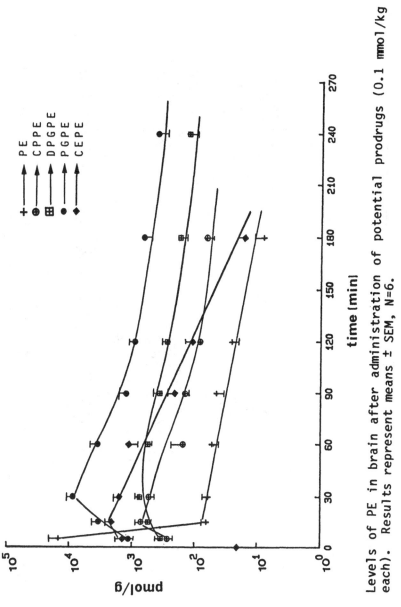

Fig. 2. Levels of PE in brain after administration of potential prodrugs (0.1 mmol/kg each). Results represent means ± SEM, N=6.

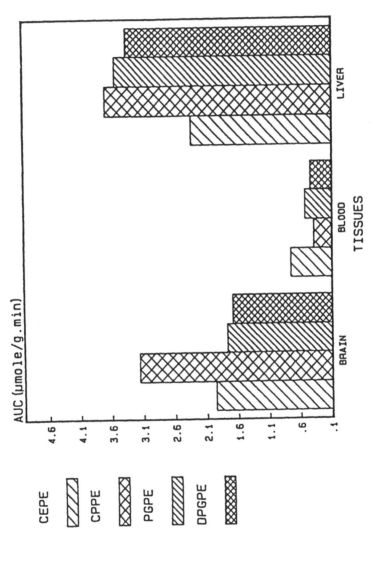

Fig. 3. Comparison of availability (AUC) of prodrugs of PE in rat brain, blood and liver after administration of 0.1 mmol/kg (i.p.) dose each.

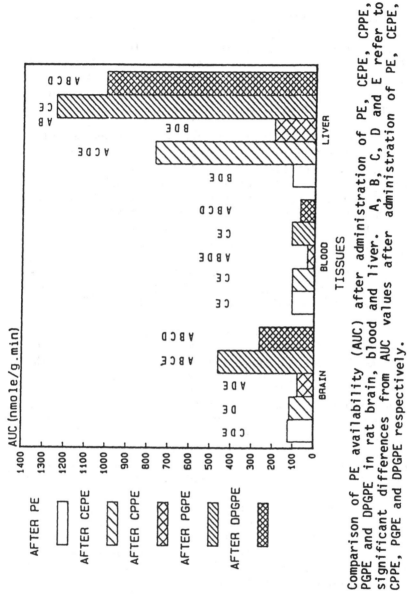

Fig. 4. Comparison of PE availability (AUC) after administration of PE, CEPE, CPPE, PGPE and DPGPE in rat brain, blood and liver. A, B, C, D and E refer to significant differences from AUC values after administration of PE, CEPE, CPPE, PGPE and DPGPE respectively.

After administration of CET, peak brain concentrations of T were observed at 15 min and they were nearly 45 times control levels; by 180 min, the levels of T were still approximately 10 times control values (Baker et al., 1987c). This was in contrast to the situation after injection of T itself, when brain concentrations of T declined rapidly after 5 min.

Unlike the MAO inhibitors, the prodrugs produce elevations of brain PE or T without causing marked prolonged increases in brain levels of the catecholamines or 5-HT (Baker et al., 1987a,b; Rao et al., 1987a,b,c). However, all of them do produce transient changes in levels of these biogenic amines and their metabolites; in several cases, these changes in biogenic amine concentrations are decreases rather than increases. These effects, particularly the decreases in brain levels of NA, are most marked with CPPE; since this amine also produces the lowest increase in brain concentrations of PE, it appears to be the least useful of the four prodrugs of PE tested.

The decreases in concentrations of the catecholamines and 5-HT combined with changes in the levels of their metabolites, suggest that the PE formed from the prodrugs or the prodrugs themselves may be having effects on the uptake and release of these putative neurotransmitter amines from nerve terminals. Such effects of PE have been demonstrated by several groups of researchers using a variety of techniques (Jonsson et al., 1966; Fuxe et al., 1967; Jackson and Smythe, 1973; Raiteri et al., 1977; McQuade and Wood, 1983; Baker and Yasensky, 1981; Philips and Robson, 1983; Nielsen et al., 1983). In addition, there have been reports of N-alkylated analogues of phenylethylamines affecting the uptake and release of biogenic amines (Koe, 1976; Baker et al., 1980; Suckling et al., 1985). Experiments are now being carried out in vitro to study the effects of prodrugs on uptake of radiolabelled neurotransmitter amines in nerve ending fractions prepared from brain.

PGPE and DPGPE, as expected from their N-propargyl structure (similar to deprenyl and pargyline), did exhibit some inhibition of MAO ex vivo, but this inhibition was considerably less than that demonstrated by the MAO inhibitors tranylcypromine and pargyline after injection of an equimolar dose (see Table 1). Since PGPE and DPGPE at this level of inhibition of MAO produced brain levels of PE considerably higher than those produced by conventional MAO inhibitors such as tranylcypromine at doses which produce almost complete inhibition of MAO, it seems likely that metabolism of PGPE and DPGPE to PE rather than inhibition of MAO and subsequent increases in concentrations of endogenous PE is primarily responsible for the PE increases seen. However, radiolabelled or stable isotope-labelled PGPE and DPGPE will have to be investigated in future experiments to delineate the contributions of these two effects.

% Inhibition

Prodrug	MAO-A	MAO-B
CEPE	0.5 ± 0.3	7.7 ± 2.6
PGPE	19.8 ± 3.9	73.3 ± 1.0
DPGPE	27.8 ± 3.3	67.9 ± 2.8
CET	9.6 ± 2.4	0
TCP	90.8 ± 1.3	94.2 ± 0.9
PAR	93.1 ± 1.1	95.6 ± 1.0

Table 1. Ex vivo inhibition of MAO activity in rat whole brain by prodrugs and by tranylcypromine (TCP) and pargyline (PAR). Each drug was injected at a dose of 0.1 mmol/kg, and the rats were killed 60 min following injection. The results represent means ± SEM (N=6).

In summary, several N-alkylated analogues of PE and T have been synthesized and have been shown, through studies in a rat model, to act as prodrugs of PE or T. Studies of effects on levels of brain catecholamines and 5-HT suggest that some of the prodrugs may affect uptake and release of the putative neurotransmitter amines, and in vitro studies in this direction are warranted. In the case of the propargylated prodrugs, future experiments with radiolabelled or stable isotope-labelled forms of these drugs may be useful in determining the extent to which metabolism of the drugs and their effects on MAO activity contribute to the PE levels observed.

The results presented here represent the findings obtained from studies at one dose of the prodrugs. It is anticipated that some of these prodrugs, particularly the N-propargyl analogues, may still be producing elevations of brain PE at lower doses without directly affecting uptake and release of catecholamines or 5-HT. Chronic studies are also planned to determine what effects these prodrugs have on brain levels of PE and on behaviours such as locomotion and intracranial self-stimulation which have been reported to be altered by antidepressant drugs. In addition, experiments are planned to investigate the propargyl analogues of T.

ACKNOWLEDGEMENTS

The research reported here was funded by the Alberta Provincial Mental Health Advisory Council (PMHAC) and the Alberta Heritage Foundation for Medical Research (AHFMR). The authors are grateful to Drs. W. G. Dewhurst and A. J. Greenshaw for useful discussion and to Miss Heather Schmidt for typing this manuscript.

REFERENCES

Baker G.B., Coutts R.T., Benderly A., Cristofoli H.A. and Dewhurst W.G. (1980) The synthesis of N-alkylated p-chlorphentermine derivatives and their effects on releasing 5-hydroxytryptamine from rat striatum in vitro. Can. J. Pharm. Sci. 15, 71-74.

Baker G.B., Coutts R.T. and Rao T.S. (1987a) Neuro-pharmacological and neurochemical properties of N-(cyanoethyl)-2-phenylethylamine, a prodrug of 2-phenylethylamine. Br. J. Pharmacol. (in press).

Baker G.B., Rao T.S. and Coutts R.T. (1986) Electron-capture gas chromatographic analysis of β-phenylethyl-amine in tissues and body fluids using pentafluoro-benzenesulfonyl chloride for derivatization. J. Chromatogr. Biomed. Appl. 381, 211-217.

Baker G.B., Rao T.S. and Coutts R.T. (1987b) Neuro-chemical and neuropharmacological investigation of N-cyanoethyltryptamine, a potential prodrug of trypt-amine. Proc. West Pharmacol. Soc. (in press).

Baker G.B. and Yasensky D.L. (1981) Interactions of trace amines with dopamine in rat striatum. Prog. Neuro-psychopharmacol. 5, 577-580.

Beckett A.H., Shenoy E.V.B. and Salmon J.A. (1972) The influence of replacement of the N-ethyl group by the cyanoethyl group on absorption, distribution and metabolism of (±)-ethylamphetamine in man. J. Pharm. Pharmacol. 24, 194-202.

Boulton A.A. (1976) Identification, distribution, metabo-lism, and function of meta- and para-tyramine, phenyl-ethyalmine, and tryptamine in brain. In, Advances in Psychopharmacology, Vol. 15 (Costa, E., Giacobini, E., and Paoletti, R., eds) pp 57-68, Raven Press, New York.

Boulton A.A. and Milward L. (1971) Separation, detection and quantitative analysis of urinary β-phenylethyl-amine. J. Chromatogr. 57, 287-296.

Coutts R.T., Baker G.B., Nazarali A.J., Rao T.S., Micetich R.G. and Hall T.W. (1985) Prodrugs of trace amines. In, Neuropsychopharmacology of the Trace Amines, (Boulton, A.A., Maitre, L., Bieck, P.R., and Riederer, P., eds) pp 175-180, Humana Press, Clifton, N.J.

Coutts R.T., Foster B.C. and Pasutto F.M. (1981) Fungal metabolism of (-)-deprenyl and pargyline. Life Sci. 29, 1951-1958.

Coutts R.T., Nazarali A.J., Baker G.B. and Pasutto F.M. (1986) Metabolism and disposition of N-(2-cyanoethyl)-amphetamine (Fenproporex) and amphetamine: study in the rat brain. Can. J. Physiol. Pharmacol. 64, 724-728.

Dewhurst W.G. (1968) New theory of cerebral amine function and its clinical application. Nature (Lond) 218, 1130-1133.

Dourish C.T. (1982) A pharmacological analysis of the hyperactivity syndrome induced by β-phenylethylamine in the mouse. Br. J. Pharmacol. 77, 129-139.

Dourish C.T. (1984) Studies on the mechanisms of action of β-phenylethylamine stereotypy in rodents: implications for a β-phenylethylamine animal model of schizophrenia. In, Neurobiology of the Trace Amines, (Boulton, A.A., Baker, G.B., Dewhurst, W.G. and Sandler, M., eds) pp 389-411, Humana Press, Clifton, N.J.

Durden D.A., Philips S.R. and Boulton, A.A. (1975) Identification and distribution of benzylamine in tissue extracts isolated from rats pretreated with pargyline. Biochem. Pharmacol. 24, 1-2.

Fischer E., Spatz H., Heller B. and Reggiani H. (1972) Phenylethylamine content of human urine and rat brain, its alterations in pathological conditions and after drug administration. Experientia 28, 307-308.

Fuxe K., Grobecker H. and Jonsson J. (1967) The effect of β-phenylethylamine on central and peripheral monoamine containing neurons. Eur. J. Pharmacol. 2, 202-207.

Gibaldi M. and Perrier D. (1982). Pharmacokinetics, Marcel Dekker, New York.

Greenshaw A.J. (1984) β-Phenylethylamine: a functional role at the behavioural level? In, Neurobiology of the Trace Amines, (Boulton, A.A., Baker, G.B., Dewhurst, W.G. and Sandler, M., eds), pp 351-373, Humana Press, Clifton, N.J.

Jackson D.M. and Smythe D.B. (1973) The distribution of β-phenylethylamine in discrete regions of the rat brain and its effect on brain noradrenaline, dopamine and 5-hydroxytryptamine levels. Neuropharmacol. 12, 663-668.

Jonsson J., Grobecker H. and Holtz P. (1966) Effects of β-phenylethylamine on content and subcellular distribution of norepinephrine in rat heart and brain. Life Sci. 5, 2235-2246.

Koe B.K. (1976) Molecular geometry of inhibitors of uptake of catecholamines and serotonin in synaptosomal preparations of rat brain. J. Pharmacol. Exp. Ther. 199, 649-661.

McKim H.R., Calverly D.G., Philips S.R., Baker G.B. and Dewhurst W.G. (1980) The effect of tranylcypromine on the levels of cerebral amines in rat diencephalon. In, Recent Advances in Canadian Neuropsychopharmacology, (Grof, P. and Saxena, B.M., eds) pp 7-13, Karger, Basel.

McQuade P.S. and Wood P.L. (1983) The effects of β-phenylethylamine on tyramine and dopamine metabolism. Progr. Neuro-Psychopharmacol. & Biol. Psychiat. 7, 755-758.

Nazarali A.J., Baker G.B., Coutts R.T. and Pasutto F.M. (1983) Amphetamine in rat brain after intraperitoneal injection of N-alkylated analogs. Progr. Neuro-Psychopharmacol. & Biol. Psychiatr. 7, 813-816.

Nazarali A.J., Baker G.B., Coutts R.T. and Wong J.T.F. (1987a) N-2-(Cyanoethyl)tranylcypromine, a potential prodrug of tranylcypromine: its disposition and interaction with catecholamine neurotransmitters in brain. Pharmaceut. Res. 4, 16-20.

Nazarali A.J., Baker G.B., Coutts R.T., Yeung J.M. and Rao T.S. (1987b) Rapid analysis of β-phenylethylamine in tissues and body fluids utilizing pentafluoro-benzoylation followed by electron-capture gas chromatography. Progr. Neuro-Psychopharmacol. & Biol. Psychiatry (in press).

Nielsen J.A., Chapin D.S. and Moore K.E. (1983) Differential effects of d-amphetamine, β-phenylethylamine, cocaine and methylphenidate on the rate of dopamine synthesis in terminals of nigrostriatal and mesolimbic neurons on the efflux of dopamine metabolites into cerebroventricular perfusates of rats. Life Sci. 33, 1899-1907.

Philips S.R. (1981) Amphetamine, p-hydroxyamphetamine and β-phenylethylamine in mouse brain and urine after (-)- and (+)-deprenyl administration. J. Pharm. Pharmacol. 6, 542-544.

Philips S.R. and Boulton A.A. (1979) The effect of monoamine oxidase inhibitors on some arylalkylamines in rat striatum. J. Neurochem. 33, 159-167.

Philips S.R. and Robson A.M. (1983) In vivo release of endogenous dopamine from rat caudate nucleus by phenylethylamine. Neuropharmacol. 22, 1297-1301.

Pirisino R., Ciottoli G.B., Buffoni F., Anselmi B. and Curradi C. (1978) N-methylbenzylamine, a metabolite of pargyline in man. Br. J. Clin. Pharmacol. 7, 595-598.

Raiteri M., Del Carmine R., Bertollini A. and Levi G. (1977) Effect of sympathomimetic amines on the synaptosomal transport of noradrenaline, dopamine and 5-hydroxytryptamine. Eur. J. Pharmacol. 41, 133-143.

Rao T.S., Baker G.B. and Coutts R.T. (1987a) Pharmacokinetic and neurochemical studies on N-propargyl-2-phenylethylamine, a prodrug of 2-phenylethylamine. Naunyn-Schmiedeberg's Arch. Pharmacol. (in press).

Rao T.S., Baker G.B. and Coutts R.T. (1987b) N-(3-Chloropropyl)phenylethylamine as a possible prodrug of β-phenylethylamine: studies in the rat brain. Progr. Neuro-Psychopharmacol. Biol. Psychiat. (in press).

Rao T.S., Baker G.B. and Coutts R.T. (1987c) N,N-Dipropargyl-2-phenylethylamine, a potential prodrug of 2-phenylethylamine: neurochemical and neuropsychopharmacological studies in rat. Brain Res. Bull. (in press).

Reynolds G.P., Elsworth J.D., Blau K., Sandler M., Lees A.J. and Stern G.M. (1978) Deprenyl is metabolized to metamphetamine and amphetamine in man. Br. J. Clin. Pharmacol. 6, 542-544.

Sabelli H.C. and Mosnaim A.D. (1974) Phenylethylamine hypothesis of affective behavior. Am. J. Psychiat. 131, 695-699.

Suckling C.J., Breckenbridge R.J., Bansal S.S., Williams L.C., Walting K.J. and Iversen L.L. (1985) Non-selective inhibition of GABA and 5-HT uptake systems in rat brain by N-n-alkylhydroxybenzylamine and N-n-alkylphenylethylamine derivatives. Biochem. Pharmacol. 34, 4173-4177.

Weli A.M. and Lindeke B. (1985) The metabolic fate of pargyline in rat liver microsomes. Biochem. Pharmacol. 34, 1993-1998.

Weli A.M. and Lindeke B. (1986) Peroxidative N-oxidation and N-dealkylation reactions of pargyline. Xenobiotica 16, 281-288.

Wurtman R.J. and Axelrod J. (1963) A sensitive and specific assay for the estimation of monoamine oxidase. Biochem. Pharmacol. 12, 1439-1440.

ALCOHOL AND BIOGENIC AMINE-DERIVED ALKALOIDAL ADDUCTS IN

THE NERVOUS SYSTEM

M.A. Collins, D.A. Pronger, B.Y. Cheng and
N.A. Ung-Chhun
Department of Biochemistry
Loyola University Medical School
2160 South First Avenue

Maywood, IL 60153

Chronic ethanol abuse causes chemical and
membrane alterations in central nervous system (CNS)
which are manifested as accelerated neuronal "aging" and,
ultimately, permanent brain damage. Clarification of the
largely unknown mechanisms for these ethanol-dependent
neuronal and perhaps glial alterations may eventually be
of clinical benefit in alcoholism. Our laboratory has
been investigating the hypothesis that trace metabolites
oxidatively derived from "alkaloidal" condensation
products of biogenic amine neurotransmitters with
acetaldehyde, the ethanol metabolite, or with related
biological carbonyls are responsible in part for
alcoholic brain damage. The biogenic amine, dopamine
(DA), reacts bimolecularly to yield the
1,2,3,4-tetrahydroisoquinolines (TIQs), salsolinol or SAL
(from acetaldehyde) or salsolinol-1-carboxylic acid or
SAL-1-CA (from pyruvic acid). In theory, cellular
oxidation (dehydrogenation) of these TIQs would yield
relatively electron-deficient 3,4-dihydroisoquinolines,
isoquinolines and possibly their N-oxides that might bind
covalently to membrane nucleophilic groups (Collins
1982). Facile covalent binding and crosslinking with
neuronal proteins is considered to be a major component
of the degenerative mechanism of (oxidized) 6-hydroxy-DA,
a neurotoxin which the oxidized TIQs resemble
structurally. Serotonin also reacts with the afore-
mentioned carbonyl compounds to produce 6-hydroxylated

1,2,3,4-tetrahydro-beta-carbolines (THBCs). Analogous to
the 6-hydroxy-DA/TIQ relationship, a potentially toxic
alkaloid resembling 5,6-dihydroxy-tryptamine (a
serotonergic toxin) could be envisioned following
hydroxylation/oxidation of the THBC moiety.

Urinary studies with detoxifying alcoholics have
generally provided support for the idea that formation
and excretion of condensation products or their
metabolites---SAL, salsoline (SALN;7-O-methyl-
salsolinol), 1-methyl-6-hydroxy-THBC (the conjugated
fraction) or harman---are promoted in vivo during ethanol
ingestion (Collins et al. 1979; Sjoquist et al. 1981a;
Beck et al. 1982; Rommelspacher et al. 1985; Adachi et
al. 1986a; Collins 1987). Some investigators also have
found increased excretion of the alkaloids by
nonalcoholic volunteers during or after controlled
ethanol administration (Sjoquist et al. 1985; Hirst et
al. 1985; Adachi et al. 1986b). However, as will be
discussed, replicable mammalian alkaloid elevations in
CNS tissues and fluids as a consequence of ethanol abuse
have not been shown in humans or animal models. Since the
conception of alcoholic brain damage due to
endogenously-generated alkaloidal toxins presupposes
ethanol-dependent CNS increases in the initial
tetrahydro-condensation products, we set out to clarify
this dilemma with two different long-term ethanol
intoxication studies.

Previously, we had obtained HPLC and GLC/mass
spectrometry (MS) results implicating, as trace CSF
constituents in untreated Rhesus monkeys, four alkaloidal
products apparently resulting from bimolecular
condensations of DA and serotonin with either
acetaldehyde or pyruvic acid, and one TIQ product, SALN,
derived from in vivo O-methylation of SAL (Collins et al.
1982a). The first study, therefore, was an examination of
the effect of prolonged ethanol administration in monkeys
on the CSF concentrations of these three TIQs and two
THBCs. The intubation technique of Ellis and Pick (1970)
which produces an ethanol-dependent monkey within several
weeks, was utilized. Secondly, we attempted to determine
the levels of these alkaloids in striatum and
hypothalamus of adult rats which had been pair-fed liquid
diets containing ethanol or isocaloric sugar for a period
of 5 months.

METHODOLOGIES

Following CSF sampling during a 6-day
pre-intoxication phase (daily intubation with dextrose),
three male Rhesus monkeys were gavaged twice daily with
20% ethanol in amounts of 4.5-6 g/kg/day for 21 days.
Liquid diet supplement (vanilla Ensure) was given with
ethanol to minimize nutritional deficits and weight loss.
Blood levels of ethanol, acetaldehyde and pyruvic acid
were monitored daily by GC and UV spectrophotometry. CSF
was drawn into EDTA/semicarbizide solutions on the 15th
day of intoxication, and 24-32 hr following the last
ethanol on day 21, which is the peak interval for
"withdrawal" symptomology (Ellis and Pick 1970). No overt
convulsions were seen by our research or animal care
staff during intoxication or withdrawal in our study,
however. CSF was assayed by a capillary GLC/MS method
developed in our laboratory for the concurrent
measurement of total (conjugated and unconjugated) TIQs
and THBCs (Pronger et al., in preparation). After
addition of five deuterated alkaloids as internal
standards, samples (5 ml CSF volumes) were treated with
sulfatase/glucuronidase to hydrolyze conjugates, total
alkaloids were isolated by alumina and/or C-18 reversed
phase cartridges, and derivatization was effected with a
mixture of trifluoroacetic anhydride and
hexafluoroisopropanol. The derivatized alkaloids were
separated on a 30 meter DB-5 fused silica column and
analyzed by selected ion monitoring (SIM) using a
Finnegan 1020B GLC/MS in the EI mode. Artifactual
formation of alkaloids was shown with labeled precursors
to be minimal or nonexistent. Similar methodology was
employed for the analysis of TIQs and THBCs in perchloric
acid supernatants of brain homogenates obtained from rats
given chronic ethanol as described. Catecholamines and
their oxidized metabolites were analyzed in aliquots of
these brain supernatants using reversed phase HPLC with
electrochemical detection (Collins and Origitano 1983).

RESULTS

Determined two hours after the morning ethanol
intubation in the monkeys, blood ethanol levels varied
between 170 and 350 mg/dl. Blood acetaldehyde levels
hovered around 1-2 uM for several days, rising to 5-7 uM

on the 10th day of intoxication. Blood pyruvic acid
levels were between 40-100 uM throughout the entire
study. The three TIQs were detectable in most, but not
all, samples from the three phases of the experiment. One
monkey demonstrated ethanol-dependent elevations in all
three TIQs, but no TIQ was increased over
pre-intoxication concentrations in all three subjects,
nor was there a consistent change between intoxication
and withdrawal. For example, SAL, discernible in 7/9
samples, appeared to be maintained in the range of 1-4 nM
throughout all phases. SALN, detectable in 8/9 samples,
increased sharply in one monkey during intoxication (from
1.6 to 19.7 nM), but decreased by 60% in the second and
became undetectable in the third. There are indications
that ethanol treatment may have had the greatest effect
on the pyruvic acid product, SAL-1-CA. Its levels became
detectable (2-5 nM) in two monkeys only during
intoxication and withdrawal; however, they were
relatively unaffected by ethanol in the third. Prolonged
intoxication appeared to have inconsistent effects on the
THBC alkaloids as well. Levels of MO-THBC were increased
2.5-fold during intoxication in one monkey, but were
reduced 80% in the second subject and were undetectable
in the intoxication phase in the third (qualitatively
similar to SALN, above). Furthermore, MO-THBC was not
detectable in any of the three withdrawal samples, and
its carboxylated analog (MO-THBC-1-CA) was detectable
(3nM) in only one of the CSF samples from the entire
study.

    Similar assays of the TIQs and THBCs were carried
out, with some methodological modifications, on two
monoaminergic brain regions from rats treated chronically
with 6.6% ethanol/ or isocaloric dextrose/liquid diets
for 5 months. The limitations of the methodology and the
GC/MS system required that 14 individual rats/group be
pooled to give an N of 5/region. As an initial surprise,
there was a clear tendency for decreases in striatal
levels of TIQs as a result of chronic ethanol
exposure---but this became understandable after DA levels
were obtained. The most readily observable alkaloid was
SALN, which with one exception (an experimental
hypothalamic sample), was always detected (>1 ng/g).
Striatal SALN levels (ng/g tissue) were significantly
reduced in the ethanol group to 1.48 ± 0.12 from the
control group value of 2.62 ± 0.42 (p <0.01);

hypothalamic SALN levels were unchanged by the ethanol
(2.5 ng/g). For either group, mean levels of SAL-1-CA in 3
control and 4 experimental samples (absent or <3 ng/g in
1 or 2 samples/group) were an order of magnitude higher
than SALN, and, in the striatum, were decreased 35%
(0.08) by ethanol treatment. The mean detectable
hypothalamic levels of SAL-1-CA were also 35% lower in
the alcohol group, but large standard errors precluded
statistical significance. Also, the primary TIQ
condensation product, SAL, discernible in only 2/5
striatal samples in either group, nevertheless appeared
to be lower as a result of ethanol (2.3 and 3.5 ng/g vs.
4.2 and 5.3 ng/g). Interestingly, SAL was undetectable
(<1.5 ng/g) in all hypothalamic samples.

In terms of THBCs, the two serotonin-derived
alkaloids were undetectable in all pooled rat brain
samples, but this may be due in part to relatively high
detection limits for these alkaloids (>5-7 ng/g).
Finally, significant ethanol-related reductions were
apparent in the striatal concentrations of DA (63% of
control; p<0.03) and DOPAC (57% of control; p<0.005), and
in the hypothalamic concentrations of 5-HIAA, the
serotonin metabolite (59% of control; p<0.02). DA, DOPAC
and norepinephrine levels were not altered significantly
in the hypothalamus, nor were serotonin levels in either
brain region.

DISCUSSION

The lack of consistent increases in the TIQs or
THBCs in monkey CSF following 10 days ethanol treatment
argues against involvement of the alkaloids in what might
be termed the early stages of alcoholism and perhaps
alcohol toxicity. We cannot draw conclusions on the
etiology of physical dependence, as withdrawal
convulsions were not observed. However, it is worth
noting that the monkey (#1) which tended to show the most
behavioral anxiety during the latter part of each day,
prior to the second ethanol gavage when blood levels had
dropped, also had apparent increases in all three TIQs in
his intoxication phase CSF sample. Nevertheless, one
monkey does not a study make, and we are left with
results which provide little support for reports of
increased or more frequently detected TIQs (esp. SALN) in
human CSF during alcoholism (Sjoquist et al. 1981a;

1981b). If CSF levels reflect the brain content, our
results are also not in accord with the finding of
increased striatal SALN in mice after 1 wk exposure to
ethanol by liquid diet or vapor chamber (Hamilton et al.
1980), nor with the large increases in SAL in striatum of
rats given ethanol as drinking fluid (Sjoquist et al.
1982)

The rat experimental results show that 5 months
exposure to ethanol/liquid diet induces decreases, not
increases, in striatal SALN and SAL-1-CA which simply may
be due to reduced precursor DA---all of which may reflect
alcohol-induced brain damage. These may be compared to
reported results of SAL, SALN and SAL-1-CA levels in the
caudate nucleus of chronic alcoholics (Sjoquist 1985). In
that case, sober alcoholics (no blood ethanol levels at
autopsy) had statistically significant reductions in SAL
levels relative to nonalcoholics. However, in alcoholics
with blood ethanol levels at death, SAL levels were
equivalent to those in nonalcoholics, which implies, if
DA is depleted in chronic alcoholism, that there was
increased SAL biosynthesis during drinking episodes
(Collins et al. 1982b). We suggest that our experimental
rats perhaps more closely resembled the sober alcoholics,
as their blood ethanol levels were relatively low at the
time of sacrifice. It is possible that elevations in one
or more of the rat brain TIQs may have occurred at
earlier periods in the ethanol liquid diet treatment when
DA levels are normal, as was reported in mice (Hamilton
et al. 1980).

ACKNOWLEDGEMENTS

Dr. Mark Perlow, Dr. Philip Hilton, Dr. George
Drucker and Ms. Carol Grandstaff are gratefully
acknowledged for their capable assistance and advice at
various stages of the experiment with Rhesus monkeys. We
also appreciate the use and attention of the Loyola
Animal Research Facility. Dr. Drucker, Mr. Rudy Abores
and Mr. K. Raikoff are recognized for their invaluable
assistance in the rat experiments. These studies were
supported by AA00266.

REFERENCES

Adachi J., Mizoi Y., Fukunaga T., Ueno Y., Imamichi H.,
Ninomiya I. and Naito T. (1986a) Individual difference in urinary excretion of salsolinol in alcoholic
patients. Alc. 3, 371-376.
Adachi J., Mizoi Y., Fukunaga T. Kogame M., Ninomiya I.
and Naito T. (1986b) Effect of acetaldehyde on
urinary salsolinol in healthy man after ethanol
intake. Alc. 3, 215-220.
Beck O., Bosin T.R., Lundman A. and Borg S. (1982)
Identification and measurement of 6-hydroxy-1-methyl-
1,2,3,4-tetrahydro-beta-carboline by GC-mass
spectrometry. Biochem. Pharmacol. 31, 2517-2521.
Collins M.A., Nijm W.P., Borge G. Teas G., and
Goldfarb C. (1979) Dopamine-related tetrahydroisoquinolines: Increased urinary excretion by alcoholics
following alcohol consumption. Sci. 206, 1184-1186.
Collins M.A. (1982) A possible neurochemical mechanism
for brain and nerve damage associated with chronic
alcoholism. Trends Pharmacol. Sci. 3, 373-375.
Collins M.A., Dahl K., Nijm W.P. and Major L.F. (1982a)
Evidence for homologous families of dopamine and
serotonin condensation products in CSF from monkeys.
Abst. Soc. Neurosci. 8, 277.
Collins M.A., Hannigan J.J., Origitano T., Moura D. and
Osswald W. (1982b) On the occurrence, assay and
metabolism of simple tetrahydroisoquinolines in
mammalian tissues. In, Beta-Carbolines and Tetrahydroisoquinolines (Bloom F., Barchas J., Sandler M.
and Usdin E., eds) pp 155-166, Alan Liss, NY.
Collins M.A. and Origitano T.C. (1983) Catecholamine-
derived tetrahydroisoquinolines: O-Methylation
patterns and regional brain distribution following
intraventricular administration in rats. J. Neurochem.
41, 1569-1575.
Collins M.A. (1987) Acetaldehyde and its amine condensation products as markers in alcoholism. In, Recent
Dev. Alcohol. Vol. 6 (Galanter M., ed) Plenum Press,
NY, in press.
Ellis, F.W. and Pick J. (1970) Experimentally induced
ethanol dependence in Rhesus monkeys. J. Pharmacol.
Exp. Ther. 175, 88-93.
Hamilton M.G., Blum K. and Hirst M. (1980) In vivo formation of isoquinoline alkaloids: Effect of time and

route of administration of ethanol. In, Biol. Effects
Alcohol (Adv. Exp. Biol. Med.) Vol. 126
(Begleiter H., ed), pp. 73-86, Plenum Press, NY.

Hirst M., Evans D. R., Gowdey C.W. and Adams M. (1985)
The influences of ethanol and other factors on the
excretion of urinary salsolinol in social drinkers.
Pharmacol. Biochem. Behav. 22, 993-1000.

Rommelspacher H., Damm H., Schmidt L. (1985) Increased
excretion of harman by alcoholics depends on events
of the life history and the state of the liver.
Psychopharmacol. 87, 64-68.

Sjoquist B., Borg S. and Kvande H. (1981a) Catecholamine-
derived compounds in urine and cerebrospinal fluid
from alcoholics during and after long-standing intoxication. Subst. Alc. Actions Misuse 2, 63-72.

Sjoquist B., Borg S. and Kvande H. (1981b) Salsolinol
and methylated salsolinol in urine and cerebrospinal
fluid from healthy volunteers. Subst. Alc. Actions
Misuse 2, 73-77.

Sjoquist B., Liljequist S. and Engel J. (1982) Increased
salsolinol levels in rat striatum and limbic forebrain
following chronic ethanol treatment. J. Neurochem.
39, 259-262.

Sjoquist B., Johnson H.A. and Borg S. (1985) The influence
of acute ethanol on the catecholamine system in man as
reflected in cerebrospinal fluid and urine. A new
condensation product, 1-carboxysalsolinol. Drug Alc.
Depend. 16, 241-249.

Sjoquist B. (1985) On the origin of salsolinol and
1-carboxysalsolinol. In, Aldehyde Adducts in Alcoholism
(Collins M.A., ed) pp. 115-124, Alan Liss, NY.

INHIBITION OF MONOAMINE OXIDASE, DEPLETION OF LUNG BIOGENIC

AMINES BY CIGARETTE SMOKE AND THE INTERACTION OF SOME

COMPONENTS OF THIS SMOKE WITH TRACE AND OTHER AMINES

P.H. Yu, D.A. Durden, B.A. Davis and A.A. Boulton

Neuropsychiatric Research Unit
CMR Building, University of Saskatchewan
Saskatoon, Saskatchewan, Canada S7N 0W0

INTRODUCTION

Low platelet monoamine oxidase (MAO, EC 1.4.3.4.) has been found to be associated with cigarette smoking (Coursey et al., 1979; Oreland et al., 1981; Norman et al., 1982; Littlewood et al., 1984; von Knorring et al., 1984). It is not quite clear, however, whether it is the smoking that produces an inhibition of MAO activity or a reduction in the synthesis of MAO, or whether individuals with low platelet MAO activity are more prone to cigarette smoking. It has been reported that low platelet MAO is related to sensation seeking personality (Murphy et al, 1977; Fowler et al., 1980). Cessation of smoking has been claimed not to change platelet MAO activity (Littlewood et al., 1984). This study is inconsistent with other reports in which platelet MAO activity of ex-smokers has been shown to be not different from that of non-smokers, but significantly higher than that of smokers (Oreland et al., 1981; von Knorring et al., 1984). It is not known whether there are compounds in the cigarette smoke that may inhibit MAO activity. Nicotine and thiocyanate produce little or no effect on MAO activity (Oreland et al., 1981; Norman et al., 1982). In order to assess the effects of cigarette smoke on MAO activity, we have examined the effects of a smoke solution and direct smoke exposure on rat lung tissue. During the course of this study we have also observed that monoamines such as β-phenylethylamine, p-tyramine and 5-hydroxytryptamine, etc. could readily interact with some compoents in the cigarette smoke non-enzymatically to produce lipophilic adducts. The formation of nitrosamines via nitrosation during cigarette smoking has been widely investigated (Hoffman and Adams, 1981; Hoffman and Brunneman, 1983) but these amine adducts were not formed via nitrosation with nitric oxides. Components in the cigarette smoke were also found to react with other primary amine groups, i.e. basic amino acids and proteins. In this paper we describe the formation and identification of these amine derivatives. The possible implications of such a novel reaction or of these new compounds in vivo are discussed.

EXPERIMENTALS

MAO Assay. MAO activity was determined by a radio-metric procedure as previously described (Yu, 1986). The enzyme preparations were incubated at 37°C for 30 min in the presence of radioactive substrate (0.1 µCi, 1 µCi=37 GBq) which had been diluted with unlabelled substrate to yield a final concentration of 5×10^{-4} M for p-tyramine and 5-hydroxytryptamine and 1×10^{-5} M for β-phenylethyl-amine in a final volume of 200 µl of 0.05 M phosphate buffer (pH 7.5). The reaction was terminated by adding 250 µl of 2 M citric acid. The oxidized products formed were extracted into 1 ml of toluene:ethyl acetate (1:1, v/v) of which 600 µl was transferred to a counting vial containing 10 ml Omnifluor fluid (New England Nuclear, Boston). The radioactivity was assessed by liquid scintillation spec-trometry (Beckman LS 7500, Fullerton, CA). Blank values were obtained by including pargyline (5×10^{-4} M) in the incubation mixtures under identical experimental conditions.

Preparation of Cigarette Smoke Solution. The cigar-ette smoke solution was prepared from filtered cigarettes with a tar and nicotine content of 16 mg and 1.2 mg, respectively. Smoking was simulated as one puff per half minute with a two second puff duration. The resultant smoke was bubbled through 0.02 M phosphate buffer solution at pH 7.5 (3 ml solution/cigarette).

Preparation of Rat Lung MAO. Rat (180 g male Wistar) lung mitochondrial MAO was prepared by homogenizing freshly dissected lung tissue in 0.32 M sucrose solution. The homogenates were then centrifuged at 600 g for 15 min; after removing the nuclear fraction sediment, the crude mitochondria in the supernatant were isolated by centrifuging at 6500 g for 20 min. The mitochondrial pellet was then suspended in chilled, distilled water. After lysis the mitochondrial membrane fragments were obtained by centrifugation at 30,000 g for 30 min. The pellet was sonicated in 0.01 M phosphate-buffer (pH 7.5).

Exposure of Lung Tissue to Cigarette Smoke. We have adopted two types of exposure of rat lung tissue to cigarette smoke: (a) The freshly dissected rat lung tissues were cut into 0.1 mm slices with a Sorvall tissue chopper and the slices then incubated at 37°C for 30 min in the above prepared cigarette smoke solution; (b) the chopped rat lung tissue (0.5 g) was spread out on a wet glass filter paper and then exposed to cigarette smoke at room temperature. The cigarette smoke was drawn through the tissues utilizing a filter funnel under light vaccum. A single cigarette, which generates a total of ten puffs of smoke (2 seconds/puff, one minute interval) was used. The tissues were then carefully rinsed with saline three times and homogenized in 0.01 M phosphate buffer (pH 7.5).

Formation and Detection of Amine-Smoke Adducts. Radio-isotopically labelled amines including 5-[2-^{14}C]-hydroxytryptamine, β-[ethyl-^{14}C]-phenylethylamine, p-[1-^{14}C]-tyramine, [2-^{14}C]-tryptamine and [8-^{3}H(N)]-dopamine (New England Nuclear, Boston, MA) were incubated separately with freshly prepared cigarette smoke solution in total volumes of 200 μL in 0.02 M phosphate buffer (pH 8.0) at 37°C for 30 min. The reaction was terminated by adding 250 μL of 2 M citric acid. The formed adducts were extracted into 1 mL of toluene: ethylacetate (1:1 v/v) of which 600 μL was transferred to a counting vial containing 10 mL Omnifluor fluid (New England Nuclear, Boston). The radioactivity was assessed by liquid scintillation spectrometry (Beckman LS 7500, Fullerton, CA).

RESULTS AND DISCUSSION

Inhibition of MAO Activity by Cigarette Smoke

As can be seen in Fig. 1, the initial rate of deamination of p-tyramine, serotonin and β-phenylethylamine, catalyzed by rat lung mitochondrial MAO, was greatly reduced when the enzyme was incubated in the presence of a cigarette smoke solution. We have included pargyline (5 x 10^{-4} M) in the assay to serve as blanks. It is interesting to note that in the presence of the cigarette smoke solution the blank values are much higher than in its absence; this probably indicates the formation of labelled substances

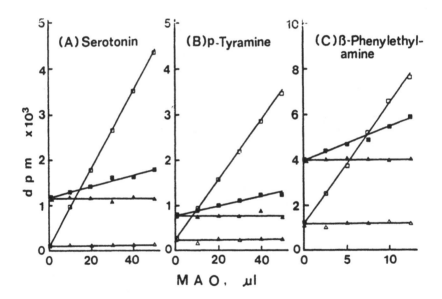

Figure 1. Effects of a cigarette smoke solution on rat lung mitochondrial MAO activity toward different substrates. Enzyme activities were measured in the presence (■) and absence (□) of cigarette smoke solution (10 µl). The blank values were obtained by including the MAO inhibitor pargyline (5 x 10^{-4} M) and preincubation for 20 min at room temperature in the presence (▲) and absence (△) of cigarette smoke solution.

from certain components of the smoke reacting with the labelled amine.

The inhibition was slightly higher for serotonin (type A substrate for MAO) than for β-phenylethylamine (type B substrate). Total inactivation of MAO activity was reached at higher concentrations of cigarette smoke solution. It is interesting to note that tobacco extract also exhibits inhibitory activity towards MAO. Unlike the cigarette smoke solution, however, the tobacco extract did not cause an increase in the blank values. It indicates that amine-cigarette smoke adducts were not formed when tobacco extract was used.

A kinetic analysis of the inhibition suggests that the inhibition of MAO activity by the component(s) in the cigarette smoke is mixed type. The inhibition was irreversible.

When the enzyme after incubation with the cigarette smoke solution (>90% MAO activity was inhibited) was subsequently dialyzed, or separated through a small PD-10 column of Sephadex G-25 (Pharmacia, Uppsala, Sweden) (Figure 2A), MAO activity could not be recovered. Such a gel filtration or dialysis is capable of removing all low molecular substances, if they are not tightly bound to the enzyme. It is, however, uncertain whether larger molecules (mol. wt. >10,000) may also be present in the cigarette smoke solution, although it is unlikely that these would inhibit MAO activity. This has been ruled out, however, because when the smoke solution was separated through the same column, MAO inhibitory activity was found only in the fractions of low molecular weight compounds (Fig. 2B). It suggests that MAO forms an inactive complex with small molecular weight components in the cigarette smoke (Fig. 2B).

It is interesting to note that some "apparent" MAO activity was observed in fractions 13-20 (Fig. 2A). This was due to non-enzymatic formation of adducts between p-tyramine and low molecular weight substances in the cigarette smole solution. This is consistent with the previous observation (Fig. 1).

When rat lung tissue slices were incubated in the cigarette smoke solution or directly exposed to cigarette smoke, the particle bound MAO activity was found to be

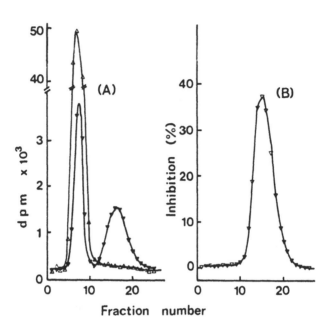

Figure 2. Sephadex G-25 gel filtration of rat lung mitochondrial MAO and the same enzyme pretreated with cigarette
smoke solution (A), and the separation of MAO inhibitory
substance(s) in the cigarette smoke solution (B). A PD-10
column (Pharmacia, Uppsala, Sweden) was used in these
experiments. (A) Rat lung MAO (250 μl) was incubated with
250 μl cigarette smoke solution at 37°C for 30 min. The
sample (▼) was eluted through the column with 0.02 M phosphate buffer (pH 7.5). Fractions of 0.5 ml were collected
and MAO activity in each fraction was determined. The same
amount of untreated MAO was incubated with water, separated
and analyzed using the same procedure (△). (B) Cigarette
smoke solution (500 μl) was eluted under the same condition. The inhibitory activity towards MAO was estimated.

drastically reduced. The inhibition is slightly in favor
of type A MAO.

MAO inhibitory activity has also been demonstrated in
human saliva collected after smoking, whilst the saliva
collected before smoking did not exhibit any effect on rat
lung MAO activity (Fig. 5). The extent of inhibition of

the MAO was however dependent on how long the cigarette smoke was held in the mouth. Inhibition of rat lung MAO by the post-smoke saliva was found to be in the range 10-70%.

The present study has demonstrated that some component(s) of cigarette smoke can irreversibly inhibit monoamine oxidase activity in vitro. It will now be important to ascertain whether these substances can cause a reduction of MAO activity in vivo and if so, whether it is related to the finding of low platelet MAO activity in cigarette smokers. When the rat lung tissue slices were exposed to cigarette smoke or incubated in saline in the presence of a cigarette smoke solution, MAO activity in these treated rat lung tissues was drastically reduced. We have also found that saliva obtained after smoking one cigarette exhibited MAO inhibiting activity, while the saliva obtained before smoking did not. In the case of human smoking, the exposure of the lung tissue to smoke could be quite extensive considering the large total surface area of the respiratory tract. We have yet to resolve whether and how the MAO inhibiting substances diffuse into the blood capillaries or other lung tissues, and consequently reduce platelet or other MAO activities. It is quite important to recognize that the inhibition of MAO activity by smoke is irreversible. This suggests that the inhibition can be progressive and cumulative. It is known that the synthesis of MAO is rather slow (Tipton and Corte, 1979; Turkish et al., 1987) and the effect caused by an acute use of an irreversible MAO inhibitor, for example, can take three weeks to disappear (Planz et al., 1972).

Our observation that cigarette smoke can directly reduce MAO activity is consistent with a report that MAO in mouse skin was found to be inhibited after exposure to cigarette smoke (Essman, 1977). The inhibition of MAO activity by cigarette smoke is by no means selective. For example, salivary amylase (Callegari and Lami, 1984) and rat lung α_1-antitrypsin activity (Janoff et al., 1979) were also found to be reduced by cigarette smoke. The reduction of platelet MAO activity in cigarette smokers (Coursey et al., Oreland et al., 1981; Norman et al., 1982; Littlewood et al., 1984; von Knorring et al., 1984) is possibly due to the direct action of the inhaled smoke being infused into the blood stream in the lung during the gas exchange process. The identity of the inhibitors is currently under

investigation. Our preliminary results indicate that the
primary amino group tends to react with volatile small
molecular weight components in the smoke. Basic amino
acids, such as lysine and arginine, are depleted after
incubation with cigarette smoke. If the free amine groups
in the enzyme molecules are blocked, an effect on the
catalytic function of some enzymes can be expected.

One of the best characterized epidemiologic findings
in Parkinson's disease is its lower incidence in cigarette
smokers compared to non-smokers (Kessler and Diamond,
1971; Baumann et al., 1980; Marttila and Rinne, 1980).
Some factors in the cigarette smoke could induce protection
from the disease. Parkinsonism does not appear to be
genetically determined and may instead involve environmental factors (Düvosin, 1984; Barbeau and Pourcher, 1982; Ward
et al., 1983). The idea that an environmental toxin is involved in idiopathic Parkinson's disease has been supported
by the recent discovery that MPTP (N-methyl-4-phenyl-
1,2,3,6-tetrahydropyridine) can cause Parkinsonism and it
is a fact that patients respond well to 1-DOPA treatment
(Langston et al., 1983). MPTP toxicity is crucially dependent on an MAO-catalyzed reaction; namely, the conversion of MPTP to MPP^+ (1-methyl-4-phenylpyridine) (Chiba et
al., 1984). Subsequently the MPP^+ produces a neuronal
degeneration, especially in the substantia nigra (Heikkila
et al., 1984a). If MAO activity is inhibited by deprenyl
or pargyline, the toxicity can be prevented (Heikkila et
al., 1984b). In fact, deprenyl has been shown to be bene-
ficial to Parkinsonian patients during 1-DOPA therapy
(Youdim et al., 1979). Our finding that cigarette smoke
can irreversibly inhibit MAO activity suggests that chronic
use of cigarettes may contribute to the prevention of oxidation of MPTP or MPTP-like compounds that may be present in
the environment.

Interaction of Primary Amines With
Some Components in Cigarette Smoke

As can be seen in Fig. 3, when ^{14}C-labelled amines
such as p-tyramine, β-phenylethylamine, tryptamine and
serotonin are incubated with cigarette smoke solution,
labelled adduct are formed in proportion to the amount of
cigarette smoke and time of incubation. Adducts also
produced from histamine and from basic amino acid, lysine

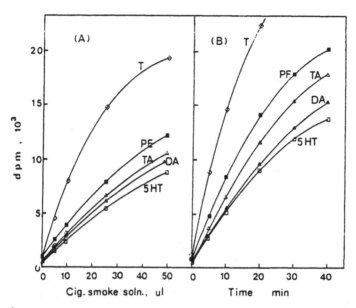

Figure 3. Reactions of biogenic amines with components of cigarette smoke. (A) Radioactively labelled amines (0.1 nmole, 0.1 µCi) [p-tyramine (p-TA), 5-hydroxytryptamine (5-HT), tryptamine (T), dopamine (DA) and β-phenylethyl-amine (PE) were incubated with increasing amounts of cigarette smoke solution at pH 8.0, 37°C for 20 min and the labelled adducts formed isolated by extraction with toluene/ethyl acetate (1:1, v/v) and counted in a liquid scintillation counter. (B) Time course of the reaction. The same amines were incubated with the cigarette smoke solution (50 µl) and the reactions terminated after different time intervals.

and arginine. The formation of the adducts occurs favorably in alkaline condition and at higher temperatures.

The smoke obtained from different brands of cigarettes interact with the amines to a different degree (Table 1). Apparently more amine adducts are formed as the tar and nicotine content increases.

TABLE 1

The formation of lipophilic adducts of β-phenylethylamine and 5-hydroxytryptamine with cigarette smoke obtained from different commercially available brands of cigarettes.

Cigarette Brand	Tar[a] Content (mg)	Nicotine Content (mg)	Adduct Formation[b] (dpm) x 10^{-3}	
			5-Hydroxytryptamine	β-phenylethylamine
1	16	1.2	48.3 ± 4.1	39.0 ± 4.4
2	15	1.2	47.7 ± 3.1	32.0 ± 1.5
3	13	1.0	43.0 ± 1.0	32.0 ± 0.3
4	13	1.0	36.7 ± 1.3	25.7 ± 0.9
5	10	0.8	34.7 ± 0.9	22.7 ± 1.3
6	9	0.8	31.0 ± 1.0	19.7 ± 0.9
7	8	0.8	23.3 ± 1.4	12.3 ± 0.3
8[c]	8	0.8	33.7 ± 2.4	22.7 ± 2.3
9	4	0.4	26.3 ± 3.2	16.7 ± 3.3

a Tar and nicotine contents were indicated on the packs of cigarettes. b ^{14}C-labelled β-phenylethylamine or 5-hydroxytryptamine (1x10^{-5} M, 0.1 µCi) dissolved in 100 µL of 0.1 M phosphate buffer (pH 8.0) were incubated at 60° for 30 min with 100 µL of cigarette smoke solution obtained from different brands of cigarettes. At the end of the incubation period 1 mL of a toluene:ethyl acetate (1:1, v/v) mixture was added to each tube and after vigorous shaking followed by centrifugation, 600 µL of the organic phase was removed for scintillation counting. The values listed are the Mean ± S.E. of three experiments. c A menthol cigarette.

The chemical nature of some of these adducts were determined (Durden et al., 1987). The primary amino group interacted with cyanide and formaldehyde and formed cyanomethyl derivatives. When 5-hydroxytryptmine and dopamine were reacted with cigarette smoke, the cyanomethyl derivatives formed were further condensed via Pictet-Spengler condensations to produce cyanomethyl tetrahydro-β-carbolines and tetrahydroisoquinolines. It is not surprising that both cyanate and formaldehyde are major toxic components present in the cigarette smoke.

It is not yet known whether such a novel chemical reaction as the one described above actually occurs in vivo. The fact that saliva after smoking a single cigarette does facilitate the formation of cyanomethylamine derivatives, however, suggests that such a reaction is quite possible, at least locally (i.e., in the lung). In the case of human smoking, the exposure of lung tissues to smoke could be quite extensive. The total surface area of the respiratory tract and the aveolar capullaries is very large. Cigarette smoke components could easily diffuse into these tissues and into the blood stream. It is known that the concentration of some primary biogenic amines (5-hydroxytryptamine and histamine, for example) are very high in the lung (Sadavongvivad, 1970). The lungs, in contrast to other organs, receive the total venous return and have been proposed to play a role in regulating the concentration of amines in venous blood before these amines reach the arterial circulation where they can exhibit profound effects (Alabaster, 1977; Juno, 1985). The level of 5-hydroxytryptamine in rat lung was substantially depleted following direct exposure to cigarette smoke in vitro. The interaction of lung amines and perhaps circulating amines with cigarette smoke components may be implicated in some clinical conditions. Cigarette smoking is a major environmental factor associated with the development of chronic obstructive pulmonary disease, cancer of the upper respiratory tract, lung, oral cavity, esophagus, pancreas, kidney and urinary bladder, and it has been implicated as a risk factor in the development of thrombosis, atherosclerosis and coronary heart diseases (Doll and Hill, 1964; U.S. Department of Health and Human Services, 1982; Greenhalgh, 1981).

On the other hand cigarette smoking seems to be related to low incidence of developing Parkinsonism and it is

much increased in patients suffering from schizophrenia and
other major psychiatric disorders (Hughes et al., 1986).
Whether these cyanomethyl or ring closed condensation
smoke-amine, smoke-amino acid or smoke protein adducts are
related to any of the above disorders or alternatively
whether such compounds exert chronic effects in such disorders or indeed exert any biological effects in their own
right remains to be established.

It is interesting to note that tetrahydro-β-carboline
and tetrahydroisoquinoline substances formed between
indoleamine and certain phenolic and catecholic amines, and
some components of cigarette smoke, are very similar to the
condensation products formed between these amines and acetaldehyde as produced in alcohol metabolism and/or alcoholism. Indeed, in addition to the compounds reported here,
we have also observed similar products formed between the
biogenic amines, acetaldehyde and cyanide, although at
somewhat lower concentrations. It is tempting to speculate
as to whether such substances might be involved in
addiction.

ACKNOWLEDGEMENTS

We thank Saskatchewan Health and the Medical Research
Council of Canada for continuing financial support and D.
Schneider, D. Young, L. Merrison, M. Mizuno and N.
Pidskalny for technical assistance.

REFERENCES

Alabaster V.A. (1977) Inactivation of endogenous amines in
the lungs. In, Metabolic Functions of the Lung (Y.S.
Bakkla and J.R. Vane, eds.), Marcel Dekker Inc., New
York, pp.
Barbeau A. and Pourcher E. (1982) New data on the genetics
of Parkinson's disease. Can. J. Neurol. Sci. 9, 53-60.
Baumann R.J., Jameson H.D., McKean H.E., Haack D.G. and
Weisberg M. (1980) Cigarette smoking and Parkinson
disease: 1. A comparison of cases with matched neigh-
bors. Neurology 30, 839-843.
Callegari C. and Lami F. (1984) Amylase activity and
cigarette smoke. Gut 25, 909.

Chiba K., Trevor A. and Castagnoli Jr. N. (1984) Metabolism of the neurotoxic tertiary amine, MPTP, by brain monoamine oxidase. Biochem. Biophys. Res. Comm. 120, 574-578.

Coursey R.D., Buchsbaum M.S. and Murphy D.L. (1979) Platelet MAO activity and evoked potentials in the identification of subjects biologically at risk for psychiatric disorders. Brit. J. Psychiat. 134, 372-381.

Doll R. and Hill A.B. (1964) Mortality in relation to smoking: Ten years' observations of British doctors. Brit. Med. J. 1, 1399-1410.

Durden D.A., Davis B.A., Yu P.H. and Boulton A.A. (1987) Formation of cyanomethyl-derivatives of biogenic amines from cigarette smoke: Their synthesis and identification. In, Proceedings of the 3rd Trace Amines Symposium, Humana Press, New Jersey, pp.

Duvoisin R. (1984) Is Parkinson's disease acquired or inherited? Can. J. Neurol. Sci. 11, 151-155.

Essman W.B. (1977) Serotonin and monoamine oxidase in mouse skin: Effects of cigarette smoke exposure. J. Med. 8, 95-102.

Fowler C.J., von Knorring L. and Oreland L. (1980) Platelet monoamine oxidase activity in sensation seekers. Psychiat. Res. 3, 273-279.

Greenhalgh R.M. (1981) Smoking and Arterial Disease. Pitman Med., Bath, United Kingdom.

Heikkila R.E., Hass A. and Duvoisin R.C. (1984) Dopaminergic neurotoxicity of 1-methyl-4-phenyl-1,2,5,6-tetrahydropyridine in mice. Science 224, 1451-1453.

Heikkila R.E., Manzino L., Cabbat F.C. and Duvoisin R.C. (1984) Protection against the dopaminergic neurotoxicity of 1-methyl-4-phenyl-1,2,5,6-tetrahydropyridine by monoamine oxidase inhibitors. Nature 311, 467-469.

Hoffman D. and Adams J.D. (1981) Carcinogenic tobacco-specific N-nitrosoamines in snuff and in the saliva of snuff dippers. Cancer Res. 41, 4305-4308.

Hoffman D. and Brunneman K.D. (1983) Endogenous formation of N-nitrosoproline in cigarette smokers. Cancer Res. 43, 5570-5574.

Hughes J.R., Hatzukami D.K., Mitchell J.E. and Dahlgren L.A. (1986) Prevalence of smoking among psychiatric outpatients. Am. J. Psychiat. 143, 993-997.

Janoff A., Carp H., Lee D.K. and Drew R.T. (1979) Cigarette smoke inhalation decreases α_1-antitrypsin activity in rat lung. Science 206, 1313-1316.

Juno A.F. (1985) 5-Hydroxytryptamine and other amines in the lung. In, Textbook of Physiology, Sec. 3, The Respiratory System, Vol. 1, Circulation and Non-respiratory Functions (A.P. Fishman, ed.) Bethesda, Maryland, pp. 337-349.

Kessler I.I. and Diamond E.L. (1971) Epidemiologic studies of Parkinson's disease: I. Smoking and Parkinson's Disease: A survey and explanatory hypothesis. Am. J. Epidemiol. 94, 16-25.

Littlewood J., Glover V., Sandler M., Langston J.W., Ballard P., Tetrud J.W. and Irwin I. (1983) Chronic Parkinsonism in humans due to a product of meperidine-analog synthesis. Science 219, 979-980.

Littlewood J., Glover V., Sandler M., Peatfield R., Petty R. and Rose F.C. (1984) Low platelet monoamine oxidase activity in headache: No correlation with phenolsulfo-transferase, succinate dehydrogenase, platelet preparation method or smoking. J. Neurol. Neurosurg. & Psychiat. 47, 338-343.

Marttila R.J. and Rinne U.K. (1980) Smoking and Parkinson's disease. Acta Neurol. Scand. 62, 322-325.

Murphy D.L., Belmaker R.H., Buchsbaum M., Martin N.F., Ciaranello R.D. and Wyatt R.J. (1977) Biogenic amine related enzymes and personality variation in normals. Psychological Med. 7, 149-157.

Norman T.R., Chamberlain K.G., French M.A. and Burrows G.D. (1982) Platelet monoamine oxidase activity and cigarette smoking. J. Affect. Disorders 4, 73-78.

Oreland L., Fowler C.J. and Schalling D. (1981) Low platelet monoamine oxidase activity in cigarette smokers. Life Science 29, 2511-2518.

Planz G., Quiring K. and Palm D. (1972) Rates of recovery of irreversibly inhibited monoamine oxidases: A measure of enzyme protein turn-over. Arch. Pharmacol. 273, 127-142.

Sadavongvivad C. (1970) Pharmacological significance of biogenic amines in the lung: 5-Hydroxytryptamine. Br. J. Pharmacol. 38, 353-365.

Tipton K.F. and Corte L.D. (1979) Problems concerning the two forms of monoamine oxidase. In, Monoamine Oxidase, Structure, Function and Altered Function (T.P. Singer, R.W. von Korff and D.L. Murphy, eds.), pp. 87-99, Academic Press, New York.

Turkish S., Yu P.H. and Greenshaw A.J. (1987) Monoamine oxidase inhibition: A comparison of in vivo and ex vivo measure of reversible effects. J. Neural Transmission (in press).

United States Department of Health and Human Services (1982) The health consequences of smoking in cancer. USPHS Publication No. 82-50179.

von Knorring L., Oreland L. and Winblad B. (1984) Personality traits related to monoamine oxidase activity in platelets. Psychiat. Res. 12, 11-16.

Ward C.D., Duvoisin R.C., Ince S.E., Eldridge R. and Calne D.B. (1983) Parkinson's disease in 65 pairs of twins and in a set of quadruplets. Neurology 33, 815-824.

Youdim M.B.H., Riederer P., Birkmayer W. and Mendelwicz J. (1979) The functional activity of monoamine oxidase: The use of deprenyl in the treatment of Parkinson's disease and depressive illness. In, Monoamine Oxidase: Structure, Function and Altered Functions (T.P. Singer, R.W. von Korff and D.L. Murphy, eds.) pp. 477-496, Academic Press, New York.

Yu P.H. (1986) Monoamine oxidase. In, Neuromethods, Vol. V: Neurotransmitter Enzymes (A.A. Boulton, G.B. Baker and P.H. Yu, eds.), Humana Press, New Jersey, pp. 235-272.

FORMATION OF CYANOMETHYL-DERIVATIVES OF BIOGENIC AMINES

FROM CIGARETTE SMOKE: THEIR SYNTHESIS AND IDENTIFICATION

D.A. Durden, B.A. Davis, P.H. Yu and A.A. Boulton

Neuropsychiatric Research Unit
CMR Building, University of Saskatchewan
Saskatoon, Saskatchewan, Canada S7N 0W0

The previous paper in this series (Yu et al., 1987) describes the finding that biogenic amines react with a component or components in an aqueous solution of cigarette smoke to produce new compounds. The compounds are readily extractable into an organic phase, are readily separable from the precursor amine by thin layer chromatography and appear to have relatively low molecular weights. These properties indicate that the unknown products would be amenable to identification by gas chromatography-mass spectrometry (GC-MS). This paper describes how these new compounds, with the aid of deuterated phenylethylamine, were identified to contain the N-cyanomethyl group, the postulated methods of formation, and the synthesis of other expected products from the other phenylalkyl, indolylalkyl and imidazolylalkyl amines.

METHODS AND MATERIALS

Materials

All solvents were of HPLC grade and water was 18 meg. ohm purity from a Barnstead Nanopure II system. The biogenic amines were obtained from Sigma (St. Louis, MO) and the deuterated amines 2,2-dideutero-phenylethylamine (PE-d_2) and 1,1-dideutero-5-hydroxytryptamine (5-HT-d_2) were synthesized in this laboratory. The derivatizing reagents, 1-dimethylaminonapthalene-5-sulphonyl chloride (dansyl

chloride and N-methyl-N-(t-butyldimethylsilyl)-trifluoro-
acetamide (MTBSTFA) were obtained from Pierce (Rockford,
IL).

Mass spectra were obtained using a VG 70-70F double
focussing mass spectrometer. It was interfaced to an
HP5700 GC which contains a megabore capillary column (J&W
DB1, 15 mx 0.53 mm i.d.) and a conventional wide bore
capillary column (J&W DBI 60 m x 0.32 mm i.d.), both using
helium as carrier gas. The MS was also equipped with a
direct insertion probe. Spectra were initially recorded at
1000 resolution and elemental compositions of significant
spectral ions at 5000 resolution.

Smoke solution. The cigarette solution was prepared
by simulated smoking (one puff of two second duration per
half minute) of cigarettes with indicated tar content of 16
mg and nicotine 1.2 mg. The smoke was bubbled through 3 ml
of 0.02 M phosphate buffer at pH 7.3. A "purified" smoke
solution was obtained by extracting the "raw" smoke solution three times with equal volumes of toluene/ethylacetate
(1:1, v/v).

Adduct formation. For the mass spectrometric identification, amine adducts were formed by reaction of approximately 10 ng of the appropriate amine with the "purified"
smoke solution of one cigarette, at 37°C for 30 min. The
adducts were then extracted into 1 ml toluene/ethylacetate
(1:1 v/v), and the volume was reduced to about 10 to 20 µl,
of which 0.5 µl was injected into the megabore column. The
analysis was then repeated using the conventional capillary
column.

Identification. For the initial identification an
equimolar mixture of phenylethylamine (PE) and 2,2-di-
deutero-phenylethylamine (PE-d$_2$) was used as substrate. In
a second experiment, an equimolar mixture of 5-hydroxy-
tryptamine (5-HT) and 1,1-dideutero-5-hydroxytryptamine
(5-HT-d$_2$) was used as reactant. Since the adduct from this
latter mixture was not amenable to GC analysis, the direct
probe was used. Ammonia chemical ionization spectra were
also obtained.

Once an initial identification had been made, the
1-dimethylaminonaphthalene-5-sulfonyl (dansyl) and t-butyl
dimethyl silyl (TBDS) derivatives were prepared. For

formation of the dansyl derivative, the organic solution of adduct was taken to dryness under a stream of nitrogen, re-dissolved in 1 ml water, saturated with sodium bicarbonate and reacted with dansyl chloride (8 mg) in acetone (2 ml) at 35° to 40° for 30 min. The solution volume was partial-ly reduced under nitrogen to remove the acetone and the derivative was extracted into toluene and its spectrum recorded after direct probe insertion. For formation of the TBDS derivative, the organic solution of adduct was taken to dryness under nitrogen, redissolved in 100 µl of acetonitrile in a Reactivial and reacted for 1 hour at 80°C with 100 µl of N-methyl-t-butyldimethylsilyltrifluoroacet-amide (MTBSTFA).

Chemical Synthesis

Adducts from PE, 5-HT and other biogenic amines were formed with the chemical reactants formaldehyde and sodium cyanide following the procedure of Winstead et al. (1978). A solution of 37% formaldehyde (1.5 eq.) and sodium bisul-phite (1.5 eq.) in 2-5 ml water was added to the amine (free base or hydrochloride) followed by sodium cyanide (1 eq.) in 2 to 5 ml water. After sitting 1 hour at room temperature the product was extracted with ether dried over anhydrous sodium sulphate and the solvent removed. In the case of non-extractable products, any precipitate was first removed by filtration and the supernatant lyophilized. At present, attempts to isolate and purify multiple products have not been made. In all cases, however, the spectra of the products were obtained after separation by GC-MS either as the free bases or as TBDS derivatives, or by selective evaporation from the direct probe.

RESULTS

Figure 1 shows the mass spectrum of the major product formed in the reaction of cigarette smoke solution with the equimolar mixture of PE and PE-d$_2$.

The ion pairs m/z 160, 162, 130, 132, 91 and 93 are due to ions containing the deuterium label whereas m/z 42 and 69 contain only the new part of the molecule. High resolution mass analysis established m/z 91 (measured at 91.0548) as C$_7$H$_7$$^+$ as expected (confirmed by m/z 93 as

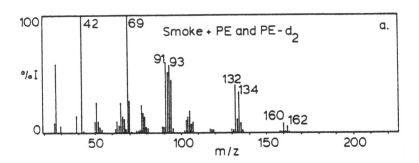

Figure 1. Mass spectrum of the extractable compound formed by reaction of PE and PE-d$_2$ with smoke solution.

C$_7$H$_5$^{2}H$_2$), m/z 69 (69.0466) as C$_3$H$_5$N$_2$$^+$, m/z 160 (160.0987) as C$_{10}$H$_{12}$N$_2$, and m/z 162 (162.1111) as C$_{10}$H$_{10}$^{2}H$_2$N$_2$. The ammonia chemical ionization spectrum and the spectrum of the dansyl derivative (M$^+$ 393.1450, C$_{22}$H$_{23}$N$_3$SO$_2$) establish-ed the molecular weight of the new compound to be 160, i.e. due to C$_{10}$H$_{12}$N$_2$. N-cyanomethylphenylethylamine, synthesized according to the procedure of Winstead et al. (1978), gave an almost identical spectrum to Figure 1 (without the deuterium ions) and the same GC retention time.

The spectrum of the compound prepared from the reac-tion of smoke solution with the equimolar mixture of 5-HT and 5-HT-d$_2$ is shown in Figure 2. The elemental

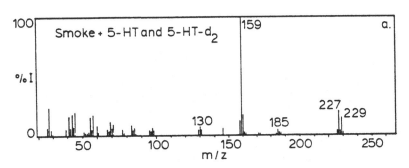

Figure 2. Mass spectrum of the extractable compound formed by reaction of 5-HT and 5-HT-d$_2$ with smoke solution.

composition was established to be ($C_{13}H_{13}N_3O$) for m/z 227 and $C_{13}H_{11}N_3O^2H_2$ for m/z 229. Synthetic 6-hydroxy-2-cyanomethyl-1,2,3,4-tetrahydro-beta-carboline was prepared from the reaction of 5-HT with the formaldehyde-cyanide solution and gave essentially identical mass spectral information. Two other compounds were also detected as formed in this reaction, 6-hydroxy-1,2,3,4-tetrahydro-β-carboline (M^+ = 188) and N-cyanomethyl-5-hydroxytryptamine (M^+ = 215). These latter compounds were, however, not as readily extractable into ether.

DISCUSSION

The biogenic amines contain many functional groups; primary and secondary amino, phenolic, alcoholic and activated carbons which could react with the initially unknown reactive components of cigarette smoke. By choosing PE as the initial substrate it was possible to limit reactions to those of the primary amino group and even though this would not necessarily apply completely to reactions of the other biogenic amines, it would give a clue to the identity of some of the newly formed compounds. PE had a second useful characteristic in that the mass spectrum of many derivatives of PE contain m/z 91 ($C_7H_7^+$) due to cleavage of the alpha-beta carbon bond. By adding to the mixture, PE labelled with the deuterium in the beta position, we would expect additional ions at m/z 93 due to $C_7H_5{}^2H_2{}^+$. Thus a computer search of the many spectra of the GC-MS analysis of the complex cigarette smoke-amine reaction for the simultaneous presence of m/z 91 and 93 would indicate only those compounds which were due to reaction with PE (and PE-d$_2$). From the initial complex mixture we were able to obtain a somewhat purified extract in which the major component from the PE/PE-d$_2$ reaction was dominant and thus identify the new compound. This compound was not listed in the mass spectrum library (of 30,000 compounds) that we used initially to find a possible identity of the compound with elemental composition $C_{10}H_{12}N_2$, but the reports by Langis et al. and Winstead et al. indicated that a N-cyanomethyl derivative of PE could be formed by reaction with formaldehyde and cyanide present in the cigarette smoke extract. Synthesis of N-cyanomethylphenylethylamine following the procedure of Winstead et al. (1978) confirmed the identity of the smoke extract compound.

A mechanism for formation of the cyanomethylamine derivatives in the presence of the reducing agent sodium bisulphite was proposed by Winstead et al. (1978) (Fig. 3).

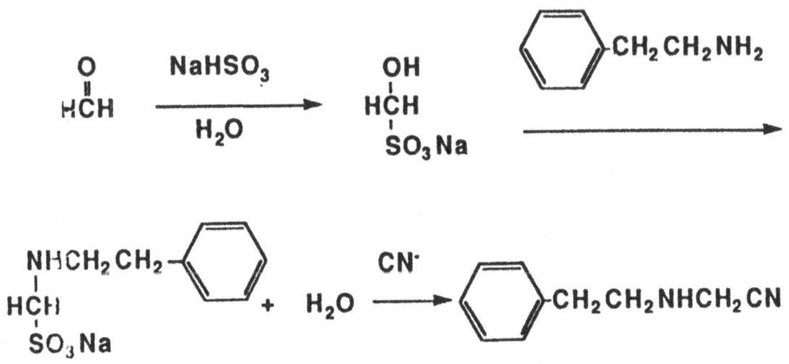

$$\text{H}\overset{\text{O}}{\overset{\|}{\text{C}}}\text{H} \xrightarrow[\text{H}_2\text{O}]{\text{NaHSO}_3} \text{H}\overset{\text{OH}}{\underset{\text{SO}_3\text{Na}}{\overset{|}{\text{C}}}}\text{H} \xrightarrow{\quad\quad} $$

with $\text{CH}_2\text{CH}_2\text{NH}_2$ substituted benzene ring above.

$$\text{H}\overset{\text{NHCH}_2\text{CH}_2-\bigcirc}{\underset{\text{SO}_3\text{Na}}{\overset{|}{\text{C}}}}\text{H} + \text{H}_2\text{O} \xrightarrow{\text{CN}^-} \bigcirc\text{-CH}_2\text{CH}_2\text{NHCH}_2\text{CN}$$

Figure 3. Proposed mechanism for formation of cyanomethyl phenylethylamine involving sodium bisulphite.

We have, however, observed that the reaction can proceed without bisulphite and a more direct mechanism may be involved as in Figure 4.

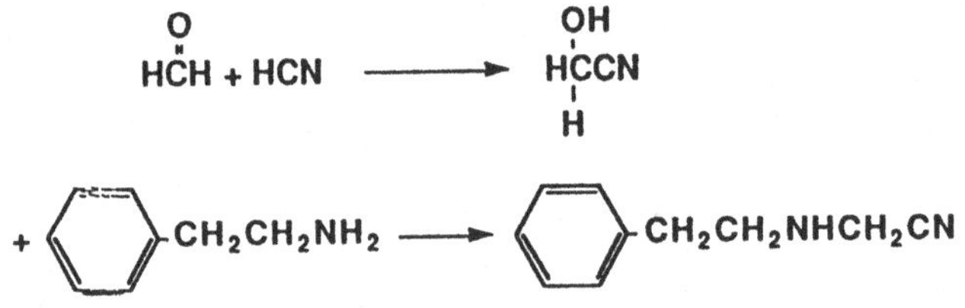

$$\text{H}\overset{\text{O}}{\overset{\|}{\text{C}}}\text{H} + \text{HCN} \longrightarrow \text{H}\overset{\text{OH}}{\underset{\text{H}}{\overset{|}{\text{C}}}}\text{CN}$$

$$+ \bigcirc\text{-CH}_2\text{CH}_2\text{NH}_2 \longrightarrow \bigcirc\text{-CH}_2\text{CH}_2\text{NHCH}_2\text{CN}$$

Figure 4. Mechanism for direct formation of cyanomethylphenylethylamine.

Reaction of formaldehyde and cyanide in cigarette smoke with 5-HT would be expected to form up to three compounds as shown in Figure 5. Formaldehyde alone would produce 6-hydroxy-1,2,3,4-tetrahydro-β-carboline via Pictet-Spengler condensation; formaldehyde and cyanide

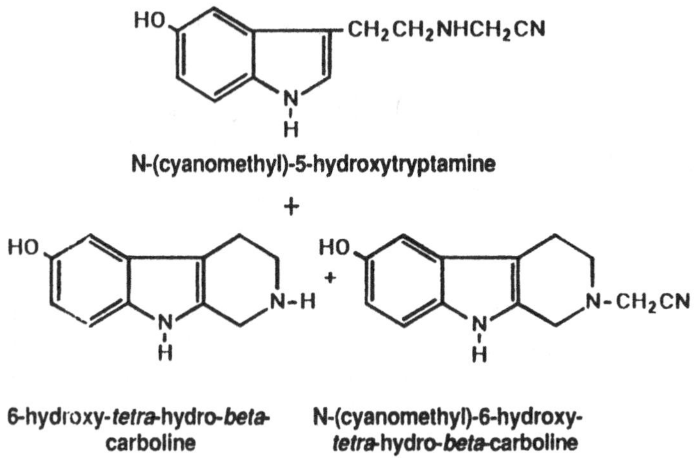

N-(cyanomethyl)-5-hydroxytryptamine

+

6-hydroxy-*tetra*-hydro-*beta*-carboline   N-(cyanomethyl)-6-hydroxy-*tetra*-hydro-*beta*-carboline

Figure 5. Products of reaction of 5-HT with smoke solution or a solution of formaldehyde and cyanide.

(1 equivalent of each) would produce 5-hydroxy-N-cyano-methyltryptamine in a reaction analagous to that involving PE, and reaction of two equivalents of HCHO and one of cyanide would be expected to form the 6-hydroxy-2-cyano-methyl-1,2, 3,4-tetrahydro-β-carboline via an initial Pictet-Spengler condensation followed by cyanomethylation of the now secondary amino group. The latter compound was then identified as the extractable product obtained from the reaction of 5-HT with cigarette smoke solution. We have not yet identified the former two compounds in the smoke solution reaction, as the method to do so is under development.

The other biogenic amines may be expected to react with cigarette smoke solution to produce a variety of com-pounds; the expected products of reaction of formaldehyde/cyanide solution are indicated in Table 1. Thus para-tyramine (p-TA) and para-octopamine (p-Oct) may be expected to produce only the N-cyanomethyl compounds as indicated. The other phenylalkylamines, i.e. meta-tyramine (m-TA), dopamine (DA), meta-octopamine (m-Oct), noradrenaline (NA), with their meta hydroxy group which activates the number 6 position on the ring would also be expected to undergo

Possible Products of Reaction of Biogenic Amines
with Cigarette Smoke Extract

Amine	N-cyanomethyl compound	Alkaloid	N-cyanomethyl alkaloid	Potential Number of Products
Phenylalkyl				
PE	X			1
p-TA	X			1
p-Oct	X			1
m-TA	X	X	X	3
m-Oct	X	X	X	3
DA	X	X	X	3
NA	X	X	X	3
Indolyalkyl				
Trypt	X	X	X	3
5-HT	X	X	X	3
Imidazolylalkyl				
Hist	X	X	X	5

Notes:
1. Alkaloid from phenylalkyls is a tetrahydroisoquinoline.
2. Alkaloid from indolylalkyls is a tetrahydro-β-carboline
3. Alkaloid from imidazolylalkyls is a tetrahydro-6-aza-benzimidazole.

Pictet-Spengler cyclization with formaldehyde. Thus this
group could form three compounds for each amine, the
N-cyanomethyl compound, the tetrahydroisoquinoline, and the
N-cyanomethyl-tetrahydroisoquinoline as in Figure 6.
Tryptamine (T) similarly forms three compounds, the
N-cyanomethyl, the tetrahydro-β-carboline and the N-cyano-
methyl-tetrahydro-β-carboline again analogous to products
of 5-HT. Histamine (Hist), the other biogenic amine, could
potentially be expected to produce many products, the
N-cyanomethyl of both the primary and secondary amino
groups (3 products), the cyclization to produce an alkaloid, tetrahydro-6-azabenzimidazole, and combined products
of the alkaloid with cyanomethylation of the amino groups
(Fig. 7). We have identified four of these products at

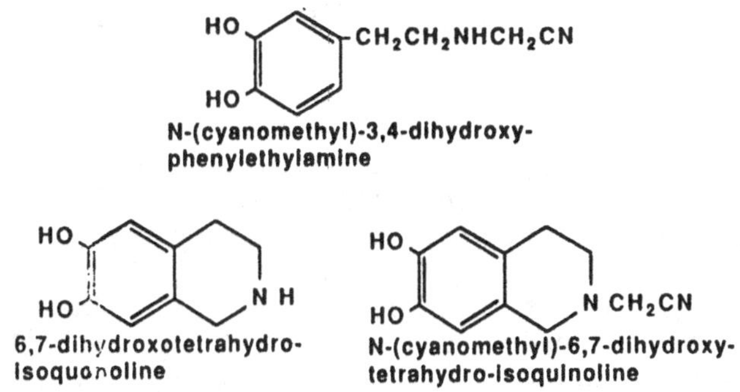

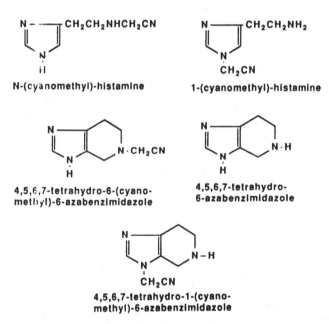

Figure 6. Possible products of reaction of dopamine with smoke solution or a solution of formaldehyde and cyanide.

Figure 7. Possible products of reaction of histamine with smoke solution or a solution of formaldehyde and cyanide.

present. We are now developing methods to test whether these compounds are formed in vivo. The 5-HT product has already been detected in human saliva after smoking even a single cigarette.

It is interesting to speculate the effect that these compounds in the lungs and circulatory system of smokers would have on their behaviour and health. The previous paper (Yu et al., 1987) indicates that they have some MAO inhibitory activity and we are synthesising the compounds in high purity to test this possibility. The alkaloids involving the tetrahydroisoquinoline and beta-carboline structures are similar to compounds associated with addiction in alcoholism and similar natural compounds (harmaline, etc.). Finally it has been observed (Winstead et al., 1978) that these compounds decompose under physiological conditions to release cyanide. Their high lipophilicites could thus facilitate transport of cyanide to parts of the body not normally available to the free cyanide ion.

ACKNOWLEDGEMENTS

We wish to thank R. Janzen and N. Pidskalny for assistance with preparation of the derivatives and with the GC-MS analysis and Saskatchewan Health for continuing financial support.

REFERENCES

Langis, A.L. and Steger, M.G.P. (1965) U.S. Patent 3202674; Chem. Abstr. 63, 14872a.

Winstead, M.B., Ciccarelli, C.A. and Winchell, H.S. (1978) Liberation of cyanide from α-aminonitriles relative to amygdalin. J. Pharm. Exp. Therap. 205, 751-756.

Yu, P.H., Durden, D.A., Davis, B.A. and Boulton, A.A. (1987) Inhibition of monoamine oxidase, depletion of lung biogenic amines by cigarette smoke and the interaction of some components of this smoke with trace and other amines. In, The Trace Amines: Their Comparative Neurobiology and Clinical Significance (eds. Boulton, Juorio and Downer) Humana Press, New Jersey, 1987.

THE TYRAMINE CONJUGATION TEST IN UNIPOLAR AND BIPOLAR

DEPRESSED PATIENTS: UTILITY AS A PREDICTOR

OF TREATMENT RESPONSE

M. Sandler[‡], A.S. Hale[*], P. Hannah[‡], P.K. Bridges[*]

[‡]Bernhard Baron Memorial Research Laboratories, Queen Charlotte's Hospital, London, W6 OXG, UK. [*]The Geoffrey Knight Unit for Affective Disorders, Brook General Hospital and Guy's Hospital Medical School, London, UK.

Over the last 12 years, evidence has accumulated that depressed patients show abnormal metabolism of orally-administered loads of tyramine, excreting significantly less tyramine-O-sulphate than normal controls (Sandler et al., 1975; Harrison et al., 1984; Hale et al., 1986). These studies have demonstrated that the abnormality is present only in that group of patients on whom the presence of characteristic biological symptoms and signs confer the label, endogenous depression. Patients without these features (variously labelled neurotic, non-endogenous, atypical, mood-reactive, minor) have a tyramine response to oral challenge indistinguishable from that of normal controls. Such tyramine excretion appears to be independent of severity of illness, age and sex. The abnormality persists after recovery from an episode of illness (Bonham Carter et al., 1978; Harrison et al., 1984) and is present in non-affected first degree relatives of patients (Hale et al., 1986). It may thus represent a trait marker for endogenous depression.

Current clinical practice suggests that patients with an endogenous clinical picture respond more readily to treatment with tricyclic antidepressants, provided that psychotic features are not present (Paykel and Hale, 1986). We wondered, therefore, whether patients showing a low tyramine excretion response would be more likely to respond to tricyclic antidepressants than those with

higher output values. The current study was designed to
test this hypothesis in a cohort of depressed inpatients.
Patients were drawn from a larger group from whom those
with psychotic symptoms were excluded. Results after
treatment with electroconvulsive therapy in the psychotic
subgroup will be presented elsewhere.

Analysis of preliminary results pointed to a strong
group effect for bipolar depressed patients which led us
to question the findings of Harrison et al. (1984) that
tyramine excretion following oral challenge in depressed
bipolar patients is similar to that of unipolars. We
report a direct comparison between carefully matched groups
of unipolar and bipolar depressed patients, the results of
which provide a rationale for excluding bipolar cases from
the main treatment response study.

UNIPOLAR-BIPOLAR COMPARISON

Method

The tyramine excretion test was performed on a
consecutive series of 14 hospitalised bipolar patients,
and a similar number of unipolar depressive patients
matched for age and sex. Patients were all diagnosed
according to the Research Diagnostic Criteria (Spitzer
et al., 1977). All patients had been free of psychoactive drugs, apart from benzodiazepine night sedation, for
at least 2 weeks prior to testing and had been hospitalised
within the previous week. Severity of depressive symptoms
at the time of testing was determined using the Hamilton
Depression Scale (17 item) (HamD) (Hamilton, 1960).

Ten of the bipolar patients were depressed at the
time of testing and four were hypomanic. Unipolar and
bipolar patients were compared as groups, both overall and
after excluding the hypomanic patients and their matched
unipolar cases. Groups were compared using the Mann-
Whitney U test.

Results

The groups were well matched for age (bipolar
mean=51.36 years + 13.43 S.D.; unipolar 51.50 years $\pm$ 13.51
S.D.) and sex (both groups 4 males, 10 females). Mean
tyramine excretion for the bipolar group was 5.70mg/3h $\pm$

2.51 S.D. This was significantly higher (p<0.001) than for the unipolar group (mean 2.33mg/3h + 0.90 S.D.). HamD scores were significantly different between the two groups of all 14 cases (p<0.05) but not for the matched groups excluding the hypomanic bipolar cases. Tyramine excretion for these smaller groups (bipolar depressive, mean 5.22mg/3h + 2.48 S.D.; unipolar depressives 2.33mg/3h + 0.98 S.D.) was significantly different (p<0.001). These results are shown in Fig.1.

Conclusions

Unipolar and bipolar depressed patients differed markedly on tyramine excretion rates after oral challenge, to the extent that it seemed justifiable to exclude bipolar cases from the treatment response study. Indeed, results for tyramine excretion for the bipolar cases were similar to those previously reported for normal controls (Hale et al., 1986).

TREATMENT RESPONSE PREDICTION STUDY

Method

Thirty unipolar depressive inpatients (23 female, 7 male) fulfilling Research Diagnostic Criteria for Major Depressive Disorder, definite or probable, with Hamilton Depression Scale (17 item) scores greater than 18, were given a tyramine excretion test within one week of admission, and prior to being given medication. All subjects had been drug free for at least two weeks at the time of the test. Patients with psychotic features were excluded as it was thought probable that they would require electroconvulsive therapy; hence, entry into the present study would not have been clinically justifiable.

Patients were treated as inpatients with 150mg of a tricyclic antidepressant (either imipramine, amitriptyline or clomipramine) for a period of 8 weeks. At the end of this period, patients were reassessed with the HamD. It was thus possible to decide on a number of criteria by which patients had improved compared with pre-treatment. Two criteria were used for the analysis, to divide patients into responder and non-responder groups; patients were considered to have responded if their final

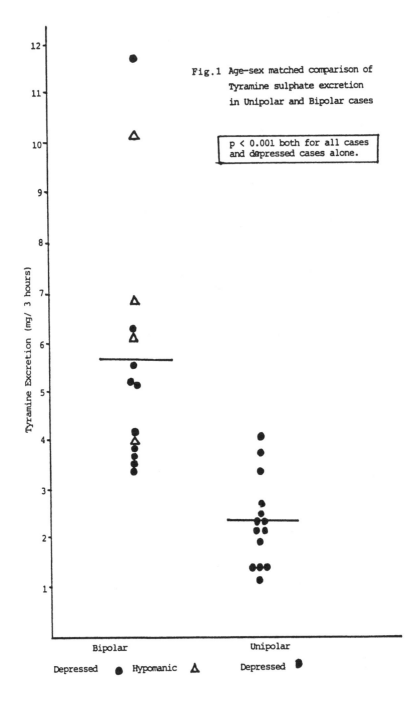

Fig.1 Age-sex matched comparison of
Tyramine sulphate excretion
in Unipolar and Bipolar cases

p < 0.001 both for all cases
and depressed cases alone.

HamD score fell below 12, and if their 8 week HamD score had fallen to less than 50% of their pre-treatment score. Groups were then compared by pre-treatment tyramine excretion response. Tyramine excretion was also correlated with HamD percentage change, initial HamD, final HamD and age.

Results

Mean age of patients was 46.87 years + 13.47 S.D. Mean initial HamD was 26.33 + 6.21 S.D. Mean pre-treatment tyramine excretion was 3.68mg/3h + 1.55 S.D. Using the percentage change in HamD criterion, 15 patients improved and 15 failed to improve, during the 8 weeks of the study. Using a final HamD score of <12, 13 patients improved and 17 failed to do so.

Mean tyramine excretion at baseline for responders (>50% change in HamD) was 2.68mg/3h + 1.12 S.D. and for non-responders 4.67mg/3h + 1.25 S.D. Excretion was significantly lower in responders (p<0.01) (Fig.2).

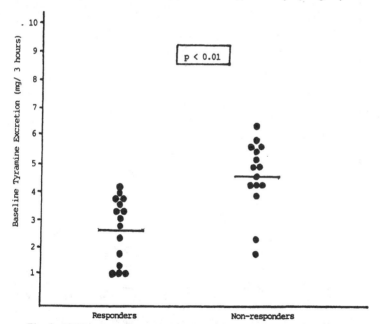

Fig.2 .TYRAMINE EXCRETION AT BASELINE IN UNIPOLAR DEPRESSIVES WHO SHOWED RESPONSE AND NON-RESPONSE AFTER 8 WEEKS TREATMENT WITH 150 mg TRICYCLIC ANTIDEPRESSANT DAILY

Tyramine excretion at baseline showed a highly significant correlation with percentage change in HamD score after
8 weeks treatment (rho= -0.62, p=0.0005) (Fig.3). There
was also a significant correlation with final HamD score
(rho= 0.51, p=0.003). There was no significant correlation
with initial HamD score or with age. Age did not correlate
with any measure of improvement. There was a weak, non-
significant relationship between initial HamD score and
percentage change. There was no significant relationship
between initial and final HamD scores.

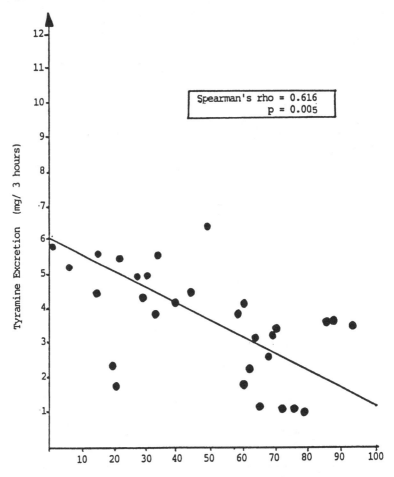

Fig.3   % CHANGE IN HamD SCORE BASELINE TO 8 WEEKS
          TREATMENT WITH TRICYCLIC ANTIDEPRESSANTS,
          150 mg DAILY.

Using the previously determined cutoff point of
4.1mg/3h to divide tyramine excretion into high and low
groups (Hale et al., 1986), there was also a significant
relationship to both measures of improvement (Table 1).

Table 1

		Final HamD score		Percentage change in HamD score from baseline to 8 weeks treatment		
		<12	>12		>50%	<50%
Tyramine Excretion (mg/3 hours)	<4.1	13	5	<4.1	1	3
	>4.1	0	12	>4.1	0	12

chi-squared=12.49   chi-squared=16.81
p=0.004   p<0.0001

Tyramine excretion measured at baseline in a group of
unipolar depressed cases predicts response to 8 weeks
treatment with 150mg of tricyclic antidepressant
(amitriptyline, imipramine or clomipramine). No relationship showed between baseline tyramine excretion and baseline
HamD (severity) score. Tyramine excretion does not change
with tricyclic treatment, in keeping with its putative
status as a trait rather than state marker for unipolar
depressive illness.

The second finding demonstrated above is that of a
highly significant relationship between pre-treatment
tyramine excretion and response to tricyclic antidepressants
in normal clinical doses, over an assessment period of
adequate length (Quitkin et al., 1984). As both initial
and final assessment of clinical state was performed blind
to tyramine-O-sulphate determination, it would seem that,
in this group of patients with major affective disorder,
tyramine excretion was a reasonable predictor of treatment
response. As all patients in the study had sufficient
biological features to satisfy RDC major depressive
disorder criteria, a trial of a tricyclic drug in the doses
given was a clinically reasonable procedure. Whilst there
was a weak relationship between initial severity and change,

this was not significant. Evidence that it is possible to
predict response to the tricyclics on the basis of such
constructs as endogenicity is scanty (Paykel, 1979),
despite widespread clinical belief to the contrary. There
was no way clinically of differentiating between the
responders and the non-responders in the sample reported
above.

Hence, in patients with unipolar depressive illness
in whom there is clinical justification for a trial of
tricyclic antidepressants, pre-treatment tyramine
excretion may predict treatment outcome. As it has been
demonstrated that tyramine excretion is robust to
treatment with tricyclics (Hale et al., 1986), it is
possible that the procedure may also have utility in
patients who have failed to respond at low dosage or short
trials of treatment. It is clear that some patients who
eventually respond show no sign of doing so for the first
four or five weeks (Quitkin et al., 1984). There is also
a group of patients who respond to tricyclic medication
only in large daily doses (Paykel and Hale, 1986). Hence,
a low tyramine excretion result in a non-responder might
suggest the need for a dosage increase and/or patience.

REFERENCES

Bonham Carter, S.M., Sandler, M., Goodwin, B.L., Sepping, P.,
    and Bridges, P.K. (1978) Decreased urinary output of
    tyramine and its metabolites in depression.
    Br. J. Psychiat. 132, 125-32.
Hale, A.S., Walker, P.L., Bridges, P.K., and Sandler, M.
    (1986) Tyramine conjugation deficit as a trait marker
    in endogenous depressive illness. J. Psychiat. Res.
    20, 251-61.
Hamilton, M. (1960) A rating scale for depression. J. Neurol.
    Neurosurg. Psychiat. 23, 56-62.
Harrison, W.M., Cooper, T.B., Stewart, J.W., Quitkin, F.M.,
    McGrath, P.J., Leibowitz, M.R., Rabkin, J.R.,
    Markowitz, J.S., and Klein, D.F. (1984) The tyramine
    challenge test as a marker for melancholia. Arch. Gen.
    Psychiat. 41, 681-85.
Paykel, E.S. (1979) Predictors of treatment response. in:
    Psychopharmacology of Affective Disorders (Paykel, E.S.
    and Coppen, A., eds), pp.193-220, Oxford University
    Press, Oxford.

Paykel, E.S., and Hale, A.S. (1986) Recent advances in the
    treatment of depression. in: The biology of
    Depression (Deakin, J.F.W., ed), pp.153-173, Gaskell,
    London.
Quitkin, F.M., Rabkin, J.G., Ross, D., and McGrath, P.J.
    (1984) Duration of antidepressant treatment. What is
    an adequate trial. Arch. Gen. Psychiat. 41, 238-45.
Sandler, M., Bonham Carter, S., Cuthbert, M.F., and
    Pare, C.M.B. (1975) Is there an increase in monoamine
    oxidase activity in depressive illness? Lancet i,
    1045-49.
Spitzer, R.L., Endicott, J., and Robbins, E. (1977)
    Research Diagnostic Criteria (RDC) for a selected
    group of functional disorders. New York Psychiatric
    Institute, New York.

DECREASED   TYRAMINE   CONJUGATION   IN   MELANCHOLIA:

A   TRAIT   MARKER

Thomas B. Cooper, Wilma M. Harrison,
Jonathan W. Stewart and Donald F. Klein

New York State Psychiatric Institute
722 West 168th Street, New York, N.Y. 10032
and Nathan Kline Institute, Orangeburg, N.Y. 10962.

Detailed accounts of the investigation of a tyramine
conjugate deficit as a trait marker for depressive illness
can be found in a review by Sandler et al. (1984) and
papers by Harrison et al. (1984) and Hale et al. (1986).

We describe herein a brief synopsis of work in this
area, a description of the details of application of the
test in our clinic and present additional evidence
suggesting that the role of MAO in this deficit warrants
re-evaluation.

The discovery of trait markers in biological
psychiatry is of crucial heuristic as well as practical
importance. State dependent variables while useful in
diagnosis and in monitoring recovery may simply depend
upon epiphenomena, whereas state independent (trait)
markers are much more likely to be closely linked to the
underlying pathophysiological predisposition to depressive
episodes. In a series of papers Sandler et al. (1975)
and Bonham Carter et al. (1978) have reported that patients
with severe chronic unipolar depressive illness excreted
significantly lower amounts of free and conjugated urinary
tyramine in response to an oral tyramine load when
compared with controls, and that this deficit remained
in patients who had completely recovered from depression.
They also reported that in a group of apparently normal
pregnant women whose psychiatric histories were unknown,

those with the lowest output of urinary tyramine after an
oral tyramine load were found to have higher lifetime
incidence of depressive illness compared with those with
the highest tyramine output (Bonham Carter et al. 1980).
The low tyramine excretors were found prospectively to have
a higher chance of developing a post-partum depression. As
none of these women were depressed when loaded with
tyramine, the decreased excretion of tyramine appears to
be associated with a lifetime vulnerability to depressive
illness. A screening test for vulnerability to depressive
illness was suggested.

Tyramine has also been implicated in the search for
trait markers by the observation of Ghose et al. (1975)
who reported that depressed patients show a hyper-
responsiveness to the infusion of tyramine by the intravenous route and that this response was state independent
i.e. a trait marker.

BACKGROUND TO AND APPLICATION OF THE TEST

The test requires that the patient observe a low
monoamine diet for 24 hours preceding the test. On the
morning of the test the patient voids urine, ingests 125 mg
tyramine hydrochloride ($\approx$ 100 mg tyramine) and collects all
urine for the next three hours. The urine volume is
recorded, and acidified aliquots are stored frozen until
analysis. The test is therefore simple to carry out, of
minor inconvenience to the patient and the initial processing of urine can be accomplished by ward personnel.

The selection of the three hour urine collection
design is based upon the observation of Bonham Carter et al.
(1970) that "most" metabolized tyramine is excreted within
three hours post loading dose. The data of Sandler et al.
(1975) showed at least twice as much conjugated tyramine is
excreted during the first three hour period compared to the
second three hour period. In our experiments we have collected blood samples over a 12 hour period in four subjects
and determined the conjugated tyramine plasma concentration.
These data are shown in Figure 1. The $AUC_{0-3 \text{ hrs.}}$
represents only 50-60% of the $AUC_{0-12 \text{ hrs.}}$ and therefore
the three hour collection period may not be optimal. A
different time interval 0 - 4, 5, 6 or 8 hours may yield
additional statistical power in the separation of the

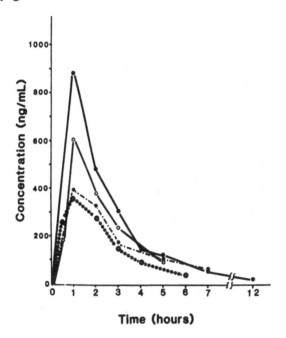

FIG. 1. Plasma tyramine-o-sulfate concentration in four subjects following 100 mg tyramine p.o.

diagnostic groups. The convenience and practicality of the three hour collection period however cannot be over-emphasized. The data in Figure 1 appear to indicate that a 6 or 8 hour collection period may represent > 80% of the AUC. These data also indicate that urine collections from 12 - 24 hours may in fact decrease statistical power by including 12 hours of endogenous conjugated tyramine production during which time little conjugated tyramine remains from the loading dose. While it does appear that a more detailed investigation of the collection time inter-val is warranted we emphasize that the longer the time interval of urine collection the less willing the patient will be to remain in an outpatient department waiting room! General experience with urine collection in an unsupervised outpatient study, i.e. the patient collecting timed inter-val samples at home or at work have proven totally unsatis-factory. We may therefore have to accept the compromise of the three hour interval because we have generated consider-able data and patients may find additional time constraints too onerous.

ASSAY PROCEDURE

The three hour urine volume is recorded and aliquots are acidified and stored frozen until assay. The assay procedure involves acid hydrolysis, liquid/liquid extraction followed by HPLC separation and amperometric detection.

A standard curve is generated with each analytical run. This consists of a range of tyramine-o-sulfate concentrations and a fixed amount of internal standard (3 methoxytyramine). This standard curve is then processed exactly as the unknown samples. The tyramine-o-sulfate used as standard was prepared in our laboratory as described by Bonham-Carter et al. (1978) with an additional chromatographic clean up step.

Samples and standard curve samples are adjusted to pH 1.0 and heated to 90°C for 30 minutes. After cooling, the tube contents are adjusted to pH 10 using a borate buffer and the sample extracted with ethyl acetate (passed through an alumina column immediately prior to use). The organic supernatant is then back extracted with 0.1 N hydrochloric acid and the extract diluted 1 : 10 in the HPLC mobile phase (0.05 M KH_2PO_4 and acetonitrile (97 : 3). Chromatographic conditions are: column 250 x 4.6 mm. i.d. Whatman Partisil 10-ODS-3, flow rate 1.5 ml/min and amperometric detector potential 1.15V with reference to an Ag/AgCl electrode. Chromatograms are free of interfering peaks with retention times for tyramine and internal standard 4.8 and 7 minutes respectively. The procedure has high precision R.S.D. $<$ 6% and accuracy (recovering 98 $\pm$ 3%). Excellent linearity over the entire concentration range expected in urine or plasma, the additional specificity of the detector and consistently clean chromatograms have made this procedure our method of choice. Plasma samples are processed in identical fashion using 2 ml plasma with appropriate reduction in concentration of internal standard and standard concentration range.

CONFIRMATION AND EXTENSION OF ORIGINAL FINDINGS

We attempted to replicate the study of Sandler et al. (1975) and to extend it to less severely depressed patients, bipolar patients and to nonendogenous subtypes

of depression (Harrison, et al. 1984). These studies
confirmed a decreased urinary conjugated tyramine excretion
in patients with melancholia (unipolar and bipolar), but
not in nonendogenous depressives when these groups were
compared to normals. This latter observation of the specificity of the measure for melancholia has recently been
confirmed by Hale et al. (1986).

An ANCOVA with tyramine sulfate excretion as the
dependent variable, the diagnostic group and sex as factors,
and age as the covariate, demonstrated that the main
effects of sex and diagnostic group were significant
(F = 8.1; p < .001) and age was not a significant covariate when sex and group were held constant. Tyramine
sulfate excretion was lower for female subjects than for
males as a main effect.

In a "post hoc" analysis of tyramine sulfate excretion
levels we investigated this sex difference and found a
natural break point in our data (females < 5 mg/3 hours and
males < 6 mg/3 hours) that appeared to differentiate
between patients with melancholia and normal controls. In
a prospective study of melancholic patients and normal
volunteers these natural break points were used and found
to separate the diagnostic groups accurately. Pooled data
from these experiments are shown in Table 1. This sex
difference had not been previously described but recently
Hale et al. (1986) reported that normal males had significantly higher tyramine conjugate excretion than normal
females (p < 0.05). However these same authors were again
unable to demonstrate a sex difference in their depressive
groups.

Table 1

Tyramine Sulfate Excretion mg/3 Hours

	Females < 5 mg Males < 6 mg	Females > 5 mg Males > 6 mg
Normals	6	27
Melancholia Unipolar and Bipolar	37	5

Chi^2 34.12 (p < 0.001)

In 12 of our patients the tyramine sulfate excretion levels were tested after recovery from episodes of melancholia. At the time of testing these patients were free of medication. All 12 had excretion values below the empirical cut-off points, thus confirming the state independent character of this marker.

POSSIBLE MECHANISM OF DEFICIT

Investigation of possible mechanisms have included altered gut motility, sulfate depletion, sulfate conjugation and MAO activity (Bonham Carter et al. 1978, 1980, 1981; Sandler et al. 1984).

The hypothesis with most intuitive appeal is that of increased MAO in this group of patients. In the most simplistic model a minor (5 - 10%) increase in gut MAO activity could decrease the availability of tyramine for sulfate conjugation by up to 50%. Recently Georgotas et al. (1986) reported a study of MAO activity in 67 patients with the DSM III diagnosis of major depression, 26 of whom met the additional criteria for melancholia. The melancholic patients had significantly higher platelet MAO activity than those without melancholia ($p < .05$), data controlled for age and sex effects on MAO activity.

In a study of platelet MAO activity in 1,129 males Cloninger et al. (1985) found that a five component model was required to adequately describe the distribution. They were able to show that the upper and lower deciles contain contributions from two extreme components that differ markedly from a much larger intermediate component with activity near the mean of the general population.

A recent observation from our laboratory (Stewart et al. 1987) has been the finding that the tyramine excretion test may predict treatment response to an MAOI (phenelzine). In a test of 84 patients randomly assigned to imipramine, phenelzine or placebo, tyramine sulfate excretion was significantly lower in responders in the phenelzine group (29 of 37, $p < 0.01$). This preliminary finding again raises the question of MAO activity in these subjects. The MAO levels at baseline from these patients did not correlate with tyramine excretion but as the phenelzine group contained only eight nonresponders statistical power is lacking.

Given these three observations (Georgotas et al., 1986; Cloninger et al., 1985; Stewart et al., 1987) the indication is clear: before we can rule out MAO activity as the biochemical lesion, a large N study of melancholic and non-melancholic subjects with and without the tyramine conjugation deficit will require detailed concomitant platelet MAO evaluation.

CONCLUSION

Essentially two research groups have reported upon clinical studies of single oral dose tyramine loading. The preponderance of experimental data has been generated by the originator of the hypothesis and recently reviewed by this group (Sandler et al., 1984). Our group has published their findings in Harrison et al. (1984) and Stewart et al. (1987). We have confirmed essentially in every detail the original findings: the laboratory data generated by completely different analytical procedures show good agreement in terms of mean and range of values and the findings have been confirmed and extended in prospective studies. It is hoped that other clinical investigators will join in further field evaluations of this promising, safe and technically simple procedure.

Acknowledgements: This work was supported, in part, by NIMH Grant No. MH 30906. The excellent technical assistance of Ms Rhoda Lonow, Ms Elaine Tricamo and the nursing staff of the Depression Evaluation Service at Pschiatric Institute is gratefully acknowledged.

REFERENCES

Bonham Carter, S., Karoum, F., Sandler, M., and Youdim, M.B.H. (1970) The effect of tyramine on phenolic acid and alcohol excretion in man. Br. J. Pharmacol. 39, 202P.

Bonham Carter, S.M., Sandler, M., Goodwin, B.L., Sepping, P., and Bridges, P.K. (1978) Decreased urinary output of tyramine and its metabolites in depression. Br. J. Psychiat. 132, 125-132.

Bonham Carter, S.M., Reveley, M.A., Sandler, M., Dewhurst,
    J., Little, B.C., Hayworth, J. and Priest, R.G. (1980)
    Decreased urinary output of conjugated tyramine is
    associated with lifetime vulnerability to depressive
    illness. Psychiat. Res. 3, 13-31.

Bonham Carter, S.M., Glover, V., Sandler, M., Gillman,
    P.K. and Bridges, P.K. (1981) Human platelet phenol-
    sulphotransferase: separate control of the two forms
    and activity range. Clin.Chim.Acta 117, 333-344.

Cloninger, C.R., vonKnorring, L. and Oreland, L. (1985)
    Pentametric distribution of platelet monoamine oxidase
    activity. Psychiat. Res. 15, 133-143.

Georgotas, A., McCue, R.E., Friedman, E., Hapworth, W.E.,
    Kim, O.M., Cooper, T.B., Chang, I. and Stokes, P.
    (1986) Relationship of platelet MAO activity to
    characteristics of major depressive illness. Psychiat.
    Res. 19, 247-256.

Ghose, K., Turner, P. and Coppen, A. (1975) Intravenous
    tyramine pressor response in depression. Lancet 1,
    1317-1318.

Hale, A.S., Walker, P.L., Bridges, P.K. and Sandler, M.
    (1986) Tyramine conjugation deficit as a trait-marker
    for endogenous depression. J. Psychiat. Res. 20, 251-
    261.

Harrison, W.M., Cooper, T.B., Stewart, J.W., Quitkin, F.M.,
    McGrath, P.J., Leibowitz, M.R., Rabkin, J.R.,
    Markowitz, J.S. and Klein, D.F. (1984) The tyramine
    challenge test as a marker for melancholia. Arch.
    Gen.Psychiat. 41, 681-685.

Sandler, M., Bonham Carter, S.M., Cuthbert, M.F. and Pare,
    C.M.B. (1975) Is there an increase in monoamine
    oxidase activity in depressive illness? Lancet, 1,
    1045-1049.

Sandler, M., Bonham Carter, S.M. and Walker, P.L. (1984)
    Tyramine and depressive illness. In: Neurobiology of
    The Trace Amines (Boulton, A.A., Baker, G.B.,
    Dewhurst, W.G. and Sandler, M. eds.) pp.487-498.
    Humana Press, Clifton, N.J. U.S.A.

Stewart, J.W., Harrison, W.M., Cooper, T.B. and Klein, D.F.
    Manuscript in preparation.

DEUTERATED TYRAMINE CHALLENGE TEST AND PLATELET MAO AND

PST ACTIVITIES IN NORMAL, MIGRAINE AND DEPRESSIVE SUBJECTS

B.A. Davis[*], S. Blackshaw[†], R. Bowen[†], S. Cebrian-
Perez[§], B. Dawson[°], N.R. Dayal[§], D.A. Durden[*],
S. Saleh[§], S. Shrikhande[§], P.H. Yu[*], D. Keegan[†]
and A.A. Boulton[*]

Neuropsychiatric Research Unit[*], Departments of
Psychiatry[†] and Family Medicine[°], University of
Saskatchewan, Saskatoon, Saskatchewan, Canada S7N 0W0 and
Mental Health Services Branch[§], Saskatchewan Health,
Regina, Saskatchewan, Canada S4S 6X6

INTRODUCTION

The trace amines have been implicated in a number of
psychiatric and other brain disorders (Davis et al., 1982).
It has been suggested that a defect in the metabolism of
p-tyramine (p-TA) may underlie a predisposition to and
attacks of migraine (Smith et al., 1971; Youdim et al.,
1971; Sandler et al., 1974; Glover et al., 1977; Glover et
al., 1981; Littlewood et al., 1982; Glover et al., 1983)
and depression (Sandler et al., 1975; Bonham Carter et al.,
1978; Agren, 1983; Harrison et al., 1983; Harrison et al.,
1984). Tyramine challenge tests in which p-TA was ingested
by normal, migraine (Smith et al., 1971; Youdim et al.,
1971) and depressed subjects (Sandler et al., 1975; Bonham
Carter et al., 1978; Harrison et al., 1983; Harrison et
al., 1984) have shown highly significant reductions in the
excretion of conjugated tyramine in the migraine and depressed patients. Consistent with these findings, the
activity of platelet phenolsulfotransferase (PST), the conjugating enzyme for p-TA, has been reported to be significantly lower in migraine (Littlewood et al., 1982; Glover
et al., 1983). There have also been conflicting claims
that monoamine oxidase (MAO) activity is lower in migraine
(Sandler et al., 1974; Glover et al., 1981; Littlewood et
al., 1982), whereas in depression there have been

indications that MAO activity may be increased (Bonham Carter et al., 1978; Agren, 1983.

A number of problems have characterized the challenge tests. The ingested and endogenous p-TA (and their metabolites) cannot be distinguished from each other and the large daily variations in the excretion of endogenous conjugated p-TA which have been reported (Huebert and Boulton, 1979) may mask or exaggerate changes in p-TA conjugation following ingestion. Furthermore, the ingestion of a large dose of p-TA may perturb the endogenous tyramine system with unforeseen effects on the observed level of excreted conjugated tyramine. Finally, in most of these earlier studies p-TA was measured by a non-specific fluorimetric method.

In this paper we report challenge tests in which normal, migraine and depressed subjects ingested p-tyramine-β, β-^{2}H$_2$ (p-TA-d$_2$). In contrast to the earlier studies, the origin of the deuterium-labelled metabolites is unambiguous and our method of quantitation, mass spectrometry with selected ion monitoring (MS-SIM), is specific and sensitive. We have also determined the activities of platelet MAO and PST to several substrates. Significant differences in one or more of these parameters could serve as biological state or trait markers in the diagnosis of migraine and depression and could possibly even lead to improvements in treatment.

MATERIALS AND METHODS

p-Tyramine-β,β-^{2}H$_2$ and Deuterated Internal Standards

p-Tyramine-β,β-^{2}H$_2$, p-tyramine-α,α,β,β-^{4}H$_2$ (p-TA-d$_4$) and p-hydroxyphenylacetic acid-^{4}H$_2$ (p-HPA-d$_4$) were synthesized by recently published procedures (Davis, 1987; Davis et al., 1986). Deuterium incorporation was ascertained by MS-SIM of the derivatized compounds as described below.

Enzyme Assays

Platelet MAO activity was determined by a radioenzymatic procedure using ^{14}C-labelled p-TA, tryptamine (T) and phenylethylamine (PE) as substrates (Yu et al., 1982). Platelet PST activity (to phenol and p-TA) was also determined radioenzymatically using ^{35}S-labelled phospho-adenosine-phosphosulfate as sulfate group donor (Yu et al. 1985).

Analytical Procedure for
Unconjugated and Conjugated p-TA-d₂

Deuterium-labelled internal standard (p-TA-d₄) was added to urine (1.0 mL); the solution was made basic by the addition of solid sodium carbonate and then derivatized with an acetone solution of 1-dimethylaminonaphthalene-5-sulfonyl chloride (dansyl chloride). The derivative was isolated by extraction into toluene-ethyl acetate and purified by thin-layer chromatography. The dansyl derivative was then evaporated from the probe in the ion source of the mass spectrometer and the protio and deutero dansyl amines were measured by selected ion monitoring. A complete description of the procedure has been published elsewhere (Philips and Boulton, 1979). Total tyramine was measured as described above but after hydrolysis of the urine with hydrochloric acid. The difference between the total and unconjugated tyramine values gave the conjugated value.

Analytical Procedure for
Unconjugated and Conjugated p-HPA-d₂

Deuterium-labelled internal standard (p-HPA-d₄) was added to urine (200 µL); the solution was acidified with sulfosalicylic acid and the acid metabolites were extracted into ethyl acetate. The solvent was evaporated and the residue was derivatized with trifluoroethanol and penta-fluoropropionic anhydride. Following a buffer wash and extraction into hexane, the p-HPA-d₂ was quantitated by capillary column gas chromatography-mass spectrometry with selected ion monitoring. Full experimental details have been published recently (Davis et al., 1986). Total p-HPA was determined as above following hydrolysis in dilute sulfuric acid. Conjugated p-HPA-d₂ equals the difference between total and unconjugated p-HPA-d₂.

Selection of Subjects

Migraine patients were diagnosed as having classical or common migraine, the duration of which was at least two years and the patients must have suffered from two to eight attacks per month. All prophylactic medication was withdrawn two weeks prior to obtaining samples, but symptomatic medication was allowed. Patients suffering from significant cardiac, renal, hepatic or metabolic disease or headaches due to hypertension, menstruation or known organic aetiology were excluded, as were pregnant or lactating women and patients with known neurological or psychiatric disorders.

Unipolar depressed patients exhibiting a Beck depression score greater than 20, a Hamilton depression score greater than 16, a Raskin depression score greater than 8, a Covi anxiety score greater than 7 and meeting the criteria of major depression outlined by DSM III and diagnosed through the application of a modified diagnostic interview schedule (DIS) were recruited. Subjects diagnosed as suffering from schizophrenia, bipolar affective disorder, anxiety or panic disorder, organic brain disorder, alcohol or drug abuse or an active medical illness as well as pregnant women or those taking monoamine oxidase inhibitors, antipsychotics, antihypertensive drugs, corticosteroids or contraceptives were excluded.

Control subjects were healthy volunteers recruited from the laboratory staff.

The patients and controls donated a morning fasting blood sample and immediately after emptying their bladder, ingested a gelatin capsule containing 125 mg of p-TA-d_2 hydrochloride (equivalent to 100 mg of free base). Urine was then collected for 3 h or 24 h.

RESULTS AND DISCUSSION

Control, migraine and depressed subjects ingested deuterated tyramine and the urinary levels of the major deuterated metabolites, unconjugated and conjugated p-TA-d_2 and unconjugated and conjugated p-HPA-d_2, were assayed by MS-SIM. A summary of the results for the completed control (Boulton and Davis, 1987) and migraine (Davis et al., 1987) studies and preliminary results for the depression study are presented in Table I.

TABLE 1

Urinary Deuterated Metabolites of Ingested p-Tyramine-β,β-$^2H_2$ in Control, Migraine and Depressed Subjects

Metabolite	Control mg/24 h (n=8)	Migraine mg/24 h (n=19)	Control mg/3 h (n=9)	Depression mg/3 h (n=12)
Unconjugated p-TA-d_2	0.41±0.15	0.06±0.01	0.34±0.13	-
Conjugated p-TA-d_2	8.70±1.11	6.99±0.75	4.46±0.70	4.66±0.80
Unconjugated p-HPA-d_2	62.3±3.6	52.4±3.9	44.1±3.2	47.3±3.0
Conjugated p-HPA-d_2	6.6±2.0	3.8±1.8	2.5±1.3	-

Values are the mean ± standard error of the mean.

No significant deuterium isotope effect should be observed for the metabolism of p-TA-d_2 since MAO acts only at the carbon atom alpha to the amino group (Yu et al., 1981). The results in Table I show that although the conjugated p-TA-d_2 was reduced in migraine the reduction did not achieve significance. No reduction in conjugated p-TA-d_2 was observed in the depressed patients. This is contrary to the results of previous challenge tests in which highly significant reductions were claimed to have been found (Table II) both in migraine and in depression.

A comparison of the results in Tables I and II reveals that the absolute values for the levels of excreted conjugated tyramine are substantially lower for the deuterated p-TA challenge test than for the unlabelled p-TA test, except for depressed patients. Four factors may be invoked in an attempt to explain these discrepancies:

1. Non-specificity of the methods of quantitation in the earlier studies.

TABLE II

Previously Reported Urinary Conjugated Tyramine Levels
Following Ingestion of Unlabelled p-Tyramine

Reference	Control	Migraine	Depression
Smith et al. (1971)	15.3±1.1	10.5±1.2 (p <0.01)	-
Youdim et al. (1971)	8.7±1.0	1.7±0.3 (p <0.001)	-
Sandler et al. (1975)	7.6±0.5	-	4.4±0.6 (p <0.005)
Bonham Carter et al. (1978)	7.6±0.5	-	3.0±0.5 (p <0.001)
Harrison et al. (1983)	6.2±0.3	-	a 3.9±0.3 (p <0.05) b 5.0±0.4 (p <0.05)
Harrison et al. (1984)	6.3±0.3	-	a 4.6±0.3 (p <0.001) b 4.6±0.3 (p <0.001)

Values are mean ± standard error of the mean [mg/3 h except for Smith et al. (1971)
which is mg/24 h]. a = unipolar depression; b = bipolar depression.

2. Inability to distinguish endogenous from exogenous tyramine in the earlier studies.

3. Possible perturbation of the endogenous tyramine system by the ingested p-tyramine.

4. Different compositions of the populations of the subjects tested.

Since most of the values obtained for conjugated tyramine in the p-TA-d_2 challenge test are markedly lower than those for the unlabelled p-TA test, one should suspect that for the latter the fluorimetric method of quantitation may include some other fluorophore along with tyramine in the measurement. Although the mean level of excretion of endogenous conjugated p-TA is only 5-10% of that excreted following a challenge test, Huebert and Boulton (1979) have shown that daily variations in the excretion of conjugated p-TA can be as great as ten-fold; in such cases, the level of endogenous conjugated p-TA could equal that produced from ingested tyramine. The ingestion of 100 mg of tyramine might also be expected to perturb the endogenous tyramine system, perhaps causing the release of stores of tyramine and therefore creating an artificially high level of excreted conjugated p-TA. It turns out, however, that the levels of excreted unlabelled conjugated p-TA before and after the p-TA-d_2 challenge are essentially the same and amount to only a few percent of the conjugated p-TA-d_2. Finally, there appear to be no obvious differences in composition of the groups of subjects investigated. In all the studies subjects were advised to avoid tyramine-containing foods and individuals known to suffer from active physical or mental illnesses were excluded from the control groups. No age or sex correlations with conjugated p-TA excretion were reported in any of the other earlier studies and none were found in the present study. Migraine and depressed subjects were drug-free for at least ten days prior to the challenge tests. Harrison et al. (1984) reported significantly lower (p <0.003) conjugated p-TA in depressed females than in depressed males; in the present study a similar trend was observed but was not statistically significant. To summarize, the discrepancy between the results from the earlier studies and the present one would appear to be due mainly to the non-specific method of quantitation and the inability to distinguish between endogenous and exogenous tyramine. No perturbation of the

endogenous tyramine system was found and the compositions
and exclusion criteria of the groups studied were similar.

The levels of unconjugated p-HPA-d$_2$, the major metabolite of p-TA-d$_2$, were found to be lower in migraine and
depression compared to controls, but the reductions were
not statistically significant. Sandler et al. (1979) have
reported a significant (p <0.005) reduction in the excretion of endogenous p-HPA in depression, whereas Smith et
al. (1971) and Youdim et al. (1971) found non-significant
increases in p-HPA in migraine subjects challenged with
p-TA. As the major metabolite of tyramine, significantly
reduced levels of excretion of p-HPA would have suggested a
deficiency in MAO activity.

In Table III the activities of platelet MAO and PST to
several substrates in control, migraine and depressed subjects are given. PST activity to phenol was found to be
reduced in migraine, confirming the observations reported

TABLE III

MAO and PST Activities to Some Substrates in Control, Migraine and Depressed Subjects

Enzyme-Substrate	Control (n=14)	Migraine (n=12)	Depression (n=16)
PST-p-TA	34.5±2.7	36.7±1.6	34.4±3.6
PST-Phenol	12.2±1.2	8.6±0.5[†]	8.6±0.9[†]
MAO-PE	10.6±1.1	12.4±0.9	10.0±1.1
MAO-p-TA	2.53±0.28	2.01±0.12[††]	2.52±0.27
MAO-T	0.90±0.08	0.79±0.06	0.47±0.05[†]

Values are mean ± standard error of the mean in pmol/mg/min
for PST and nmol/mg/h for MAO. [†] p <0.01, Student's
t-test. [††] p <0.05, Student's t-test.

previously by Littlewood et al. (1982) and Glover et al. (1983). A similar reduction of PST (phenol) in depression is reported here for the first time. The phenol-inactivating variant of PST has no known endogenous substrate at the present time so the significance of a reduction in its activity in migraine and depression remains unclear. Orally administered p-TA-d$_2$ is probably metabolized by MAO and PST in the gut. Since access to the gut variants of these enzymes was not practicable, however, platelet MAO and PST were measured instead.

Early results (Sandler et al., 1974) indicated very significant reductions in MAO activity to PE, p-TA, dopamine and serotonin. In later papers (Glover et al., 1977 and 1981) these results could not be reproduced except for very narrowly-defined sub-groups of the migraine patients. Sandler et al. (1975) also predicted that MAO activity should be increased somewhat in depression. Agren (1983) confirmed this, finding a significant correlation between depression severity scores (Hamilton Scale and the Global Assessment Scale) and MAO activity -- the deeper the depression the higher the MAO activity. In the present study, these results could not be reproduced. MAO activity to p-TA (but not to PE or T) was reduced in migraine, whereas in depression, MAO activity to T (but not to p-TA or PE) was reduced.

In conclusion, our results do not confirm in a statistically significant manner earlier findings of reduced conjugation of p-TA in migraine and depression.

The significance of reduced activity of PST to phenol is not clear. These results do not provide any support for the claim that p-tyramine has an important role in migraine or depression and so claims that urinary conjugated p-tyramine may serve as biological state or trait markers in these disorders should be viewed with caution at this time.

ACKNOWLEDGEMENTS

We thank Saskatchewan Health, the Medical Research Council of Canada, Boehringer Ingelheim (Canada) Ltd. and Eli Lilly (Canada) Ltd. for financial support.

REFERENCES

Agren H. (1983) Depression and altered transmission states, traits and interactions, in The Origins of Depression: Current Concepts and Approaches (Angst J., ed.), pp. 197-311. Springer-Verlag, New York.

Bonham Carter S., Sandler M., Goodwin B.L., Seeping P. and Bridges P.K. (1978) Decreased urinary output of tyramine and its metabolites in depression. Brit. J. Psychiat. 132, 125-132.

Boulton A.A. and Davis B.A. (1987) The metabolism of ingested deuterium-labelled p-tyramine in normal subjects. Biomed. Environ. Mass Spectrom. 14, 207-211.

Davis B.A. (1987) Crown ether catalyzed deuterium exchange in the synthesis of benzyl cyanides. J. Lab. Compds. & Radiopharmac. 24, 199-204.

Davis B.A., Dawson B., Boulton A.A., Yu P.H. and Durden D.A. (1987) Investigation of some biological trait markers in migraine: Deuterated tyramine challenge test, monoamine oxidase, phenolsulfotransferase and plasma and urinary biogenic amine and acid metabolite levels. Headache (in press).

Davis B.A., Durden D.A. and Boulton A.A. (1986) Simultaneous analysis of twelve biogenic amine metabolites in plasma, cerebrospinal fluid and urine by capillary column gas chromatography-high resolution mass spectrometry with selected ion monitoring. J. Chromatogr. Biomed. Appl. 374, 227-238.

Davis B.A., Yu P.H., Carlson K., O'Sullivan K. and Boulton A.A. (1982) Plasma levels of phenylacetic acid, m- and p-hydroxyphenylacetic acid and platelet monoamine oxidase activity in schizophrenic and other patients. Psychiat. Res. 6, 97-105.

Glover M.A., Littlewood J., Sandler M., Peatfield R., Petty R. and Rose F.C. (1983) Biochemical predisposition to dietary migraine: The role of phenolsulfotransferase. Headache 23, 53-58.

Glover V., Peatfield R., Zammit-Pace K., Littlewood J., Garvel M., Rose F.C. and Sandler M. (1981) Platelet monoamine oxidase activity and headache. J. Neurol. Neurosurg. & Psychiat. 44, 786-790.

Glover V., Sandler M., Grant E., Rose F.C., Orton D., Wilkinson M. and Stevens D. (1977) Transitory decrease in platelet monoamine-oxidase activity during migraine attacks. The Lancet 391-393.

Harrison W.M., Cooper T.B., Quitkin F.M., Liebowitz M.R., McGrath M.D., Stewart J.W. and Klein D.F. (1983) Tyramine sulfate excretion test in depressive illness. Psychopharmacol. Bull. 19, 503-504.

Harrison W.M., Cooper T.B., Stewart J.W., Quitkin F.M., McGrath P.J., Liebowitz M.R., Rabkin J.R., Markowitz J.S. and Klein D.F. (1984) The tyramine challenge test as a marker for melancholia. Arch. Gen. Psychiatry 41, 681-685.

Huebert N.D. and Boulton A.A. (1979) Longitudinal urinary trace amine excretion in a human male. J. Chromatogr. Biomed. Appl. 162, 169-176.

Littlewood J., Glover V., Sandler M., Petty R., Peatfield R. and Rose F.C. (1982) Platelet phenolsulfotransferase deficiency in dietary migraine. The Lancet, 983-985.

Philips S.R. and Boulton A.A. (1979) The effect of mono-amine oxidase inhibitors on some arylalkylamines in rat striatum. J. Neurochem. 33, 159-167.

Sandler M., Bonham Carter S., Cuthbert M.F. and Pare C.M.B. (1975) Is there an increase in monoamine-oxidase activity in depressive illness? The Lancet, 1045-1049.

Sandler M., Ruthven C.R.J., Goodwin B.L., Reynolds G.P., Rao V.A.R. and Coppen A. (1979) Deficient production of tyramine and octopamine in cases of depression. Nature 278, 357-358.

Sandler M., Youdim M.B.H. and Hanington E. (1974) A phenyl-ethylamine oxidizing defect in migraine. Nature 250, 335-337.

Smith I., Kellow A.H., Mullen P.E. and Hanington E. (1971) Dietary migraine and tyramine metabolism. Nature 230, 246-248.

Youdim M.B.H., Bonham Carter S., Sandler M., Hanington E. and Wilkinson M. (1971) Conjugation defect in tyramine-sensitive migraine. Nature 230, 127-128.

Yu P.H., Barclay S., Davis B. and Boulton A.A. (1981) Deuterium isotope effects on the enzymatic oxidative deamination of trace amines. Biochem. Pharmacol. 30, 3089-3094.

Yu P.H., Bowen R., Carlson K., O'Sullivan K. and Boulton A.A. (1982) Comparison of some biochemical properties of platelet monoamine oxidase in some mentally disordered and healthy individuals. Psychiat. Res. 6, 107-121.

Yu P.H., Rozdilsky B. and Boulton A.A. (1985) Sulfate con-jugation of monoamines in human brain: Purification and some properties of an arylamine sulfotransferase from cerebral cortex. J. Neurochem. 45, 836-843.

TRACE AMINE/CATECHOLAMINE RELATIONSHIPS: URINARY RATE OF EXCRETION PATTERNS OF HUMAN SUBJECTS RESPONDING TO ACUTE STRESS

J. Harris[1] and G. S. Krahenbuhl[2]
Chemistry Department[1] and Exercise & Sport
Research Institute, Arizona State
University, Tempe, AZ 85287, USA

INTRODUCTION

Trace amines and catecholamines along with their respective metabolites have been implicated in stress as well as in other specific behavioral paradigms. Further, trace amines have been found to release and potentiate the effect of catecholamines (Jones 1984), to possess a neuromodulatory role, and to interact with selected catecholamines neurotransmitters (Jones 1982, Langer et al. 1985).

We have determined that the determination of biogenic amines and their metabolites in a timed collected sample of urine can indicate changes during acute events of stress and thus may serve as a noninvasive index of acute stress response. We are stimulated by previous observations and the early suggestion by previous observations and the early suggestion by Boulton and others (Boulton 1976) of a neuromodulatory and/or interactive role of trace amines with catecholamines to review the dynamics of change in the urinary excretion rates of the biogenic amines/ metabolites in response to several types of stressors.

The several types of stressors, involved in this study include exercise, flight emergencies, sport shooting competition, and various flight training

lessons. The results subserving our discussion involve
analysis via HPLC-EC of the catecholamines, NA, A, DA,
and 5-HT with their respective metabolites MHPG, VMA,
HVA, DOPAC, and 5-HIAA. "Trace" amines PE, m- & p-TA & T
with their respective metabolites, PAA, m- & p-HPAA,
p-HMA & IAA were determined with GC-MS.

   Data collected (Harris et al. 1984, 1985a, 1985b;
Krahenbuhl et al. 1985a) reflect a change in excretion
rate, i.e., from a basal (resting) state to one in
response to a stressor. These data suggest a pattern of
differing excretion rates of the catecholamines/
metabolites and the "trace" amines/metabolites in
response to the several stressors. A relationship would
be anticipated if the "trace" amines played a
neuromodulatory or neurotransmitter role as previous
observations by Boulton and others have suggested
(Sabelli et al. 1978; Boulton 1976, 1982; Jones, 1982;
Wolf et al. 1983).

                        METHODS

   Human Subjects. The subjects were all males, ages
21-30 years, from whom informed consent was obtained; the
research was conducted in conformance with the principles
embodied in the Helsinski Declaration. The human stress
response was measured after exposure to each of the
following. a) flight-training lessons which involved a
power-on-stall and spin recovery (SPIN); and instrument
check ride (CHECK); and instrument flight with take-off
and landing with darkened canopy (BLIND); b) non-flight
laboratory simulator landings on aircraft carriers (SIM);
c) flight emergencies involving events variously grouped
as low order (electrical malfunctions, control problems),
medium order (fumes, smoke, mechanical problems), or high
order emergencies (engine failure, fire, loss of
hydraulics); d) exercise, physical workload on treadmill
to maximal aerobic power and to 60% submaximal power; e)
sport skill competition - elite shooting, a sports
activity demanding total concentration and fine physical
control on the part of the individual. Each subject
contributed a basal urine, prior to testing then was
subjected to a particular stressor, then submitted a
urine collection; the exact duration of time and total
volume were noted. Collection times were matched in time

of day so as to control for diurnal variation. The normal subjects, drank at least 250 ml of water to provide adequate amounts of urine. A 100 ml aliquot of each sample was stabilized with 1 ml of 10% EDTA- 4% thioglycolic acid mixture and stored at -90°C until analyzed.

Analyses. The analysis of the trace amines and their acid components was conducted by Dr. Bruce Davis and the group at the Psychiatric Research Center, University of Saskatchewan using a high resolution integrated mass spectrometry of the derivatized amines and acid metabolites, incorporating appropriate internal standards as originally described by Durden et al. (1973), and Davis et al. (1981, 1982).

The analysis of the catecholamines and their metabolites was carried out using HPLC with a μ-Bondapak C_{18} 10 μm reverse phase columns and on-line electrochemical detection according to the methods detailed previously (Harris et al. 1984, 1985a).

RESULTS AND DISCUSSION

In general the pattern of response of the amines and their metabolites to the several stressors would enable the stressor to be separated into two distinct groups, Group I, with short duration and less intensive stress, include the in-flight training lessons and exercise; Group II, with prolonged and greater intensity of stress, include emergency events, blind flying and shooter competition.

As shown in Table I, both groups of stressors produce an increase in the rate of excretion of A and NA. Group I, further, had an excretion rate pattern of a decrease in 5-HT, no change, or a tendency to increase in DA, and a decrease in all of the catecholamine metabolites and 5-HIAA. In Group II, the excretion rate pattern exhibited no change or a tendency to decrease in DA and 5-HT, a decrease in 5-HIAA, an increase in VMA and HMPG, and variable results pattern HVA and DOPAC. Group I stressors are distinguished from those of Group II with a tendency to increase DA, with a definite decrease in 5-HT and a uniform pattern of decrease in excretion rates of the metabolites VMA, MHPG, DOPAC, and

TABLE 1

Pattern of change in rate of excretion of trace amines and
catecholamines and metabolites over basal rates

Trace Amine	Stress Group I	II	Catechol-amine	Stress Group I	II
PE	↓48%	↑22%	A	↑100–300%	↑100–525%
P-TA	↓20%	↓10%	NA	↑50–100%	↑35–140%
m-TA	↓20%	↓10%	DA	NC↑12%	NC↓14%
T	↓40%	↓42%	5-HT	↓25%	NC↓8%
Metabolite			**Metabolite**		
PAA	↓48%	↓15%	VMA	↓21%	↑18%
P-HPA	↓25%	NC↑5%	MHPG	↓40%	↑50%
m-HPA	↓42%	↑38%	DUPAC	↓18%	Variable
IAA	NC↑10%	↑44%	HVA	↓20%	Variable
			5-HIAA	↓21%	↓20%

↑, increase; ↓, decrease; ↑̣, tendency to increase; NC, no change.
Data from Harris et al., 1985. Data from Harris et al., 1984, 1985a,b.

TABLE 2

Correlations between changes in trace amines and catecholamines

Condition	DA	HVA	DOPAC	p-TA	m-TA	5-HT	5-HIAA	PE
Cold stress	↑	↑	—	↓	↑	—	—	—
Drug receptor against[2]	NC ↓	—	↓	↑	NC	—	—	—
Stressor I[‡]	NC	↓	↓	↓	↓	↓	↓	↓
Stressor II[‡]	NC	Variable	Variable	↓	↓	↓	↓	↑
Lesion[3]	NC	↑	↑	↓	↑	↓	↓	—
Reserpine[4]	↓	—	—	↓	↓	—	—	NC

↑, increase; ↓, decrease; ⇡, tendency to increase; ⇣, tendency to decrease; NC, no charge.

[1] Juorio, A. V., 1979a.   [2] Juorio, A. V., 1979b.   [‡] Harris et al., this study.

[3] Juorio, A. V., 1979c.   [4] Juorio et al., 1985.

HVA. Group II stressors in contrast have a uniform
pattern of an increase in metabolites VMA, MHPG, variable
HVA & DOPAC.

Trace amine excretion rates also had distinguishable
patterns. Group I stressors uniformly had decreased
rates of excretion in the amines PE, m- and p-TA, and
T. Similarly there is a decrement in the metabolites
PAA, m- and p-HPA and a tendency to decrease or no change
in IAA. Group II stressors produce an increase in PE in
contrast to Group I stressors, while yielding decreases
in the other trace amines. For the trace amine
metabolites of Group II stressors, the rates of excretion
increased for IAA and m-HPA, but did not change or tended
to increase for p-HPA. There was no change or a tendency
to decrease in PAA.

There is a persistent pattern in the excretion rate
of biogenic amines and metabolites among the diversity of
stress paradigms. The measurement reflects a rate of
total response from peripheral and CNS contributions.
Nevertheless, the patterns we have obtained in the rates
of excretion tend to follow data reported for tissue
changes under various conditions. A release of a
catecholamine from a tissue causing a depletion in there
would result in an increase in excretion of the amine or
its metabolites in the urine.

We have found that the rate of excretion of p-TA is
considerably higher than the meta isomer, about five
times greater. This is in keeping with the observation
that tissue concentration of p-TA is always significantly
higher than m-TA, with m=TA exhibiting a turnover in
human brain only 16% of that of p-TA (Young et al.
1982). Albeit in basal conditions the rate of excretion
of 5-HT is higher than T, and 5-HIAA is of the same
magnitude as IAA, the ratio of IAA/T is almost twice that
of the 5-HIAA/5-HT. Under acute stress conditions, the
ratios increase 15-25% with the hydroxyindole system and
50% in the indole system. This observation supports the
contention that T has a role in 5-HT action. Indeed, the
relationship between T and 5-HT has variously been
reported as T modifying the actions of 5-HT (Jones &
Boulton 1980) or T may act as a 5-HT agonist (Mounsey
et al. 1982). Studies by Juorio (1984) have shown in
general a reciprocal relation between p & m-TA

concentrations and DA turnover in striatal tissue, albeit Huebert & Boulton (1978) could not correlate brain changes with urinary levels of p or m-TA.

Table 2 lists a few of the changes in trace amines and catecholamines reported over the years by Juorio and co-workers (1979, 1985). Our data gave a pattern suggestive of some relationship between PE and DA; p-TA and DA and T and 5-HT, a relation which is in keeping with reported drug and/or electrophysiological studies that "trace" amines possess a role in neural transmission; this role may be a "regulatory" effect on the action of the classical transmitters, or as neurotransmitters in some pathways. Some PE effects are mediated by serotonergic mechanisms (Sloviter et al. 1980) but generally is considered to exert behaviorial effects through DA - mediated mechanisms (Broestrup & Randrup 1978). Our data of a reciprocal relation between PE & NA are consistent with several studies showing PE leading to reduction in NA in tissue and no change or slight reduction in DA and 5-HT. Depletion of a catecholamine in tissue could result in an increase of the amine or its metabolites in excretion in urine.

Our pattern of response is in keeping with the possibility of simultaneous release of catecholamine and of "trace" amine as suggested by Dyck et al. (1982). The concept of co-transmission is gaining evidence for a means of receptor modulation by a "modulatory" transmitter regulating the effect of a "major" transmitter.

In this study, an attempt was made to determine whether an interrelationship between catecholamines and "trace" amines exist as possible complimentary transmitter markers of stress. The pattern of the rate of change we have obtained did indicate relationships of the catecholamines and "trace" amines which tend to follow some of the animal tissue data. One would not anticipate a mirror reflection of CNS activity in a situation involving both CNS and peripheral systems. We do, nevertheless, see that the trace amines and catecholamines interact in a role to be delineated.

ACKNOWLEDGEMENTS

We thank Mr. R. D. Malchow, Dr. J. R. Stern, Mrs. G. Wheatley for excellent technical assistance, Mr. W. Pedshalny, Mr. M. Mizuno and Dr. D. H. Durden and Dr. B. A. Davis for the mass spectrometric analyses, and Dr. A. A. Boulton for collaboration and stimulus. We also acknowledge with appreciation the financial support from several sources which enabled us to generate the fundamental data subserving the present study, namely, Arizona State University Research Fund, the U.S. Air Force, the Saskatchewan Department of Health and the Medical Research Council of Canada.

REFERENCES

Boulton, A. A. (1976) Cerebral aryl alkyl aminergic mechanisms, in Trace Amines and the Brain, (Usdin, E. and Sandler, M., eds.), pp. 22-39, M. Dekker Inc., New York.

Boulton, A. A. (1982) Some aspects of basic psychopharmacology; the trace amines. Prog. Neuropsychopharmacol. Biol. Psychiat., 6, 563-570.

Braestrup, C. and Randrup, C. (1978) Stereotyped behavior in rats induced by phenylethylamine, dependence on dopamine, and noradrenaline and possible relation to psychosis, in Noncatecholic Phenylethylamines Part I. Phenylethylamine: Biological Mechanism and Clinical Aspects (Mosnaim, A. and Wolf, M., eds.) pp. 245-269, Marcel Dekker, New York.

Davis, B. A. and Boulton, A. A. (1981) Longitudinal urinary excretion of "trace" acids in a human male. J. Chromatog., 222, 161-169.

Davis, B. A., Durden, D. A. and Boulton, A. A. (1982) Plasma concentrations of p- and m-hydroxyphenylacetic acid and phenylacetic acid in humans: gas chromatographic-high resolution mass spectrometric analysis. J. Chromatog. & Biomed. Appl., 230, 219-230.

Durden, D. A., Philips, S. R. and Boulton, A. A. (1973) Identification and distribution of β-phenylethylamine in the rat. Can. J. Biochem., 51, 995-1002.

Dyck, L. E., Juorio, A. V. and Boulton, A. A. (1982) The in vitro release of endogeneous m-tyramine, p-tyramine and dopamine from rat striatum. Neurochem. Res., 7, 705-716.

Harris, J., Krahenbuhl, G. S., Malchow, R. D. and Stern, J. R. (1984) Stress response in pilots: urinary biogenic amine/metabolite excretion, in Stress: Role of Catecholamines and Other Neurotransmitters (Usdin, E., Kuetnansky, A. and Axelrod, J., eds.), pp. 991-10001, Gordon & Breach Science Publ., N.Y.

Harris, J., Krahenbuhl, G. S., Malchow, R. D. and Stern, J. R. (1985a) Neurochemistry of stress: urinary biogenic amine/metabolite excretion rates in exercise. Biogenic Amines, 2, 261-267.

Harris, J., Davis, B. A., Krahenbuhl, G. S. and Boulton, A. A. (1985b) Trace amines/metabolite responses in stress, in Neuropharmacology of the Trace Amines (Boulton, A. A., Maitre, L., Bieck, P. R. and Riederer, P., eds.) pp. 395-406, Humana Press, Clifton, N.J.

Huebert, N. D. and Boulton, A. A. (1978) The effects of some stimulants and anti-psychotic drugs on urinary tyramine levels in the rat. Res. Comm. Chem. Path. Pharmacol., 22, 73-82.

Jones R. S. G. (1982) Tryptamine: a neuromodulator or a neurotransmitter in mammalian brain? Progress in Neurobiology, 19, 117-139.

Jones, R. S. G. (1984) Electrophysiological studies of the possible role of trace amines in synaptic function, in Neurobiology of the Trace Amines (Boulton, A. A. and Buher, G. B., Deuhurst, W. G. and Lundler, M., eds.), pp. 205-223, Humana Press, Clifton, N.J.

Jones, R. S. G. and Boulton, A. A. (1980) Tryptamine and 5-hydroxytryptamine: Actions and interactions on cortical neurons in the rat. Life Sci., 27, 1949-1956.

Juorio, A. V. (1979a) Effect of stress and L-DOPA administration on mouse striatal tyramine and HVA levels. Brain Res., 179, 186-189.

Juorio, A. V. (1979b) Drug-induced changes in the formation, storage and metabolism of tyramine in the mouse. Brit. J. Pharmacol., 66, 377-384.

Juorio, A. V. (1979c) Drug induced changes in the central metabolism of tyramine and other trace amines: Their possible role in brain functions, in Neurobiology of Trace Amines (Boulton, A. A., Baker, A. B., Deuhurst, W. G., and Sandler, M., eds.), pp. 145-

162, Humana Press, Clifton, N.J.

Juorio, A. V., Greenshaw, A. J. and Boulton, A. A. (1985) Possible pathways for some trace amine-containing neurons, in Neuropharmacology of the Trace Amines (Boulton, A. A., Maitre, L., Bieck, P. R. and Riederer, P., eds.), pp. 87-99, Humana Press, Clifton, N.J.

Krahenbuhl, G. S., Harris, J., Malchow, R. D. and Stern, J. R. (1985a) Biogenic amine/metabolite response during in-flight emergencies. Airation, Space and Environmental Medicine, June, 576-580.

Krahenbuhl, G. S. and Harris J. (1985b) Biochemistry of psychological stress in elite shooters. Unpublished data.

Langer, S. Z., Arbilla, S., Niddam, R., Benhirane, S. and P. Baud (1985) Pharmacological profile of β-phenethylamine on dopaminergic and noradrenergic neurotransmission in rat cerebral slices, in Neuropsychopharmacology of the Trace Amines (Boulton, A. A., Maitre, L., Biech, P. R. and Riederer, P., eds.), pp. 27-38, Humana Press, Clifton, N.J.

Mounsey, I., Brady, K. A., Carroll, J., Fisher, R. and Middlemen, D. N. (1982) K^+-Evoked 5-HT release from rat frontal cortex slices: The effect of 5-HT agonists and antagonists. Biochem. Pharmacol., 31, 49-53.

Sabelli, H. C., Borison, R. L., Diamond, B. I., Havdala, H. S. and Narasimhachari (1978) Phenylethylamine and brain function. Biochem. Pharmacol., 27, 1707-1711.

Sloviter, R. S. Connor, J. D. and Drust, E. G. (1980), Serotenergic properties of β-phenylethylamine in rats. Neuropharmacol., 19, 1071-1076.

Wolf, M. E. and Mosnaim, A. D. (1982) Phenyethylamine in neuropsychiatric disorders. Gen. Pharmac., 14, 385-390.

Young, S. N., Davis, B. A. and S. Gauthier (1982) Precursors and metabolites of phenylethylamine, m- and p-tyramine and tryptamine in human lumbar and cisternal cerebrospinal fluid. J. Neurol. Neurosurg. Psychiat., 45, 633-639.

PHENYLALANINE, PHENYLETHYLAMINE, AND TARDIVE DYSKINESIA

IN PSYCHIATRIC PATIENTS

M.A. Richardson,(1) R. Suckow,(2) R. Whittaker,(3)
A. Perumal,(2) W. Boggiano,(3) I. Sziraki,(1) H. Kushner(1)

(1) Nathan Kline Institute for Psychiatric Research
    Orangeburg, New York
(2) New York State Psychiatric Institute
    New York, New York
(3) Rockland Psychiatric Center
    Orangeburg, New York
New York State Office of Mental Health

INTRODUCTION

Tardive dyskinesia (TD), an abnormal movement disorder seen subsequent to treatment in patients taking neuroleptic drugs, afflicts large numbers of chronically ill psychiatric patients. Postsynaptic dopamine receptor supersensitivity, pre-synaptic dopaminergic overactivity and/or noradrenergic hyperactivity have been postulated as the pathophysiological mechanisms for the symptoms of the disorder (Jeste and Wyatt, 1981; Tarsy and Baldessarini, 1977).

Phenylketonurics (PKU) have been demonstrated to be at particular risk for developing TD (Richardson et al., 1986). In PKU the normal oxidation of the amino acid phenylalanine (Phe) to tyrosine (Tyr) in the liver is almost completely stopped (Knox, 1966). This shift away from Tyr leads to an excessive production of phenylethylamine (PE) which is the decarboxylated product of Phe. Abnormally high levels of PE have been observed in the urine of PKU patients (Oates et al., 1963).

Young schizophrenic males with a manic symptom profile (Richardson et al., 1985), schizophrenics with hypomanic

behaviors (Wilson et al., 1983) and patients with bipolar
disorders (Rosenbaum et al., 1977; Rush et al., 1982) have
all been shown to be at risk for TD. Abnormally high levels of PE have been reported in patients with manic symptoms (Fischer, 1972; Karoum et al., 1982) and in three
bipolar women with dyskinesias (Linnoila, 1983).

In view of the above reported associations of TD with
PKU and manic symptoms, the known disorder of Phe metabolism in PKU and the suggestion of abnormalities of the same
metabolic pathway in patients with mania--this study was
undertaken to test a hypothesis that an altered Phe metabolism may be associated with TD.

METHODS

Subjects

All male inpatients of a state psychiatric center, for
the period of December 1984 to March 1986, ages 18-44 and
carrying a chart diagnosis of schizophrenia were screened
for study inclusion if they had no less than 3 years nor
more than 20 years since their first neuroleptic treatment.
Patients were excluded if they were incompetent to give
informed consent, refused consent, were non-English speak-
ing, mute, deaf, failed to satisfy criteria (DSM-III or
RDC) for schizophrenia after a lifetime symptom screening
and diagnostic interview by the first author, or were dia-
betic. Diabetics were also excluded from the normals
group.

The study population consisted of 53 male schizophrenics and 10 normal males. The mean age for patients was
30.6; for the normals the mean age was 35.7.

Procedure

Patients and normals were subjected to the identical
experimental procedure. All subjects arrived at the Institute at 6:30 a.m. having fasted from 10 p.m. the prior eve-
ning. Fasting bloods were collected for plasma amino acid
and PE determinations. A breakfast was then served as a
Phe loading. The protein, fat, amino acid and carbohydrate

breakdown of the meal was as follows: 80.8 gm--Protein,
60.8 gm--Fat, 3551 mg--Phe, 906 mg--Tryptophan (Try), 3345
mg--Threonine (Thr), 4359 mg--Isoleucine (Ile), 6812 mg--
Leucine (Leu), 6314 mg--Lysine (Lys), 2065 mg--Methionine
(Met), 460 mg--Cystine (Cys), 1476 mg--Tyrosine (Tyr), 4718
mg--Valine (Val), 1614 mg--Arginine (Arg), 916 mg--Histi-
dine (His), 42.1 gm--Carbohydrates (Pennington and Church,
1985). Bloods were drawn 2 hours subsequent to each subject's meal completion for post-loading estimations of
amino acids and PE.

Patients were further evaluated at subsequent rating
sessions for neuroleptic side effects and psychiatric status. The Simpson Abbreviated Dyskinesia Scale (ADS) (Simpson et al., 1979) was used to evaluate patients for TD.
The TD designation was based on a criterion measure of a
mild-severe global rating on a subset of ADS items. A
severity score was also calculated from the sum of the same
subset of ADS items. Psychiatric status was evaluated by
use of the Brief Psychiatric Rating Scale (BPRS) (Overall
and Gorham, 1962). These data were analyzed in total score
form (items 1-18) and in the scale's five factors: anergia,
anxiety-depression, activation, hostility-suspiciousness,
and thought disturbance. In addition to the side effect
and clinical status several other variables were quantified
for study; these were, age at experimental session, age at
first neuroleptic treatment, length of time since first
neuroleptic treatment (Duration), length of the current
admission, neuroleptic dosage at the experimental session
converted to chlorpromazine equivalents, average daily neuroleptic dosage for the two weeks prior to the session converted to chlorpromazine equivalents (Davis, 1985), amount
of Phe ingested in the 24 hours prior to the session, and
body surface.

Plasma Amino Acid Analysis

Sample Preparation. Analysis of amino acids in plasma
was accomplished by a recently developed technique involving high-performance liquid chromatography preceded by pre-
column derivatization of the samples by phenylisothiocya-
nate (PITC) (Bidlingmeyer et al., 1984).

Plasma samples (0.25 ml) were deproteinized by mixing
with equal volumes of acetonitrile followed by

centrifugation. The clear protein-free supernatant was
mixed with two-volumes of methyl-t-butyl ether, centrifuged
and the ether layer discarded. The aqueous layer was evaporated to dryness under vacuum. The residue was redissolved in a mixture of ethanol: triethylamine: water
(2:1:2)(40 ul) and again redried. Derivatization was initiated by adding (50 ul) freshly prepared reagent consisting of ethanol: triethylamine: water and PITC in the ratio
of 7:1:1:1. After 20 minutes at room temperature, the
entire reagent mix was removed under vacuum. The residue
may be stored at -20° C for several weeks without any significant degradation of the amino acids. This residue was
redissolved in a phosphate buffer containing 5% acetonitrile prior to chromatographic analysis. Standards of free
amino acids in 5% albumin bovine were carried through the
entire procedure.

Chromatography. The separation of the amino acids was
performed by liquid chromatography. The solvent delivery
system consisted of two Model 6000A pumps and a Model 440
fixed wavelength UV detector (254 nm). The solvent gradient was controlled with a Model 660 Gradient Programmer and
the samples were injected by a Model 710B WISP auto injec-
tor (all equipment by Waters Assoc.). The column temperature was controlled at 39° C by an aluminum column block
which is connected to a Haake Circulating Water Bath. The
column was an application-specified reversed-phase PICO-TAG
column, 3.9 x 150 mm.

The solvent system consisted of two eluents, (a) an
aqueous buffer (A), and (b) 60% acetonitrile in water (B).
The typical buffer (A) was 0.14 M sodium acetate containing
0.5 ml of triethylamine per liter and titrated to pH 6.35
with glacial acetic acid with 6% acetonitrile. A gradient
used for the separation consisted of 3% B tranversing to
51% B at a flow rate of 1.1 ml/min. for 13 minutes using
convex curve #5 on the Model 660 Gradient Programmer. An
additional step at 100% B was used to wash the column prior
to returning to the initial conditions for re-equilibra-
tion.

Peak heights of the amino acids on the chromatogram
were compared to the peak heights of the amino acid standards processed simultaneously using norleucine as the
internal standard. A standard calibration curve developed
from the standards was used to quantitate each amino acid.

Ratios of plasma levels of Phe, Tyr and Try to the other large neutral amino acids (LNAA) were calculated to estimate the rates at which each amino acid entered the brain (Fernstrom and Wurtman, 1977; Pardridge and Choi, 1986). As an example, the Phe ratio was calculated as follows:

$$\text{Phe ratio} = \frac{\text{Phe plasma level}}{(\text{Plasma levels of His} + \text{Thr} + \text{Tyr} + \text{Val} + \text{Ile} + \text{Leu} + \text{Try})}$$

Plasma PE Analysis

PE was quantified using a GC/MS negative chemical ionization procedure. Deuterium labeled PE was used as an internal standard. The PE was separated from plasma or tissue by liquid/liquid extraction, then derivatized using pentafluorobenzoyl chloride. After removal of excess reagent the derivatized material is injected onto a 30 meter DB-17 megabore column at 210° C with methane as carrier and ionization gas. Negative chemical ionization was performed with the source temperature at 150° C and pressure of 1 torr. Fragmentation was minimal with the derivative with the major ion at amu 295 and 298 for PE and deuterated PE respectively. The method is highly sensitive with a limit of detection of 20 pg/ml and linear through 2 ng/ml. In fifty drug-free patients separate from this study with this procedure, the mean plasma PE level was 85.3 pg/ml $\pm$ 27.8.

Statistical Analysis

Each individual variable (risk factor) was screened for possible significance with respect to TD by successive univariate t-tests or the Behrens-Fischer t-test depending on which was appropriate. Also, possible dependencies among the variables were tested by correlational analysis. As is well known, such a procedure may produce "significant" p values by chance alone. The p values from these successive t-tests must, therefore, be treated with caution.

Logistic models with TD status as the dependent

variable were prepared to simultaneously test variables
(risk factors) determined to be individually significantly
associated to TD in the t-test analyses. The forced entry
approach was used for the logistic regression analyses.
The number of variables used closely corresponded to 1/10
the number of observations. It was believed that the
forced entry procedure and the limitation of variables to 4
or 5 would maximize the rigor of the logistic analyses and
enhance the interpretation of the risk factors determined
from these models. Two models were used. The first logistic model (A) analyzed significant variables pertinent to
the fasting condition (Phe ratio-fasting, duration, age at
first neuroleptic treatment, BPRS Total Score), and the
second model (B) employed significant variables appropriate
for the post-Phe loading condition (Phe ratio loading, Phe
level loading, duration, age at first neuroleptic treatment, BPRS Total Score).

RESULTS

Subjects were classified as patients or normals. The
t-test statistic was used to detect differences between
these groups with respect to the variables in the study.
No significant differences were seen between patients and
normals for either Phe or PE levels; fasting (Phe-1, PE-1)
or post-loading (Phe-2, PE-2), as follows: Phe-1 (nor-
mals=54.6; patients=53.7, t=0.17, df=61, N.S.), Phe-2 (nor-
mals=63.9; patients=65.1, t=-0.19, df=60, N.S.), PE-1 (nor-
mals=74.8; patients=84.1, t=-1.05, df=56, N.S.), PE-2
(normals=83.2; patients=90.1, t=-0.93, df=59, N.S.). A
significant difference was seen between patients and normals for the means of the fasting Tyr/LNAA ratio (Nor-
mals=.0896; Patients=.0974, t=-2.13, df=15.3, p<.05). No
differences were seen, however, for Tyr plasma levels
either fasting (Tyr-1) (normals=60.2; patients=59.1, t=.19,
df=10.2, N.S.), or loading (Tyr-2) (normals=62.7;
patients=60.9, t=.36, df=57, N.S.), or the loading ratio
(normals=.0901; patients=.0955, t=-.68, df=51, N.S.).
Also, no significant differences were detected between
patients and normals with respect to Try fasting (Try-1)
(normals=68.9; patients=66.8, t=.39, df=61, N.S.), and Try
loading (Try-2) (normals=84.9; patients=83.3, t=.32,
df=23.3, N.S.) plasma levels. No significant differences
were also seen for the following plasma/LNAA ratios: Phe-1
(normals=.0804; patients=.0763, t=.71, df=54, N.S.), Phe-2

(normals=.0768; patients=.0741, t=.42, df=51, N.S.), Try-1
(normals=.0872; patients=.0857, t=.22, df=54, N.S.), Try-2
(normals=.0708; patients=.0698, t=.15, df=51, N.S.).

A correlational analysis performed on the patient data
(see down 51) shows a significant positive association
between loading levels of PE and Phe. PE also shows some
significant correlations with Tyr and Try. Phe, however,
shows a stronger picture than PE of associations with Tyr
and Try. No significant associations were seen for the
Phe/LNAA ratio with either the Tyr or Try/LNAA ratios.

Table 1.   SIGNIFICANT CORRELATIONS

PLASMA LEVELS
Male Schizophrenic Patients

	PE-2	Phe-1	Phe-2	Tyr-1	Tyr-2	Try-1	Try-2
PE-1							
r	.74		.31			.35	.40
p	$\leq$.0001		$\leq$.05			$\leq$.05	$\leq$.01
PE-2							
r		.35	.43		.31		.39
p		$\leq$.05	$\leq$.005		$\leq$.05		$\leq$.01
Phe-1							
r			.66	.63	.34	.35	.40
p			$\leq$.0001	$\leq$.0001	$\leq$.01	$\leq$.01	$\leq$.01
Phe-2							
r				.37	.69		.56
p				$\leq$.01	$\leq$.0001		$\leq$.0001
Tyr-1							
r					.59	.49	.47
p					$\leq$.0001	$\leq$.001	$\leq$.001
Tyr-2							
r						.37	.69
p						$\leq$.01	$\leq$.0001
Try-1							
r							.65
p							$\leq$.0001

Patients were also classified by TD status, Yes
(N=34)/No (N=19). T-tests were used to test differences
between the two TD status groups on the study's continuous
variables. The demographic variables of age, age at first
neuroleptic treatment and the length of the current admission, show significant differences between the TD-Yes/No
groups with the TD-Yes group showing higher mean values
(age--t=2.66, df=51, p<.01, age at first neuroleptic treat-
ment--t=2.04, df=51, p<.05, current admission--t=2.87,
df=34.4, p<.01). The BPRS total score (t=3.21, df=28.6,
p<.01) and the BPRS factors of Activation (t=3.83, df=51,
p<.001), Hostility-Suspiciousness (t=3.51, df=51, p<.001)
and Thought Disturbance (t=2.19, df=51, p<.05) also showed
significant between-group (TD-Yes/TD-No) differences again
with the TD-Yes group showing higher mean values. Tables 2
and 3 demonstrate that the only plasma values to significantly differentiate TD status were Phe-2 and the two LNAA
ratios for Phe-1 and 2. In addition to being associated
with TD status, Phe-2 was also weakly but significantly
correlated with TD severity (r=.27, p<.05) as were Phe ratios 1 and 2 (Phe ratio 1--r=.28, p<.05, Phe ratio 2--r=.33,
p<.05). A trend for an association with TD severity was
seen for PE-2 (r=.23, p=.10) though no association was demonstrated for PE with TD status.

Table 2. PLASMA LEVELS

MALE SCHIZOPHRENIC PATIENTS

	Mean Values		t-test		
	TD-YES	TD-NO	t	df	p
	(N=34)	(N=19)			
Fasting					
PE $(10^{-12}$ gm/ml)	85.4	81.9	0.45	47	N.S.
Phe (nM/ml)	55.4	50.7	1.17	51	N.S.
Tyr (nM/ml)	67.5	65.7	0.40	51	N.S.
Try (nM/ml)	60.0	57.6	0.83	48	N.S.
Post-Loading					
PE $(10^{-12}$ gm/ml)	93.1	84.7	1.29	49	N.S.
Phe (nM/ml)	68.7	58.6	2.20	51	<.05
Tyr (nM/ml)	85.7	79.0	1.01	51	N.S.
Try (nM/ml)	61.4	59.9	0.37	47	N.S.

Table 3. PLASMA/LNAA RATIO VALUES

MALE SCHIZOPHRENIC PATIENTS

	Mean Values		t-test		
	TD-YES	TD-NO	t	df	p
	(N=34)	(N=19)			
Fasting					
Phe	.0792	.0711	2.17	48	<.05
Tyr	.0990	.0945	0.87	48	N.S.
Try	.0869	.0834	0.71	48	N.S.
Post-Loading					
Phe	.0782	.0658	3.20	46	<.01
Tyr	.0984	.0897	1.69	46	N.S.
Try	.0698	.0698	- 0.01	21.4*	N.S.

*Behrens-Fisher t

    The risk factors (variables) found to be significant
in logistic regression Model A (see Methods) thereby demonstrating significant independent association with TD, were:
Phe ratio 1, age at first neuroleptic treatment and BPRS
total score. Logistic regression Model B (see Methods)
performed on the post-loading condition found Phe ratio 2,
age at first neuroleptic treatment and the BPRS total score

to be independently significant for TD status.

DISCUSSION

This study represents the first attempt to study amino acid metabolism in relationship to TD status or severity. The significant positive association of the Phe post-loading plasma levels with TD presence and the independent ability of the Phe/LNAA ratio to explain TD status with higher Phe ratios for TD patients in both the fasting and post-Phe loading conditions can have implications for the balance of brain amino acids, and chronic neurotransmitter balances in TD. The fact that Phe levels and ratios did not significantly differ between patients and normals and that these values bore similarity between the TD-Yes group and normals raises questions about the nature of those implications that cannot be answered with the study data. While Phe also showed a significant correlation with TD severity, PE levels only showed a trend in that direction. The fact that PE levels did not differentiate TD status while Phe did so suggests some disassociation between the amine and the amino acid as regards the pathophysiology of TD. The plasma levels of Phe and PE had, however, shown significant, albeit modest, correlations. The possibility cannot be discounted from the study data, however, that metabolites of PE such as phenylacetic acid would have shown an association with TD status.

Other investigators have studied interrelationships between amino acid levels and between amino acids and amines in varied patient populations. Studies conducted on older, treated PKU patients, measuring plasma Phe and urinary dopamine, found that in 9 out of 10 patients an inverse relationship existed between plasma Phe and urinary dopamine. In the same study, urinary serotonin fell during Phe loading in six patients (Krause et al., 1985). Other work in a schizophrenic population with patients after a 1-2 week neuroleptic-free period found high plasma Phe to be associated with decreased CSF Tyr, Try, HVA and 5HIAA which was not the case for the normals studied. Plasma Tyr and Try were not, however, significantly correlated to HVA or 5HIAA in either the patients or normals (Bjerkenstedt et al., 1985). These authors suggest that raised plasma levels of competing amino acids to Tyr such as Phe may in turn lead to lower levels of dopamine and then supersensitive

dopamine receptors. In a study conducted among alcoholic
patients, increased Phe and Tyr resulted in a lowered Try
ratio for depressed alcoholics vs. non-depressed patients
and controls (Branchey et al., 1984).

Our findings for TD may not be consistent with the
Bjerkenstedt et al. (1985) suggestion or the above Branchey
et al. (1984) findings because the significant positive
association of the Phe ratio with TD did not manifest
itself in a relationship for the disorder with a lower Tyr
or Try ratio. Additionally, though in the study data substantial significant positive correlations had been seen
for plasma levels of Tyr and Phe and no correlations for
their ratios, no relatioship with TD was seen for either
Tyr plasma levels or the Tyr/LNAA.

The clues to the meaning of the study's association of
Phe and TD may be found by going back to the original stimulus for the work, which was the significant discrimination
found by the first author for TD status by PKU status
(Richardson et al., 1986). It has been suggested that the
heavy saturation of the blood-brain barrier amino acid
transport system by Phe is the major causative factor in
the pathogenesis of PKU (Choi and Pardridge, 1986). This
suggestion, however, cannot, it seems, explain the findings
of this study due to the lack of significant difference
between normals and patients in Phe values. Further work
will be needed to determine both the genesis of the Phe
association with TD, and the impact of that association on
TD vulnerability.

REFERENCES

Bidlingmeyer, B.A., Cohen, S.A., and Farvin, T.L. (1984)
    Rapid analysis of amino acids using pre-column derivati-
    zation. J. Chromatogr. 336, 93-104.
Bjerkenstedt, L., Edman, G., Hagenfeldt, L., Sedvall, G.,
    and Wiesel, F. A. (1985) Plasma amino acids in relation
    to cerebrospinal fluid monoamine metabolites in schizo-
    phrenic patients and healthy controls. Br. J. Psychia-
    try 147, 276-282.
Branchey, L., Branchey, M., Shaw, S., and Lieber, C.S.
    (1984) Relationship between changes in plasma amino
    acids and depression in alcoholic patients. Am. J. Psy-
    chiatry 141, 1212-1215.

Choi, T.B. and Pardridge, W.M. (1986) Phenylalanine transport at the human blood-brain barrier. Studies with isolated human brain capillaries. J. Biol. Chem. 261, 6536-6541.

Davis, J.M. (1985) Organic therapies, in Comprehensive Textbook of Psychiatry, 4th edit. (Kaplan, H.I. and Sadock, B.J., eds), pp. 1481-1569. Williams and Wilkins, Baltimore, Maryland.

Fernstrom, J.D. and Wurtman, R.J. (1972) Brain serotonin content: Physiological regulation by plasma neutral amino acids. Science 178, 414-416.

Fischer, E. and Heller, B. (1972) Phenylethylamine as a neurohumoral agent in brain. Behav. Neuropsychiatry 4, 8-11.

Jeste, D.V. and Wyatt, R.J. (1981) Dogma disputed: Is tardive dyskinesia due to postsynaptic dopamine receptor supersensitivity? J. Clin. Psychiatry 42, 455-457.

Karoum, F., Linnoila, M., Potter, W.Z., et al. (1982) Fluctuating high urinary phenylethylamine excretion rate in some bipolar affective disorder patients. Psychiatry Res. 6, 215-222.

Knox, L.E. (1966) Phenylketonuria, in The Metabolic Basis of Inherited Disease, (Stanbury, J.B., Wyngaarden, J.B., and Fredickson, D.S., eds), pp. 258-294. McGraw Hill Book Co., New York.

Krause, W., Halminski, M., McDonald, L., Dembure, P., Salvo, R., Freides, D., and Elsas, L. (1985) Biochemical and neuropsychological effects of elevated plasma phenylalanine in patients with treated phenylketonuria. J. Clin. Invest. 75, 40-48.

Linnoila, M., Karoum, F., Cutler, N.P., et al. (1983) Temporal association between depression-dependent dyskinesias and high urinary phenylethylamine output. Biol. Psychiatry 18, 513-517.

Oates, J.A., Nirenberg, P.Z., Jepson, J.B., et al. (1963) Conversion of phenylalanine to phenylethylamine in patients with PKU. Proc. Soc. Exp. Biol. Med. 112, 1078-1081.

Overall, J.E. and Gorham, D.R. (1962) The brief psychiatric rating scale. Psychol. Rep. 10, 799-812.

Pardridge, W.M. and Choi, T.B. (1986) Neutral amino acid transport at the human blood-brain barrier. Federation Proc. 45, 2073-2078.

Pennington, J.A.T. and Church, H.N. (eds) (1985) Food Values of Portions Commonly Used, 14th edit., Harper & Row, New York.

Richardson, M.A., Pass, R., Bregman, Z., and Craig, T.J. (1985) Tardive dyskinesia and depressive symptoms in schizophrenics. Psychopharmacol. Bull. 21, 130-135.

Richardson, M.A., Haugland, G., Pass, R., and Craig, T.J. (1986) The prevalence of tardive dyskinesia in a mentally retarded population. Psychopharmacol. Bull. 22, 243-249.

Rosenbaum, A.H., Niven, R.G., Hanson, N.P., et al. (1977) Tardive dyskinesia: Relationship with a primary affective disorder. Dis. Nerv. Syst. 38, 423-427.

Rush, M., Diamond, F., Alpert, M., et al. (1982) Depression as a risk factor in tardive dyskinesia. Biol. Psychiatry 17, 387-392.

Simpson, G.M., Lee, J.H., Zoubok, B., and Gardos, G. (1979) A rating scale for tardive dyskinesia. Psychopharmacol. (Berlin) 64, 171-179.

Tarsy, D. and Baldessarini, P.J. (1977) The pathophysiologic basis of tardive dyskinesia. Biol. Psychiatry 12, 431-450.

Wilson, I.C., Garbutt, J.C., Lanier, C.F., et al. (1983) Is there a tardive dysmentia? Schizophr. Bull. 9, 187-192.

Acknowledgment: This study was funded in part by the National Institute of Mental Health, Grant #R03 MH40629.

METABOLISM OF TYRAMINE IN SUBJECTS TREATED WITH DIFFERENT MONOAMINE OXIDASE INHIBITORS

P.R. Bieck, G. Aichele, C. Schick,
E. Hoffman and E. Nilsson

Human Pharmacology Institute
CIBA-GEIGY GmbH, D-74 Tübingen,
Federal Republic of Germany

Abstract: In the studies presented, the relative extent of tyramine (TYR) deamination to p-hydroxyphenylacidic acid (HPAA) and of TYR conjugation was determined before and during treatment with 5 different MAO inhibitor (MAOI) drugs. Healthy ambulatory subjects were treated p.o. for 2 - 4 weeks. Data were evaluated from (n) subjects. MAOI drugs were: Brofaromine (Brof): 100 - 150 mg/d (9); clorgyline (Clor): 5, 10, 15 mg/d (3); selegiline (Sel): 5, 20 mg/d (6); tranylcypromine (TCP): 20 mg/d (4) and phenelzine (Phen): 30, 45 mg/d (1).
After oral doses of TYR causing systolic blood pressure (BP) increases of 30 mm Hg, plasma concentrations of HPAA, conjugated (conj) TYR (after enzymatic deconjugation) and of unconjugated (unc) TYR were determined by HPLC. The area under the plasma level-time curve (AUC) was calculated considering the different TYR doses given (AUCspec).
The following results were obtained:
1. In 16 untreated subjects unc TYR and conj TYR in plasma amount to $0.4 \pm 0.2\%$ and $12.5 \pm 4.3\%$, of the sum of total TYR + HPAA.
2. During subchronic MAO inhibition, the AUCspec of conj TYR after oral TYR administration increases in varying degrees from 12.5% to 92% of the total AUCspec (total TYR + HPAA). This is accompanied by

a maximal decrease of AUCspec of HPAA from
87.1% to 8%.
3. The following order of increasing potency
   for inhibiting gastrointestinal MAO could
   be established: Sel (MAO-B, irreversible),
   Brof (MAO-A, reversible), Phen (MAO-A+B,
   irreversible), Clor (MAO-A, irreversible)
   and TCP (MAO-A+B, partially irreversible).
The results suggest that during inhibition of
deamination in the gastrointestinal tract,
inactivation by conjugation protects the
organism against the pharmacological action of
the symphatomimetic amine tyramine.

INTRODUCTION

The development of specific and sensitive
assays for TYR (p-hydroxyphenylethylamine) has
resulted in a better knowledge on the kinetics
and metabolism of this substance in man.
Deamination by MAO is the major route for TYR
metabolism. HPAA accounts for 44 - 81% of the
urinary excretion of an intravenous or oral
dose of TYR (Tacker et al., 1972; Jones and
Pollitt, 1976). **Sulfate conjugation** to
TYR-0-sulfate accounts for 10 - 15% of the
total dose in urine after oral TYR (Mullen and
Smith, 1971; Smith and Mitchell, 1974). Only
about 3.5% of the total radioactive dose is
excreted in the urine as **unconjugated TYR**
(Tacker et al., 1972). During subchronic
treatment with the specific reversible MAO-A
inhibitor brofaromine (Waldmeier et al., 1983)
the relative amount of unc TYR excreted into
the urine is elevated and the clearance of unc
TYR from plasma is diminished. The mean relative
bioavailability (BAVrel) of unc TYR is
increased up to 3-fold (Bieck et al., 1985).
Except studies about sulfation of TYR in
relation to depressive illness (Sandler et
al., 1985) or to migraine (Smith et al., 1971;
Youdim et al., 1971), not much attention has
been paid to the inactivation of oral TYR by
conjugation. To evaluate the contribution of
this metabolic step, the amounts of unc and of

conj TYR including the metabolite HPAA were
measured after oral TYR load. Metabolisation
before and during MAO inhibition with several
pharmacologically different drugs was compared.

MATERIALS AND METHODS

No objections against the design and
procedures of these studies were raised by the
Ethical Committee of the Human Pharmacology
Institute CIBA-GEIGY, Tübingen.

Subjects

Sixteen healthy volunteers, 4 females and
12 males, aged between 24 and 29 years,
participated in five studies. None of the
volunteers took any other medication for at
least one week before and during the studies.
During the experiments, no foodstuffs rich in
tyramine or alcohol were permitted. All
volunteers were ambulatory and carried out
their usual activities.

Dosage

In the studies, there were first **control
periods** followed by an oral **MAOI treatment
period** for 2 to 4 weeks. MAOI drugs were given
in following dosages. Brofaromine (Brof),
selective and reversible inhibitor of type A:
100 or 150 mg/d; selegiline (Sel), selective,
irreversible inhibitor of type B: 5 and 20
mg/d; clorgyline (Clor), selective,
irreversible inhibitor of type A: 5, 10, and
15 mg/d; phenelzine (Phen): 30 and 45 mg/d and
tranylcypromine (TCP): 20 mg/d, both
irreversible, nonselective inhibitors of both
type A and B.

Oral Tyramine Pressor Test

Control period: On day 1, the untreated
volunteers took tyramine capsules with 100 ml
water in hourly intervals to estimate the

amount of amine (mg TYR) to increase the
systolic BP by at least 30 mm Hg (= PD30). On
day 2, the volunteers took the PD30 of TYR as
single dose to measure the kinetics of TYR in
untreated subjects.

MAOI period: The estimation of the PD30
in the studies with MAOI drugs was repeated at
the end of each treatment period. The PD30 of
TYR found was given as single dose to get
kinetic data on the next day.

Tyramine Kinetics

Blood samples were obtained before and at
frequent intervals (for details see Fig. 1)
after oral administration of TYR. The blood
was centrifuged immediately, plasma was
separated and kept frozen at -20°C until
analysis.

Analytical Methods

Unc TYR and HPAA were measured using HPLC
with electrochemical detection (Bieck et al.,
1985). To estimate conj TYR, plasma was
enzymatically treated with
beta-glucuronidase/arylsulfatase (SIGMA G
0876) for 24 h. Incubation of plasma sample
aliquots with beta-glucuronidase/arylsulfatase
and arylsulfatase resulted in equal amounts of
liberated TYR. This shows that conj TYR in
plasma represents exclusively TYR-sulfate.

Calculations

From plasma concentrations of TYR and
HPAA, the area under the plasma level-time
curve (AUC(0-7); ng/ml * h) was calculated
using the trapezoidal rule. The AUC was
normalized by dividing by the TYR dose given
(AUCspec; ng/ml * h * mg^{-1}TYR). HPAA is
expressed in TYR equivalents by multiplication
with the ratio mol.wt of TYR/mol.wt of HPAA.
Means $\pm$ SD are given. The paired t-test of
Student was applied to test for statistical
differences.

RESULTS

Plasma Concentrations of TYR and HPAA
Unmedicated subjects

Mean plasma concentration-time curves of HPAA, conj TYR and unc TYR after administration of an oral TYR load (mean dose 532 mg) to six untreated subjects are shown in Fig. 1. Maximal concentrations of 8000, 800 and 90 ng/ml are reached within 1 - 2 h (HPAA), 2 h (conj TYR) and 0.5 h (unc TYR). Three h after intake of TYR only HPAA and conj TYR are measurable in plasma. The AUCs differ widely according to the different concentration ranges measured. The **ratio** is about **100** (AUC HPAA: 22 300) : **10** (AUC conj TYR: 2.240) : **0.5** (AUC unc TYR: 100). The AUCspec values are 42, 4.2 and 0.19.

In Tab. 1, the AUCspec values for 16 untreated subjects are summarized. They are in the same order of magnitude shown for the 6 subjects (Fig. 1). It can be seen that all three substances show a 3- to 4-fold range of AUCspec values. Expressed as % of the total AUCspec the contributions of unc TYR, conj TYR and HPAA are 0.4%, 12.5% and 87.1%.

	AUCspec (0-7) ng/ml*h*mg^{-1}TYR	AUCspec % of total
TYR unc (range)	0.17 ± 0.07 (0.08 - 0.32)	0.4 ± 0.2 (0.2 - 0.8)
TYR conj (range)	5.3 ± 1.7 (2.5 - 8.6)	12.5 ± 4.3 (3.5 - 20.0)
HPAA (range)	39.3 ± 12.7 (21.8 - 69.1)	87.1 ± 4.4 (79.8 - 96.4)

Tab. 1 TYR and its metabolites in plasma of 16 untreated subjects after oral administration of the PD30 of TYR

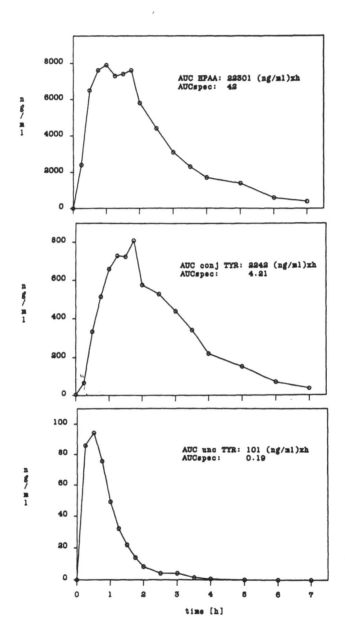

<u>Fig. 1</u>. Mean plasma concentrations of unc TYR,
conj TYR and HPAA in 6 untreated subjects
after oral administration of the PD30 of TYR
(mean dose 532 ± 166 mg).

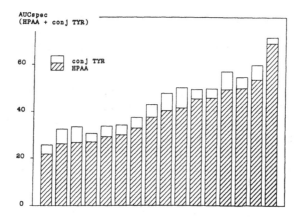

Fig. 2. Distribution of HPAA and conj TYR in plasma of 16 untreated subjects after oral administration of the PD30 of TYR. AUCspec (HPAA + conj TYR).

Fig. 2 depicts the wide range of variation for both metabolic products in these 16 subjects. There is no clear correlation between deamination and conjugation. The lowest and highest AUCspec values of HPAA, 20 and 70, are associated with AUCspec values of conj TYR of 4 and 2.5.

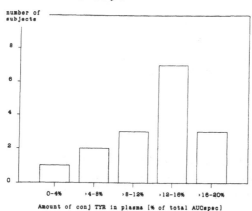

Fig. 3. Histogram of the relative amount of conj TYR in plasma (% of total AUCspec) of 16 untreated subjects after oral administration of the PD30 of TYR.

The distribution of conj TYR is shown on
a histogram (Fig. 3). In seven out of 16
subjects, 12 to 16% of the total AUCspec is
conjugated.

In one untreated volunteer, the effect of
different amounts of oral TYR on its plasma
metabolites was explored. Increasing the oral
dose of TYR from 50 mg to 100 mg enhances the
relative amount of conj TYR from 5.6% to
22.5%. However, doubling of the dose to 200 mg
does not change this ratio any further
(19.8%).

Plasma Concentrations of TYR and HPAA
MAOI treatment

Fig. 4 shows the effect of subchronic
treatment with the MAO-A inhibitor Brof on the
amounts of TYR and TYR metabolites in plasma.
An increase of the AUCspec of **unc TYR** is not
always demonstrable (Fig. 4, right). The mean
change from 0.18 to 0.38 represents a
statistically non-significant 2.1-fold
increase of systemic availability. The AUCspec
of **HPAA** shows no consistent change (Fig. 4,
center). Again, the mean decrease from 42 to
36 does not reach statistical significance.
Only the AUCspec of **conj TYR** is persistently
elevated after MAOI treatment (Fig. 4, left).
The mean change from 5.0 to 35.1 represents a
significant 7-fold increase of systematic
bioavailability (p < 0.0001).
The mean change of plasma metabolites of
TYR after administration of oral TYR before,
during and after treatment with the
irreversible MAO-A inhibitor Clor is depicted
in Fig. 5. There is a dose-dependent relative
increase of the AUCspec of conj TYR from 13%
to 53% and to 74%. During the third week of
treatment with 15 mg/d, there is no further
change (68%). Two to three weeks after
stopping Clor initial values were reached.

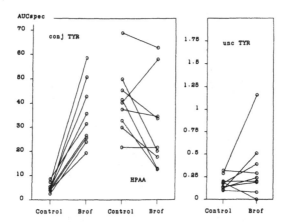

Fig. 4. Effect of Brof on amounts of conj TYR,
HPAA and unc TYR in plasma after oral
administration of the PD30 of TYR. AUCspec.

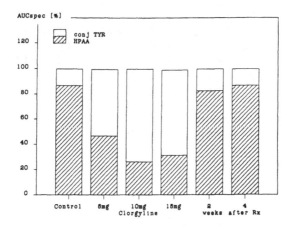

Fig. 5. Plasma TYR metabolites after oral
administration of the PD30 of TYR before,
during and after Clor treatment. Mean of 2 - 3
subjects. AUCspec (%).

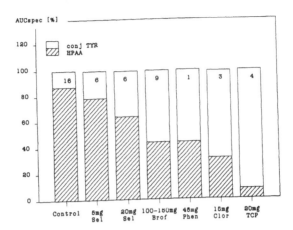

<u>Fig. 6.</u> Plasma TYR metabolites after oral
administration of the PD30 of TYR during
treatment with different MAOI drugs.
AUCspec (%). n = number of subjects.

Fig. 6 shows the relative amounts of
plasma TYR metabolites after oral TYR during
treatment with 5 different MAOI drugs. The
higher dose of 20 mg/d Sel leads to an about
3-fold increase of conj TYR, compared with
controls. Five mg/d have only a small effect.
Brof given for 2 weeks causes a 5-fold
increase of the relative amount of conj TYR,
which is similar to the effect of Phen
administered for one week (to one subject
only!). The irreversible MAOI Clor and TCP
show the largest effects. During TCP, the
initial relation between HPAA and conj TYR is
reversed.

DISCUSSION

Comparison of the time-courses of the 3
substances in plasma suggests prolonged
gastrointestinal metabolism of oral TYR mainly
by **deamination** but also by **conjugation** in a
ratio of 10 : 1. Our results on TYR metabolism
using **plasma** concentration measurements agree

well with data in humans obtained earlier by
Mullen and Smith, 1971, Smith and Mitchell,
1974 and by Jones and Pollitt, 1976. After
oral TYR these investigators determined that
60% and 10 - 15% of the dose is excreted in
urine in form of the deamination product and
the sulfate conjugate. We found in plasma of
16 untreated subjects relative amounts of 87%
and 13% for HPAA and conj TYR. The mean
HPAA/total TYR ratio of 7.2 corresponds well
with the ratio of 5 found in venous plasma of
an in situ dog preparation, using isolated
loops of intestine (Ilett et al., 1980).

Our data, showing a 0.4% contribution of
unconj TYR for the total amount of TYR
metabolites do not agree, however, with the
values given by Tacker et al., 1972. They
found 3.5% of a radioactive dose excreted in
urine in form of unc TYR. The dissimilarity
could be resulting from their non-specific
method or from the different material
analysed. But, varations between plasma
kinetics and renal excretion are not a
probable explanation since clearance of unc
TYR calculated from plasma and urinary data
are significantly correlated (Bieck et al.,
1985).

The relative amount of **conj TYR** in
plasma, appearing for 7 h after oral
administration ranges between 3.5% and 20%.
Half of the 16 subjects showed values around
13%. The reason for this large interindividual
variability in untreated subjects is unclear,
but the capacity of the sulfate conjugation
process could play a role (Smith and Mitchell,
1974; Morris and Levy, 1983). The preliminary
finding in one subject, showing an enhancement
of TYR conjugation up to a fraction of 20% by
increasing the exogenous TYR dose supports
this notion. Anyhow, the 3- to 4-fold
variation of the AUCspec of unc TYR from 2.5
to 8.6 in healthy subjects cautions against
the inconsiderate use of a reduced excretion
of conj TYR to diagnose patients with migraine
(Mullen and Smith, 1971) or with depression
(Sandler et al., 1985).

The **A-form** of MAO contributes to over 70% of the total MAO activity present in the human intestine (Hasan et al., 1987). Therefore, specific inhibition of MAO-A in the gastrointestinal tract should significantly decrease the amount of the **deamination product** HPAA in plasma. However, the AUCspec of HPAA shows no consistent change after the reversible MAOI drug Brof. The mean decrease of the AUCspec of HPAA is 26%. Only the absolute amount of conj TYR is statistically significant and persistently elevated (mean: 7-fold).

In contrast, the irreversible inhibitor of MAO-A Clor markedly inhibits deamination already after treatment for one week with 5 mg/d which becomes maximal in the second week with 10 mg/d. The AUCspec of HPAA decreases by 44% (5mg/d) and 70% (10 mg/d). In parallel, the absolute amounts of conj TYR in plasma increase 4- to 6-fold.

These differing results using two selective MAO-A inhibitors in clinically effective doses could be explained by their irreversible or reversible binding to the enzyme. Our finding that Brof protects TYR from deamination to a significantly lesser extent than Clor suggests that the substrate might have displaced the inhibitor from the enzyme. Such an assumption is in accordance with results obtained in animals (Waldmeier, 1985).

As expected, from the smaller contribution of the **B-form** of MAO in the intestine to deamination of TYR there is almost no decrease of the AUCspec of HPAA during treatment with 5 mg/d of the irreversible MAO-B inhibitor Sel. However, conj TYR in plasma is increased twofold. The higher dose of 20 mg/d Sel causes a 20% decrease of plasma HPAA and a threefold increase of conj TYR. It is unclear if this effect is related to additional inhibition of MAO-A, what has been postulated (Elsworth et al., 1978).

After inhibition of both **forms A and B** of

MAO by the (partially) irreversible nonspecific inhibitor TCP, deamination of TYR is so much impaired (decrease of AUCspec HPAA > 70%) that almost all TYR is metabolised by conjugation.

Our results suggest that during subchronic inhibition of MAO the second essential pathway of TYR inactivation, e.g. **conjugation** is used increasingly to protect the body from the systemic effects of oral TYR. It appears that changes of TYR conjugation are a more sensitive index for the assessment of gastrointestinal MAO inhibition than those of HPAA. Its measurement, therefore, can be used to establish an order of potency for pharmacologically different MAOI drugs during subchronic doses applied clinically.

Acknowledgement: We thank for the help of M. Hügler and G. Vees.

REFERENCES

BIECK, P.R., SCHICK, CH., ANTONIN, K.H., REIMANN, I., MOERIKE, K. (1985) Tyramine kinetics before and after MAO inhibition with the reversible MAO inhibitor brofaromine (CGP 11 305 A). In: **Neuropsychopharmacology of the Trace Amines**, (BOULTON, A.A., BIECK, P.R., MAITRE, L., RIEDERER, P.), pp 411 - 426. Humana Press Inc

ELSWORTH, J.D., GLOVER, V., REYNOLDS, G.P., SANDLER, M., LEES, A.J., PHUAPRADIT, P., SHAW, K.M., STERN, G.M., KUMAR, P. (1978) Deprenyl administration in man: a selective monoamine oxidase B inhibitor without the "cheese effect". **Psychopharmacology** <u>57</u>, 33 - 38

HASAN, F., McRODDEN, J.M., KENNEDY, N.P., TIPTON, K.F. (1987) The involvement of intestinal monoamine oxidase in the transport and metabolism of tyramine. **J. Neurotransmission** (in press)

ILETT, K.F., GEORGE, C.F., DAVIES, D.S. (1980) The effect of monoamine oxidase inhibitors on "first-pass" metabolism of tyramine in dog intestine. **Biochem. Pharmacol.** **29**, 2551 - 2556

MORRIS, M.E., LEVY, G. (1983) Serum concentration and renal excretion by normal adults of inorganic sulfate after acetaminophen, ascorbic acid, or sodium sulfate. **Clin. Pharmacol. Ther.** **33**, 529 - 536

MULLEN, P.E., SMITH, I. (1971) Tyramine metabolism and migraine: a metabolic defect. **Brit. J. Pharmacol.** **41**, 413P - 414P

SANDLER, M., HALE, A.S., WALKER, P.L., BRIDGES, P.K. (1985) Tyramine-conjugation deficit as a traitmarker in endogenous depressive illness. In: **Neuropsychopharmacology of the Trace Amines**, (BOULTON, A. A., BIECK, P., MAITRE, L., RIEDERER, P.), pp 427 - 432. Humana Press Inc

SMITH, I., MARCH, S.E., GORDON, A.J. (1972) A method for the quantitative extraction of radioactive tyramine and its metabolites from urine. **Clin. Chim. Acta** **40**, 415 - 419

SMITH, I., MITCHELL, P.D. (1974) The effect of oral inorganic sulphate on the metabolism of 4-hydroxyphenylethylamine (tyramine) in man: tyramine O-sulphate measurement in human urine. **Biochem. J.** **142**, 189 - 191

TACKER, M., CREAVEN, P.J., McISAAK, W.M. (1972) Preliminary observations on the metabolism of (1-^{14}C) Tyramine in man. **J. Pharm. Pharmac.** <u>24</u>, 247 - 249

WALDMEIER, P.C., FELNER, A.E., TIPTON, K.F. (1983) The monoamine oxidase inhibiting properties of CGP 11 305 A. **Eur. J. Pharmacol.** <u>94</u>, 73 - 83

WALDMEIER, P.C. (1985) On the reversibility of reversible MAO inhibitors. **Naunyn-Schmiedeberg's Arch. Pharmacol.** <u>329</u>, 305 - 310

YOUDIM, M.B.H., BONHAM CARTER, S., SANDLER, M., HANINGTON, E., WILKINSON, M. (1971) Conjugation defect in tyramine-sensitive migraine. **Nature** <u>230</u>, 127 - 128

ADRENERGIC ACTIVITIES OF OCTOPAMINE AND SYNEPHRINE STEREOISOMERS

C. Mohan Thonoor[*], Margaret W. Couch, Clyde M. Williams and John M. Midgley[*], Veterans Administration Medical Center and Department of Radiology, University of Florida, Gainesville, Florida, U.S.A., 32602 and [*]Department of Pharmacy, University of Strathclyde, Glasgow, U. K., G1 1XW

INTRODUCTION

Both m-OA and p-OA are known to occur in the mammalian sympathetic nervous system where they are taken up in NA nerve terminals, accumulated in storage vesicles and released together with NA (Ibrahim et al, 1985). It is probable that both amines are released with NA as co-transmitters as proposed for p-OA by Axelrod and Saavedra (1977). Although all nerves probably contain two or more co-transmitters (O'Donohue et al, 1985), little is known about the mechanism of neuromodulation produced by the release of multiple co-transmitters. m-OA and p-OA are structurally so similar to NA that it is reasonable to suppose that their actions might be mediated by one or more of the well-characterized adrenoceptors. The physiological effects of p-OA and m-OA were first determined with racemates, and later some investigations were performed with poorly characterized enantiomers. The latter experiments were carried out on selected in vivo responses before the different subtypes of adrenoceptors were recognized. The activities of the pure (-) and (+) forms of m- and p-OA on α-1, α-2, β-1 and β-2 adrenoceptors have now been determined (Brown et al, 1987; Jordan et al, 1987).

METHODS

Resolution of Racemates into the Corresponding Enantiomers

Salts of racemic m- and p-OA and p-synephrine (p-SYN) were converted to the free bases by ion exchange chromatography. Appropriate diastereoisomeric salts were prepared by reacting them with one equivalent of an enantiomer of a suitable organic acid (eg. (+) or (-)-dibenzoyltartaric acid) and the diastereoisomers were separated by fractional crystallization. Each of the resultant diastereoisomeric salts was dissolved in water and afforded the corresponding optically active enantiomeric HCl salt using ion-exchange chromatography. (-) and (+)m-SYN were obtained commercially.

Pharmacological Protocol

The activities of the (-) and (+) forms of m-OA, p-OA, m-SYN and p-SYN were evaluated using well established pharmacological in vitro test systems. a) α-1 adrenoceptors from rat aorta - contraction of thoracic aorta ring (with endothelium removed); b) α-2 adrenoceptors from rabbit saphenous vein - contraction of vein rings pretreated with cocaine; c) β-1 adrenoceptors from guinea pig atria - chronotropic response of isolated atria, pretreated with phenoxybenzamine, and d) β-2 adrenoceptors from guinea pig trachea-helical strips pretreated successively with phenoxybenzamine and carbachol.

RESULTS

We have recently found that the circular dichroism (CD) curves of (-)p-OA and (-)p-SYN are superimposable, indicating that these compounds have the same absolute configuration (Midgley et al, 1987). Our recent X-ray crystallographic analysis of (-)p-SYN has defined this unequivocally as (R), in agreement with previous conclusions based upon chemical correlations (Pratesi et al, 1958). CD curves of the (+)p-isomers are also superimposable; as expected, they are of opposite sign to those of the corresponding (-)-isomers. A similar pattern is observed for CD curves of the enantiomorphs of m-OA and m-SYN but the curves of the (-)m-isomers are of opposite sign to those of the corresponding (-)p-isomers, indicating that the former have the (S) configuration. This conclusion is in contrast to that proposed (on the basis of CD evidence) for (-)m-OA (Kametani et al, 1973) and it also reverses the (R) configuration generally accepted for (-)m-SYN ("phenylephrine") on the basis of ORD correlations (Lyle, 1960).

Activities on α-1 and α-2 Adrenoceptors

The rank order of potency of the (-) forms on α-1 adrenoceptors from rat aorta was NA=A>m-OA=m-SYN>p-OA=p-SYN (Figure 1). The two m-compounds were 6 fold less active and the two p-compounds were 1,000 fold less active than NA at α-1 adrenoceptors. These values are in good agreement with an earlier investigation on the rat aorta in which racemic m-OA and (-)m-SYN were 3-4 fold less active and racemic p-OA was 1,000 fold less active than NA (Ress et al, 1980). The (+) isomers were 1-3 orders of magnitude weaker than their (-) counterparts.

The rank order of potency of the (-) forms on α-2 adrenoceptors in rabbit saphenous vein was the same as for α-1, i.e. NA=A>m-OA=m-SYN>p-OA=p-SYN (Figure 2). The two m- compounds were 150 fold less active and the two p-compounds were 1,000 fold less active than NA. The (+) forms were 1-3 orders of magnitude less active than their (-) counterparts.

Activities on β-1 and β-2 Adrenoceptors

The rank order of potency of the (-) forms on β-1 adrenoceptors in the isolated guinea pig atria was NA>m-SYN>m-OA=p-OA>p-SYN; m-SYN was 100 fold less active; m- and p-OA were 6,000 fold less active, and p-SYN was about 40,000 fold less active than NA (Figure 3). The (+) forms were 1-2 orders of magnitude less active than their counterparts.

The four (-) compounds were more than 4 orders of magnitude less active than NA on β-2 adrenoceptors from guinea pig trachea (Figure 4) and the (+) forms had no detectable activity in concentrations as high as 10^{-4}M.

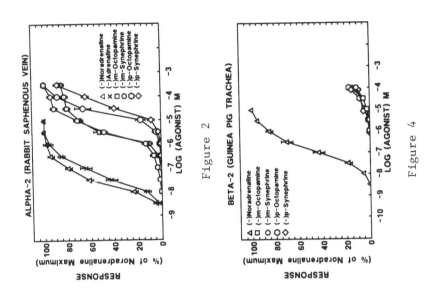

Figure 1

Figure 2

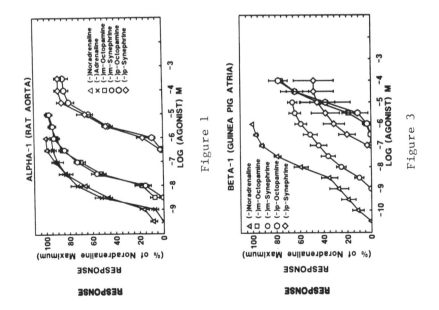

Figure 3

Figure 4

These findings are in good agreement with earlier observations that m- and p-OA had no detectable effect on the
in vivo β-adrenergic responses of initiation of thirst or
increase in tail skin temperature in the rat (Fregly et
al, 1979).

Binding Affinities on α-1 Adrenoceptors

Ligand binding data for the OA and SYN enantiomers
at α-1 and α-2 binding sites obtained from rat cerebral
cortex showed that (-) forms were more active than (+)
forms and that the rank order of affinity of (-) forms
for both binding sites was NA>m-OA=m-SYN>p-SYN>p-OA.
Relative affinities of α-1 binding sites were very similar to their pharmacological activity in rat aorta, but
affinities in both series relative to NA were greater
at α-2 binding sites than the pharmacological activity in
rabbit saphenous vein, suggesting a low receptor reserve
in the latter tissue.

DISCUSSION

The concentration of NA in the rat heart is approximately 1,000 ng/g whilst that of m-OA and p-OA are 4.0
and 3.3 ng/g, respectively, less than 1% that of NA
(Ibrahim et al, 1985). However, the half life of p-OA
(2.1 h) was found to be significantly lower than that of
NA (13-15 h) resulting in a 6 fold greater turnover of p-
OA than NA in rat heart (Molinoff and Axelrod, 1972).

It is therefore clear that if m-OA and p-OA are coreleased with NA in amounts proportional to their concentrations, it must be concluded that their activities at
adrenergic receptors (α-1, α-2, β-1 and β-2) are too low
to be physiologically significant. It follows as a cor-
ollary to this conclusion that if m-OA and p-OA have a
physiological function in vertebrates, then these functions are probably mediated by specific octopaminergic
receptors, as in the case for p-OA in the invertebrates.

REFERENCES

Axelrod J. and Saavedra J.M. (1977) Octopamine. Nature 265, 501–504.

Brown C.M., McGrath J.C., Midgley J.M., Muir A.G.B., O'Brien J.W., Thonoor C.M., Williams C.M. and Wilson V.G. (1987) Alpha-adrenergic activities of octopamine and synephrine stereoisomers. (Submitted for publication).

Fregley M.J., Kelleher D.L. and Williams C.M. (1979). Adrenergic activity of ortho- meta- and para-octopamine. Pharmacol. 18, 180–187.

Ibrahim K.E., Couch M.W., Williams C.M., Fregly M.J. and Midgley J.M. (1985) m-Octopamine: Normal occurrence with p-octopamine in mammalian sympathetic nerves. J. Neurochem. 44, 1862–1866.

Jordan R., Midgley J.M., Thonoor C.M. and Williams C.M. (1987) Beta-adrenergic activities of octopamine and synephrine stereoisomers. J. Pharm. Pharmacol. (In press, 1987).

Kametani T., Shibuya S., Sugi H. and Fukumoto K. (1973) Studies on the syntheses of heterocyclic compounds. Part DXXVII. J. Heterocyc. Chem. 10, 451–453.

Lyle G.G. (1960) Rotatory dispersion studies. I. Aralkylamines and alcohols. J. Org. Chem. 25, 1779–1784.

Midgley J.M., Thonoor C.M., Drake A.F., Williams, C.M., Koziol A.E. and Palenik G. (1987) The resolution and absolute configuration of isomeric octopamines and synephrines (Submitted for publication).

Molinoff P.B. and Axelrod J. (1972) The normal distribution and turnover of octopamine in tissues. J. Neurochem. 19, 157–163.

O'Donohue T.L., Millinton W.R., Handelmann G.E., Contreras P.C. and Chronwall B.M. (1985) On the 50th anniversary of Dale's law: multiple neurotransmitter neurons. Trends in Pharmacol. Sci. 6, 305–308.

Pratesi P., La Manna A., Campiglio A. and Ghislandi V. (1958) The configuration of adrenaline and of its p-hydroxyphenyl analogue. J. Chem. Soc., 2069–2074.

Ress R.J., Rahmani M.A., Fregly M.J., Field F.P. and Williams C.M. (1980) Effect of isomers of octopamine on in vitro reactivity of vascular smooth muscle in rats. Pharmacol. 21, 342–347.

TYRAMINE-INDUCED BRAIN INJURY
IN PHENELZINE-TREATED DOGS:
AN ANIMAL MODEL FOR CEREBRAL EDEMA

B.A. Faraj, R. Sarper, M. Camp,
E. Malveaux and Y. Tarcan,
Department of Radiology,
Emory University School of Medicine,
Atlanta, Georgia, USA

We have recently demonstrated that p-tyramine (p-TA) induced behavioral side-effects and precipitated coma in chronically treated dogs with the monoamine oxidase inhibitor (MAOI) phenelzine (PEH) (Faraj et al, 1983). The objective of the present investigation was to determine whether p-TA-induced neurotoxicity in PEH-treated dogs is a reflection of the development of brain edema and whether the edema is associated with increased permeability into the brain from peripheral circulation of diffusion-limited substances. We chose to examine the influence of p-TA in dogs before and after treatment with PEH on the first-pass brain image of technetium-99m-diethylenetriamine pentaacetic acid (Tc-99m-DTPA). Furthermore, kinetics of brain Tc-99m-DTPA were correlated with cerebrospinal fluid (CSF) pressure measurements and determination of CSF concentrations of catecholamines.

METHODS

Animals: Ten conditioned mongrel dogs were used (15-20 kg).

MAO Inhibition: MAO enzyme was inhibited in these animals after the daily oral administration of PEH (4.0 mg/kg) as tablets over a period of seven days.

p-TA and CSF Pressure: p-TA-HCl (50 mg/kg) was

administered orally in 5 ml distilled water. One hour
after the oral administration of p-TA, the dogs were
anesthetized with sodium pentobarbital (25-30 mg/kg). A
20-gauge intracath needle was inserted into cisternal
space for sampling of CSF specimens and continuous
monitoring of CSF pressure.

CSF Catecholamines:

     Dopamine (D), norepinephrine (NA) and epinephrine
(A) levels in CSF samples were measured by radioenzymatic
assay according to the method of Peuler and Johnson
(1977).

Tc-99m-DTPA Brain Imaging: Brain angiography was
performed within 10-20 min following the completion of
CSF pressure measurements and collection of CSF specimens
(for details see Sarper et al, this symposium.)

RESULTS

Clinical Observations: The oral administration of p-TA
induced behavioral side-effects in PEH-treated animals.
This was characterized by hyperexcitability and repeated
episodes of convulsive seizures. This was followed by
salivation, hyperventilation, ataxia, coma and death.
There was a significant (p<0.05) elevation in CSF
pressure (31.2 ±10.6 mmHg) in these comatose animals as
compared to controls. Contrastingly, the oral
administration of p-TA to control animals did not induce
behavioral changes and had no effect on CSF pressure
(Table 1).

CSF Catecholamines:

     The concentration of NA, A, and D increased
significantly in the CSF after the oral administration of
p-TA to PEH-treated dogs as compared to controls
(Table 1). There was a significant association between
elevation in CSF catecholamines and development of
intracranial hypertension in these animals.

Tc-99m-DTPA Brain Imaging:

     This study showed that the administration of p-TA to
PEH-treated dogs resulted in preferential uptake of Tc-

Table 1:

Cerebrospinal fluid (CSF) levels of D, NA, and A (average ± SD) and mean CSF pressure in pre-phenelzine (serving as controls n=10) and post-PEH treated dogs after the administration of an oral dose of p-TA (50 mg/kg).

Parameter	Status			
	Control		Phenelzine-treated	
	BT*	AT*	BT	AT
CSF (pg/ml)				
NA	92.4±41.4	138±50	102±68.2	4679±4343
A	0	0	5.4±12.5	77.8±61.3
D	11.42±16.2	0	27.7±22.4	117±95.6
CSF pressure (mmHg)	6.1±2.1	7.0±2.5	11.5±5.1	31.2±10.6

*BT = before p-TA and AT = after p-TA

99m-DTPA by the hemispheric brain regions vs non-brain regions as compared to control animals (for details see Sarper et al, this symposium).

DISCUSSION

The oral administration of p-TA to PEH-treated animals had a profound effect on the cerebrovascular permeability and retention of the diffusion-limited tracer Tc-99m-DTPA. These changes occurred in comatose animals concomitantly with marked and persistent elevation in CSF pressure that exceeded 25 mmHg and elevation of CSF concentration of catecholamines. These preliminary data support the hypothesis that p-TA may induce brain edema in PEH-treated animals. This neurotoxic effect of p-TA may be mediated by a central noradrenergic mechanism, since experimentally it has been demonstrated that a prompt increase in water permeability and reduction of the cerebral blood flow followed stimulation of the locus coeruleus (Hartman et al, 1980). Derangements of such a system may have obvious implications in the development of cerebral edema of Reye's syndrome, a fatal illness in children characterized by rapid swelling of the brain, seizures, and loss of consciousness accompanied by hepatic dysfunction and decreased mitochondrial MAO activity (Faraj et al, 1985).

REFERENCES

Faraj B.A., Caplan D.B., Malveaux E.J., Camp V.M., and Ali F.M. (1983). Similarity between tyramine-induced neurotoxicity and the coma of Reye's syndrome. J. Pharmacol. Exptl. Ther. 226, 608–615.

Faraj B.A., Newman S.L., Caplan D.B., Ahmann P.A., Kutner M., Ali F.M., and Lindahl J.A. (1985). Platelet monoamine oxidase activity in Reye's syndrome. J. Ped. Gastroenterol. Nutr. 4, 532–536.

Hartman B.K., Swanson L.W., Raichle M.E., Preskorn, S.H., and Clark H.B. (1980). Central adrenergic regulation of cerebral microvascular permeability and blood flow; anatomic and physiologic evidence. Adv. Exp. Med. Biol. 131, 113–126.

Peuler J.D., and Johnson G.A. (1977). Simultaneous simple isotope radioenzymatic assay of plasma norepinephrine, epinephrine and dopamine. Life Sci. 21, 625–636.

Acknowledgement

This work was funded in part by U.S. Public Health Service Grant NS17555.

A QUANTITATIVE RADIONUCLIDE
IMAGING TECHNIQUE TO STUDY
CEREBRAL EDEMA IN DOGS WITH
IMPAIRED MONOAMINE OXIDASE

R. Sarper, B.A. Faraj,
E. Malveaux, and Y.A. Tarcan,
Department of Radiology,
Emory University School of
Medicine, Atlanta, Georgia, USA

Under normal circumstances, the brain vasculature and astrocyte cells effectively exclude most substances that are not highly lipid soluble or are not used directly in brain cellular metabolism. However, in the presence of cerebral pathology such as neoplasm, infarction, Reye's syndrome and hydrocephalus, sufficient alterations in the blood-brain barrier occur to allow passage of diffusion-limited radioactive tracers into the region of abnormality. The most commonly employed diffusion-limited tracer for cerebral imaging is Tc-99m-DTPA (Russell et al, 1980). The tracer Tc-99m-diethylenetriamine pentaacetic acid (Tc-99m-DTPA) is a low molecular weight, negatively charged radiopharmaceutical.

In view of the increased cerebrovascular permeability of diffusion-limited radiopharmaceutical associated with neural trauma, we took the initiative to develop a rapid radionuclide imaging method for the quantitative assessment of brain edema _in vivo_ which is applicable to a variety of experimental conditions. The method is based on the quantitative measurement of the washout of Tc-99m-DTPA during its first passage through the brain. The method was utilized to study the development of cerebral edema in an animal model of Reye's syndrome (Faraj et al, 1983).

449

METHODS

Animals: Ten conditioned mongrel dogs were selected for this study.

Imaging Studies:

   (a) Equipment: A large-field-of-view gamma camera interfaced to a clinical nuclear medicine computer with standard nuclear medicine software library was used for the imaging studies.

   (b) Data Acquisition: An anesthetized dog was placed in a supine position over the detector of the camera, which was positioned to include the head. A catheter was inserted into the jugular vein for administration of the tracer. Data acquisition was started simultaneously with the rapid intravenous injection of Tc-99m-DTPA (25 mCi, 0.5 ml), and sixty 0.5-second images of the head were taken and stored on a disk for analysis.

   (c) Data Analysis: Using the area of interest that was indicative of the accumulation of radioactivity within brain hemispheric regions, time-activity curves were generated from the 60 successive images obtained at 0.5-second intervals. The slope of the curve representative of the first-pass Tc-99m-DTPA washout was calculated by fitting a straight line to nine data points, starting with the peak value. The slope values were normalized by dividing the slope obtained in each experiment by the corresponding number of counts obtained for the curve maximum.

Brain Edema: Tyramine induced coma in phenelzine-treated dogs (for details, see Faraj et al this symposium).

RESULTS

Brain Imaging:

   Examples of comparative composite brain images in control and comatose dogs are presented in Figure 1. The results of this study showed that in comatose animals there was a preferential uptake of the tracer by the brain regions versus non-brain regions as compared with controls following the rapid intravenous injection of Tc-99m-DTPA. This preferential brain uptake was accompanied by markedly enhanced retention of the isotope within the brain.

Slope Analysis: In comatose animals, there was a 6 to 50% reduction in brain washout slopes of Tc-99m-DTPA (Figure 2).

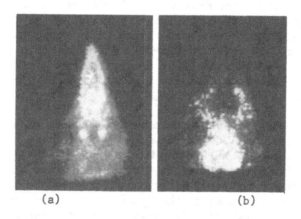

(a)   (b)

Figure 1: Whole brain image following the intravenous injection of Tc-99m-DTPA in a dog before (a) and after (b) the development of comatose behavior.

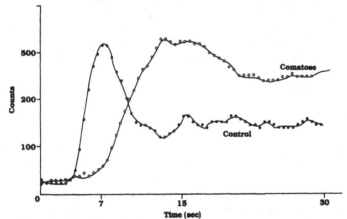

Figure 2: A time/activity curve obtained of a region of interest representative of hemispheric brain region. Each data point represents the number of counts in the region of interest as a function of time, at 0.5-s intervals.

DISCUSSION

Utilizing a scintillation camera interfaced to a
digital computer, sequential images were obtained to
generate first-pass radioactivity versus time curves for
the brain region following the rapid intravenous
injection of Tc-99m-DTPA. Analysis was based on the
measurement of slopes during the washout phase. A rapid
clearance of the tracer from the brain was observed in
control dogs as depicted by high slope value and minimal
brain accumulation. The rapid egress of an appreciable
fraction of the injected Tc-99m-DTPA from the brain
observed in this study could be interpreted as a result
of brain-capillary permeability limitation of Tc-99m-DTPA
(Rollo et al, 1977; Gates et al, 1978).

It is probable that the changes in washout slopes of
cerebral hemispheric Tc-99m-DTPA during first pass may
represent a reliable method to detect changes in
cerebrovascular permeability of diffusion-limited
substances under a variety of brain disorders. To test
our hypothesis, we assessed the dynamics of brain Tc-99m-
DTPA in dogs treated with the irreversible MAO inhibitor
phenelzine (PEH) and following the oral administration of
p-tyramine (p-TA) (for details, see Faraj et al this
symposium). These comatose animals had abnormal Tc-99m-
DTPA brain scans. The abnormalities in images were
associated with appreciable reduction in brain washout
slopes during first pass following the rapid intravenous
injection of Tc-99m-DTPA. It is interesting to note that
the development of intracranial hypertension in PEH-
treated animals following p-TA administration triggered a
prominent shift in the concentration of Tc-99m-DTPA from
other regions in the head into the brain of these
comatose animals.

In summary, the present study demonstrated that
measurement of normalized brain washout slopes of Tc-99m-
DTPA during first pass following the rapid i.v. injection
may represent a useful clinical modality to assess the
development of brain edema in a variety of neurological
disorders. Furthermore, it may also be used to assess
alteration in blood-brain permeability associated with
long-term treatment with central nervous system drugs
including antidepressants, neuroleptics, and sedatives.

REFERENCES

Faraj B.A., Caplan D.B., Malveaux E.J., Camp V.M., and Ali F.M. (1983). Similarity between tyramine-induced neurotoxicity and the coma of Reye's syndrome. J. Pharmacol. Exptl. Ther. 226, 608–615.

Gates G.F., Fishman L.S., and Segall H.D. (1978). Scintigraphic detection of congenital intracranial vascular malformations. J. Nucl. Med. 19, 235–244.

Rollo F.D., Cavlieri R.R., Born M., Blei L., and Chew M. (1977). Comparative evaluation of Tc-99m-GH, Tc-99mO$_4$ and Tc-99m-DTPA as brain imaging agents. Radiology 123, 379–383.

Acknowledgement

This work was funded in part by U.S. Public Health Service Grant NS17555.

URINARY DIPEPTIDE WITH MONOAMINE

OXIDASE INHIBITORY PROPERTIES

A. Hitri, L. Hendry, D. Ewing, and B. Diamond

Department of Psychiatry, Medicine and Endocrinology

Medical College of Georgia and Stereochemical Genetics, Inc.
Augusta, Georgia 30912, USA

INTRODUCTION

The enzyme MAO occurs in two forms,   A and B,
which differ in substrate specificity and inhibitor
specificities (Johnston, 1968; Knoll and Magyar, 1972)
apparent molecular weights (Brown et al,   1980) and
immunological properties (Denny et al,   1982). It has been
demonstrated that MAO A preferentially deaminates 5-HT and
it is more sensitive than type B to clorgyline inhibition.
Type B preferentially deaminates PE and Bz and it is more
sensitive to deprenyl inhibition than type A.

MAOI drugs are used as antidepressant agents.   The
clinical utility of these drugs is however hindered by
their side effects which are mainly precipitated by their
MAO A inhibiting properties. (Illett, 1980) and therefore
a selective MAO B inhibitor would be highly desired. 3-N-
Phenylacetyl-amino-2,6-piperidinedione (PAAP) belongs to a
class of antineoplastons, which are small to medium size
peptides and amino acid derivatives isolated initially
from blood and later from urine (Burzynski, 1969, 1973).
PAAP is presumed to be the endogenous end product of
phenylalanine metabolism (Burzynski et al,   1985).
Stereochemical modeling studies have predicted PAAP to
have psychotropic properties with minimal toxicity (Hendry
et al, 1981).   In our present study we investigated _in
vitro_ and _vivo_ this dipeptide for its MAOI potential.

455

METHODS

For the in vitro biochemical experiments human platelet preparations, as a source of MAO B, were used. The rat and human brain caudate nucleus preparations as a source of both MAO A and B enzyme type also were used.

MAO Assay: MAO B activity was determined with 14C PE as a substrate using a modified method of Wurtmon and Axelrod (1963). MAO A activity was measured in brain homogenates using 14C-5-HT as a substrate.

For in vivo animal studies male Sprague Dawley rats were used. PE stereotype behavior (SB) was assessed with varying doses of either saline, deprenyl, clorgyline and PAAP. Pretreatment usually was one hour prior to 50 mg/kg of PE. The number of animals showing forepaw padding and sniffing for at least one observation period during the 20 minute test period following PE was counted. Also SB was quantitated using a rating scale (Diamond and Borison, 1983). The rotational behavior in unilateral substantia nigra 6-OHDA lesioned rats was also used.

Twenty-one days following the lesion. Rats were administered PE (50 mg/kg) and baseline turning to the lesioned side quantitated. During the following days rats were given PAAP (50-500 mg/kg, i.p.) to ascertain if it itself produced any turning. Turning was counted for one minute every ten minutes during the half hour following drug administration.

RESULTS

Table 1 illustrates the MAO inhibitory capacity of PAAP in various tissues. The inhibitory constant for the MAO B enzyme was 63.8 + 24 in the platelets, 19.9 + 3.8 in the human brain and 8.0 + 2.1 for the rat brain. The IC50 for MAO A in the rat brain was 93.3 + 6 and in the human brain was 138.6 + 15.1. The data indicate that PAAP is about ten fold more potent in inhibiting type B than type A MAO.

In comparison to the classical MAO inhibitors PAAP appears to be two or three orders of magnitude less potent than phenelzine or pargyline. In kinetic experiments in which varying concentrations of the substrate (14C-PE) were incubated with constant amount of brain homogenate in the presence of IC50 concentrations of PAAP and phenelzine. The data revealed that PAAP exhibited a competitive type of inhibition, whereas phenelzine exhibited a noncompetitive type of inhibition.

PE administration to rats produced no SB per se. However, pretreatment with deprenyl produced forepaw padding with continuous sniffing and headweaving that lasted 60 minutes. The 5 and 10 mg/kg dose produced this behavior in 50 and 100% of the animals respectively. Pretreatment with clorgyline (5 mg/kg) did not produce this syndrome in PE treated rats. It did cause headbobbing which lasted 30 minutes. Only doses above 200 mg/kg of PAAP produced SB. The SB was identical to that of Deprenyl and lasted 50 minutes in 100% of animals. After 24 hours 60% of animals still displayed this behavior when challenged with PE. PE induced ipsilateral turning in 6-OHDA unilateral nigral lesioned rats. Duration of these rotations lasted for 30 minutes. Treatment with deprenyl doubled the number of ipsilateral turns induced by PE and increased the duration of PE rotations in these animals.

TABLE 1

Inhibition of MAO type A and B enzymes by PAAP

Enzyme Source	Enzyme	Type
	MAO A IC50 1 x 10-6M	MAO B IC50 1x10-6M
Human Platelet	-	63.8 + 24
Rat Brain Caudate Nucleus	93.3 + 6.0	8.0 + 2.1
Human Brain Caudate Nucleus	138.6 + 15.1	19.9 + 3.8

*each number represents the mean + SD of 4 determinations.

The inhibitory potency of PAAP was determined in MAO B enzyme assays by incubating a constant amount (15 microM) of 14C-PEA with aliquots of platelet or brain homogenates corresponding to 0.1 mg protein, in the presence of increasing concentrations of PAAP (10-500 microM). MAO A assays were carried out the same way with 8 microM 14C-5HT as a substrate. IC50 values were determined by computer analysis.

DISCUSSION

These results demonstrate that the dipeptide PAAP is an effective MAO B inhibitor in both animal and human tissues. This naturally occuring substance has been isolated from both blood and urine of normal human subjects and subsequently purified and synthesized (Bruzynski et al 1985). This present study has demonstrated that this endogenous dipeptide has selective MAO B inhibitory properties. Moreover unlike the classical MAOI which all act by irreversibly inactivating the enzyme (Paech et al 1979) PAAP appears to exert a competitive type of inhibition on brain and platelet MAOB.

Modeling studies in our laboratories have demonstrated that a variety of biologically active compounds possess stereochemical complementarity with double stranded DNA (Hendry, 1986). This technology can be used to predict the biological activity of candidate molecules. These predictions are based upon the type and degree of complementary fits of neurotransmitters into specific sites in DNA. PAAP fits into two sites in DNA (Hendry, 1987). These sites accommodate the neurotransmitters PE and 5-HT which are the substrates for MAO B and A, respectively. Modeling analyses in our laboratories have demonstrated that PAAP fits better into the site which is complementary to PE. On the basis of this observation, we predicted that PAAP would selectively inhibit MAO type B activity (further details of these relationships will be published elsewhere). In vitro animal and human studies and in vivo animal studies confirm these predictions.

REFERENCES

Brown, G.K., Powell, J. F. and Craig, I.W. (1980) Molecular Weight Differences Bwtween Human Platelet and Placental Monoamine Oxidase. <u>Biochem.</u> <u>Pharmacol.</u> <u>29</u>, 2595 - 2603.
Burzynski, S.R. (1969) Investigations on Unknown

Ninhydrous reacting substances in human blood serum. Experientia 25, 490-491.

Burzynski, S.R. (1973) Biologically active peptides in human urine. Isolation of medium sized peptides. Physiol. Chem. Phys. 5, 437-447.

Burzynski, S.R. and Hai, T.T. (1985) Antineoplaston A10. Drugs of the Future 10, 103-105.

Denny, R.M., Fritz, R.R. and Patel, N.T., et al (1982) Human liver MAO A and MAO B separated by immuno affinity chromatography with MAO B specific antibody. Science 215, 1400-1403.

Hendry L.B., Bransome E.D. Jr, Lehner A.F., Muldoon T.G., Hutson M.S. and Mahesh V.B. (1986) The stereochemical complementarity of DNA and reproductive steroid hormones correlates with biological activity. J. Steroid Biochem. 24. 843-852.

Hendry, L.B., Bransome, E.D., Jr., Hutson, M.S., and Campbell, L.K. (1981) First approximation of a stereochemical rationale for the genetic code based on the topography and physicochemical properties of "cavities" constructed from models of DNA. Proc. Natn. Acad. Sci. U.S.A., 78, 7440-7444.

Illett, F., George, G.F. and Davies, D.S. (1980) The effect of monoamine oxidase inhibition of "First Pass" metabolism of tyramine in dog intestine. Biochem. Pharmacol. 29, 2551.

Johnson, J.P. (1968) Some observations upon a new inhibitor of monoamine oxidase in brain tissue. Biochem Pharmacol. 17, 1285-1297.

Knoll, J. and Magyar, K. (1972) Some puzzling pharmacological effects of monoamine oxidase inhibition. In: Advances in Biochemical Psychopharmacology. (Costa, E., Sandler, M., eds.) Vol. 5, Raven Press, NY.

Paech, P. Salach, J.I., and Singer, T.P. (1979) Suicide Inactivation of Monoamine Oxidase by Tranylcypromine. In: Monoamine Oxidase Structure Function and Altered

Functions. (Singer, T.P., Von Korft, R.W., and Murphy,
D.L., eds.) Academic Press, N.Y, 1979.

Wurtman, R.J. and Axelrod, J. (1963) A sensitive and
specific assay for the estimation of monoamine oxidase.
Biochem Pharmacol. 1439-1440.

SOME INTERESTING ASPECTS OF CLINICAL APPLICATION OF THE 24

HOUR HUMAN URINE QUANTIFICATION OF PE. A PRELIMINARY STUDY.

J.Ciprian-Ollivier*,O.Boullosa,A.Lopez Mato*,
M.Cetkovich-Bakmas and J.Spatz. Centro de Inv.
Diag.y Trat.en Neurociencias.Francisco de Vitto
ria 2324 (1425)Bs.As.,Argentina.*First Profes-
sorship of Psychiatry,Faculty of Medicine,UBA.

ABSTRACT
    After one and a half follow-up study, 24 hour urinary
PE excretion quantification appears to be a state marker for
endogenous depression. This tendency is also observed in
bipolar patients but not in other pathologies.

Key Words. Bipolar (Bi); Major Depressive (MD); Minor depressive (md); Schizoaffective (Szaf); Schizophrenic (Sz).

Introduction

    Following the relation found by Fischer[5][6] between
the thymic state of the patient and urinary PE excretion,we
have widely used PE quantification as an auxiliary diagnostic tool. Spatz's Gas Chromatography Technique[8] was used,
taking special care with urine collection and the standariza
tion of the diet, since, apart from methodological differences, these two factors may be the cause of the discrepancies found between the results obtained in different laboratories [1][2][7]. Considering values under 80 mcg/24 h. to
be statistically significant[3], our method is 72% sensitive
and 81% specific for the diagnosis of endogenous depressions.
    Aggressive patients, obsessive-compulsive and phobic,
paranoid schizophrenic, non-paranoid schizophrenic, depressive and non depressive schizophrenic patients were studied

461

and significant differences were found between these groups and the controls(4). These findings led to the present follow-up study in order to determine the existence of clinical correlation and elucidate whether this is a trait or a state sign, according to the changes in the pathology and evolution of the patient.

Material and Methods

97 out psychiatric patients diagnosed following DSM III as Major Depressives, minor depressives (adaptative or neurotic), Bipolars, Schizoaffectives and Schizophrenics, were selected and compared with a control group of 14 healthy subjects. See following Table:

TABLE I

GROUP		AGE (X +/- s.d.)
	Control	34.8 +/- 10.5
DSM III	309.00//300.40	41.5 +/- 14.3
DSM III	296.23//296.33	45.4 +/- 13.0
DSM III	295.XX	28.8 +/- 10.0
DSM III	295.70	35.7 +/- 9.1
DSM III	296.6X//296.70	39.5 +/- 13.1

PE was measured using Spatz's method in patients without previous medication, after 6 months of pharmacological treatment, with or without psychotherapeutic support and after 18 months' treatment if the patients were not discharged before. Two determinations were made in the unmedicated control group with an 18 months interval. Samples were submitted with the consent of the patients and laboratory was blind with respect of their origin.

Results

Depression intensity, using Beck's score, decreased over time in all groups of patients as can be seen in TABLE II. This phenomenon is corroborated if we take as a point of reference the parameters established for the Beck:

A: Not depressed (0 to 10.9);

B: Mildly depressed (11 to 18.7);

C: Moderately depressed (18.8 to 25.4);

D: Severely depressed (25.5 or more).

TABLE II.

EVOLUTION OF BECK VALUES					
GROUP	N	0	MONTH 6	18	S.D.
Control	14 M: 7 F: 7	3.96	-	2.86	1.861
md	13 M: 4 F: 9	15.08	10.50**	-	7.152
MD	49 M:13 F:36	27.87	10.18**	-	7.705
Sz	12 M: 7 F: 5	25.75	20.00	15.17**	8.313
Szaf	11 M: 5 F: 6	27.00	17.68*	16.45*	8.435
Bi	12 M: 7 F: 5	27.50	17.75	12.00**	11.280

TABLE III.

EVOLUTION OF PE(U) VALUES					
GROUP	N	0	6	18	S.D.
Control	14	158.36 (28.57)		153.93 (14.29)	63.282
md	13	124.00 (30.77)	145.85 (30.77)	-	62.798
MD	49	78.31 (75.51)	153.65** (32.65)	-	101.835
Sz	12	129.50 (33.33)	130.17 (50.00)	139.50 (25.00)	107.784
Szaf	11	96.18 (45.45)	168.55 (27.27)	136.55 (36.36)	83.790
Bi	12	103.08 (58.33)	150.75 (33.33)	182.08 (16.67)	124.183

N: Number of patients
S.D.: Root square of the mean square of the error of
      the ANOVA with randomized blocks
  *: $p < 0.05$; **: $p < 0.01$
  (): percent of patients with abnormal values
      [PE(U) 80 ug/24 h.]

DEPRESSION (BECK SCORE) CONTROL GROUP

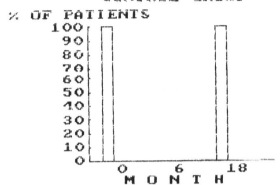

Md GROUP

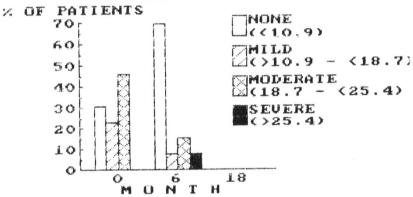

MD GROUP

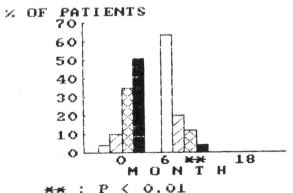

** : P < 0.01

FIG, 1.-

DEPRESSION (BECK SCORE)

SF GROUP

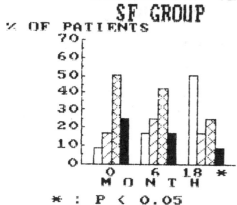

* : P < 0.05

SFAF GROUP

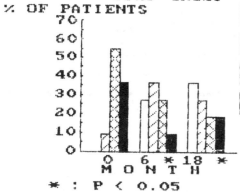

* : P < 0.05

Bi GROUP

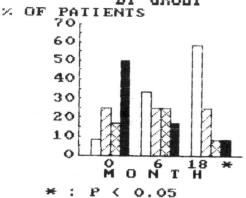

* : P < 0.05

FIG. 2.-

CORRELATION BETWEEN
PE{U} AND BECK VALUES

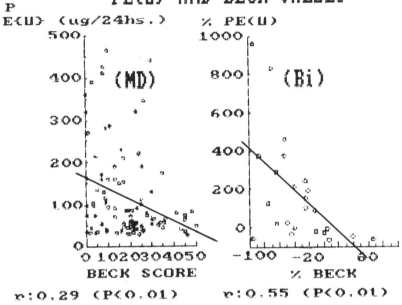

r:0.29 (P<0.01)   r:0.55 (P<0.01)

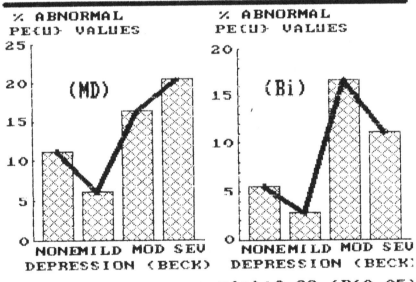

r(s):0.36 (P<0.01)   r(s):0.38 (P<0.05)

FIG.3.-

The changes in the distribution of the different groups in respective evaluations can be seen in FIGURE 1 and 2.

The study of urinary PE excretion, the other parameter studied in this trial, showed a significant increase only in the group of MD. We can also corroborate these results if we take the normal threshold. In this case there is also a strong tendency in the bipolar group, although it does not reach statistical significance due to the size of the sample. (See TABLE III).

If both parameters are correlated, positive results are obtained for the group of MD (FIGURE 3 up). If the variation percentages of the said parameters are used, there appears a correlation in the Bi group. In both MD and Bi a correlation appears if the 80 mcg. PE excretion threshold is considered. (See FIGURE 3 down).

Conclusions

Our follow-up statistical study reports on urinary PE excretion quantification as it has been practised in our country for some years. It shows that PE is an interesting and sensitive marker for endogenous (major) depressions. As was seen, there is a clear correlation in these cases between the value obtained and the clinical evaluation of the depression, both before and after treatment has been initiated. This correlation is maintained using percentual increases or reference to 80 mcg. threshold, which we consider normal. This becomes even more interesting if we consider that in non-endogenous depressive patients (adaptative or neurotic) the values remain above the threshold level without significant variations. This also occurs in the control group. In the case of bipolar patients, there is a strong tendency to find correlation between clinical improvement and percentual increase of PE; the values for depression or euthymia being relative for each patient, which would suggest a greater biological individuality. Due to the size of the sample further studies are required to confirm this impression. In schizophrenic and schizoaffective psychoses, a tendency to increase PE excretion is observed when thymia improves, but the correlation does not reach significant levels. We may then infer that we have a state marker for major and bipolar depressions, bearing in mind that the latter must be considered as percentual variations. The biochemical controversy on methodology may not yet be resolved. The fact that PE crosses the blood brain barrier makes it possible for a great part of its action on the C.N.S. to depend on its systemic production, and on account of this, we are not surprised by the values we have obtained. This study makes it possible then to confirm that the technique used in our country makes it a pioneer in the clinical use of biochemical tests in biological psychiatry. We have made hundreds of tests which give our physicians a more reliable diagnostic precision, a better follow-up and consequently a greater therapeutic certainty.

Bibliography

1. Boulton A.A. and Juorio A.V.(1982) In: Handbook of Neuro-chemistry, Vol.1 (2nd.Ed.) Edited by Abel Lujtha (Plenum Publishing Co. 1982).
2. Boulton A.A., Baker G.B., Denhurst N.G., Sandler M.(1984) (Editors) "Neurobiology of the Trace Amines". Humana Press. Clifton, New Jersey.
3. Ciprian-Ollivier J., Fernandez Labriola R., Spatz H., Gallardo F.(1978). Daimon 6, 15, 35-37
4. Ciprian-Ollivier J., Boullosa O., Cetkovich-Bakmas M. (1985) In: Biological Psychiatry 1985. Shagass C. et al. Editors. Elsevier Science Pub. Co. New York.
5. Fischer E., Heller D., Miró A.N. (1968) Arzneim-Forsch 18: 1486.
6. Fischer E. (1975). Neuropsiquiatría (Arg.) VI, 1-96.
7. Mosnaim A.D., Inwang E.E., Sugerman J.H., De Martin W.J., Sabelli H.C.(1973) In: Biol. Psychiat. 6, 235-257.
8. Spatz H., Spatz N. (1971) Biochem. Med. 6: 1.

Subject Index